Fluid Mechanics and Its Applications

Founding Editor

René Moreau

Volume 126

Series Editor

André Thess, German Aerospace Center, Institute of Engineering Thermodynamics, Stuttgart, Germany

The purpose of this series is to focus on subjects in which fluid mechanics plays a fundamental role. As well as the more traditional applications of aeronautics, hydraulics, heat and mass transfer etc., books will be published dealing with topics, which are currently in a state of rapid development, such as turbulence, suspensions and multiphase fluids, super and hypersonic flows and numerical modelling techniques. It is a widely held view that it is the interdisciplinary subjects that will receive intense scientific attention, bringing them to the forefront of technological advancement. Fluids have the ability to transport matter and its properties as well as transmit force, therefore fluid mechanics is a subject that is particulary open to cross fertilisation with other sciences and disciplines of engineering. The subject of fluid mechanics will be highly relevant in such domains as chemical, metallurgical, biological and ecological engineering. This series is particularly open to such new multidisciplinary domains. The median level of presentation is the first year graduate student. Some texts are monographs defining the current state of a field; others are accessible to final year undergraduates; but essentially the emphasis is on readability and clarity.

Springer and Professor Thess welcome book ideas from authors. Potential authors who wish to submit a book proposal should contact Dr. Mayra Castro, Senior Editor, Springer Heidelberg, e-mail: mayra.castro@springer.com

Indexed by SCOPUS, EBSCO Discovery Service, OCLC, ProQuest Summon, Google Scholar and SpringerLink

More information about this series at http://www.springer.com/series/5980

Fernando Concha A. · Juan Luis Bouso A.

Fluid Mechanics Fundamentals of Hydrocyclones and Its Applications in the Mining Industry

Springer

Fernando Concha A.
Water Research Center for Agriculture and Mining
Metallurgical Engineering, University of Concepción
Concepción, Chile

Juan Luis Bouso A.
ERAL CHILE, S. A.
Santiago de Chile, Chile

Advanced Mineral Processing S. L. (AMP)
Madrid, Spain

ISSN 0926-5112 ISSN 2215-0056 (electronic)
Fluid Mechanics and Its Applications
ISBN 978-3-030-67912-5 ISBN 978-3-030-67913-2 (eBook)
https://doi.org/10.1007/978-3-030-67913-2

This Springer imprint is published by the registered company Springer Nature Switzerland AG
The registered company address is: Gewerbestrasse 11, 6330 Cham, Switzerland

Preface

Classification in mineral processing is the operation of separating particle mixture components into two or more fractions according to size, with each resulting group more uniform in this property than the original mixture. There are two types of equipment for this purpose, mechanical classifier used in all mineral processing plant in the early 1900 and hydraulic classifiers. To cope with the immense amount of ore treated in a concentration plant, mechanical classifiers were increased in size and number, forming vast sections in the plant. The major investment in to house these numerous circuits in concentrators forced companies to introduce classifiers that occupied less floor space, the hydrocyclones that have simple construction and are ease of use. Once several hydrocyclones were installed in mining companies, it became, along with centrifugal slurry pumps, the most popular device in mineral processing plants.

Hydrocyclones are equipment with a very simple construction. They consist of an empty cylinder, or cylinder and cone, with a tangential inlet at the top of the cylinder, where the suspension to be classified comes in a discharge of the coarse material at the bottom and an overflow of the fine material at the top. This equipment, very cheap compared to mechanical classifier, was invented in the late 1800s and has suffered modest improvement thereafter.

However, the simplicity of hydrocyclones construction hides a very complicated operational behavior of the mechanism in which the particles are separated inside the equipment. This has prevented, until today, a complete knowledge of its physical fundamentals. Dyakowki and Williams in 1997 described the difficulties encountered to achieve an understanding of the physical mechanism of the hydrocyclone operation. Its three-dimensional, multiphase-multiple particle flow and its rheological behavior make it very difficult to establish a complete and fundamental physical model of the equipment operation.

Several simplified models have been presented, some of which present acceptable description of the fluid and particle movement in the equipment. Some models use experimental values to solve for pressure and velocity distributions with different physical approaches; inviscid models combining a free vortex with a sink, rotational inviscid models, models based on the exact solution of the Reynolds equation for highly idealized conditions and solutions using different mathematical techniques such as asymptotic expansions, similarity solutions and boundary layer flows and numerical solutions. These solutions have not allowed to obtain a complete picture nor the practical information for the design and operation of hydrocyclones in industrial plants. These problems lead to the postulation of several empirical models.

The aim of this work is to bring together all this information, which is scattered in a vast literature, in one book including fundamentals, applications, industrial experiences and case studies. The book is divided into 12 chapters. Chapter 1 introduces the incorporation of the hydrocyclone to the mining industry and its several applications of classification, densification and desliming. Chapter 2 deals with the fundamentals of classification, the variables involved, the material and operational variables and the analysis of the selectivity and classification function.

Sedimentation as the mechanism of the hydrocyclone classification is the subject of Chapter 3 were the fundamentals of the hydrodynamics of solid–fluid interaction is discussed. Explicit equation for the settling velocity in different types of flow is presented. An example shows the method of calculation. Chapter 4 deals with the flow of fluids in circular pipelines, as the basis for the feed, underflow and overflow piping necessary and the important off the prevention of sedimentation in the tubes.

Chapter 5 discusses the flow pattern in hydrocyclones from a fluid dynamics point of view. Experimental observations are presented, and the basic equations are laid down. The study incorporates exact and extreme solutions to these equations in Euler's, Prandtl's and Stokes flows to obtain axis–symmetrical solutions of the three velocity components. The generation and quantitative description of the air core is discussed.

Chapter 6. The problem associated with the mathematical modeling of flow patterns in hydrocyclones involves the solution of strongly coupled, nonlinear partial differential equations of mass and momentum balances, which lies well beyond any analytical approach, except when major simplifications are assumed. This chapter discuses solutions to the fundamental fluid dynamic equations using finite elements and finite difference methods comparing the several turbulence descriptions.

Chapter 7 presents a phenomenological model and a numerical simulator of a hydrocyclone developed by one of the authors in 1991–1992, based on the experimental observations of Hsieh and Rajamani in 1988 and the works of Bloor and Ingham in 1975–1976. The interior of the hydrocyclone was divided in six zones, and the fluid flow of each of one these is described by fluid mechanics equations that predict the operation of a cylindrical–conical hydrocyclones using the characteristics of feed material, the geometry of the equipment and the operating conditions. The model solves the equations that describe the fluid velocity field within the hydrocyclone as well as the particle trajectories suffering classification. From these equations, it was possible to determine the short circuit, the classification function and the size distribution of the overflow and underflow of the hydrocyclone.

Chapter 8 presents the concept of the hydrocyclone as turbomachine for its mechanical design. Several industrial designs are presented, and the effect of the equipment parts, cylinder and cone sections dimensions, overflow to underflow diameters and cone angle is discussed.

Chapter 9 studies the various empirical models of hydrocyclone selection. It presents Arterbourn method of hydrocyclone selection and the most frequently used simulation models of Lynch and Rao's, Plitt, Molycop tools and Nagerwararao's.

Chapter 10 presents applications of hydrocyclones system design, including feeding tanks geometry and volume, centrifugal pumps, hydrocyclone geometry and construction materials, hydrocyclones connecting pipes and pulp distributors.

Chapter 11 deals with instrumentation and control in a hydrocyclone control for feed and underflow concentration, particle size separation and type of underflow, spray or rope. Instruments based on electrical capacitance, image analysis, vibrating sensing and weight measurement are presented. Finally, Chap. 12 presents six study cases of application of hydrocyclones in the mining industry.

Some parts of this book, for example sedimentation of particulate systems, flow pattern in hydrocyclones and hydrocyclone system design, are the result of research by the authors and their research group at the Department of Metallurgical Engineering University of Concepcion, Chile, and at ERAL, Spain. We had important collaboration of Oscar Castro Soto, presently Process Manager at FLSmith in Chile, Dr. Arturo Barrientos and Dr. Fernando Betancourt, part of the department of Metallurgical Engineering and Dr. Pablo Cornejo from the Mechanical Department of the University of Concepción.

Many thanks to Graduate and Undergraduate Students at the University of Concepción who collaborated with their doctor's thesis: Oswaldo Bustamante, Elias Ovalle, Jaime Collantes, Julio Miranda; master's thesis: Bruno Chine, Juan Daza and engineering thesis: César Vidal, Jorge Montero, Elizabeth Alarcon, Jorge Etcheberry, M. A. Rojas, P. F., Parga, G. A., Katherin and Daniel Perez.

Especially, thanks to Barret Buchholz who did an excellent job in correcting my English.

Finally, I would like to thank several institutions that gave financial support to the research work related to the subject of this book, especially the University of Concepción; Conicyt Fondecyt Projects: 1.097/83; 841/82; Fondef Project MI-08, ANID/FONDAP/15130015.

Concepción, Chile
Santiago de Chile, Chile/Madrid, Spain

Fernando Concha A.
Juan Luis Bouso A.

Contents

Chapter 1
Introduction

Abstract Until recent centuries, nature provided ores rich in minerals that could be melted directly without any dressing, but when the grades of these ores began to decline due to extensive mining, it became necessary to develop concentration processes to achieve the grades required for melting. In all cases it was necessary to grind the ores to liberate their mineral contents so that particles rich in minerals could be separated from gangue for further processing. Screens were used to separate coarse particles from small ones after crushing, since gravimetric and flotation processes require the separation of very small particles using hydraulic devices. These operations are called *classification* and the overall size reduction-classification process is known as *close grinding circuit.*

Industrial expansion brought about by the development of nations during the last century through the present has led to an increase in the demand for primary metals such as iron, copper, lead and zinc, resulting in an increased capacity of mineral processing plants, which, in turn, have required increases in their closed-grinding circuits. To cope with the immense amount of ore to treat, crushers, mills and classifiers increased in size and number, forming various sections of open and closed size reduction circuits. A good example of these massive plants was that of the Braden Copper Company in Chile in the early twentieth century. Embedded in the side of the mountain containing the ore deposit, it made use of gravity transport of materials from the crushers to the mills, from the mills to the classifiers and then to flotation and, finally, tailings treatment. See Fig. 1.1. A flowsheet of the process is shown in Fig. 1.2.

The Sewell concentrator had 28 sections of closed grinding classification circuits, as shown in Fig. 1.3. A large mechanical spiral classifier accompanies each of the grinding sections shown in Fig. 1.2. See Figs. 1.3 and 1.4.

F. Concha A. and J. L. Bouso A., *Fluid Mechanics Fundamentals of Hydrocyclones and its Applications in the Mining Industry*, Fluid Mechanics and Its Applications 126,
https://doi.org/10.1007/978-3-030-67913-2_1

Fig. 1.1 The Braden copper Company's Sewell concentrator [4]

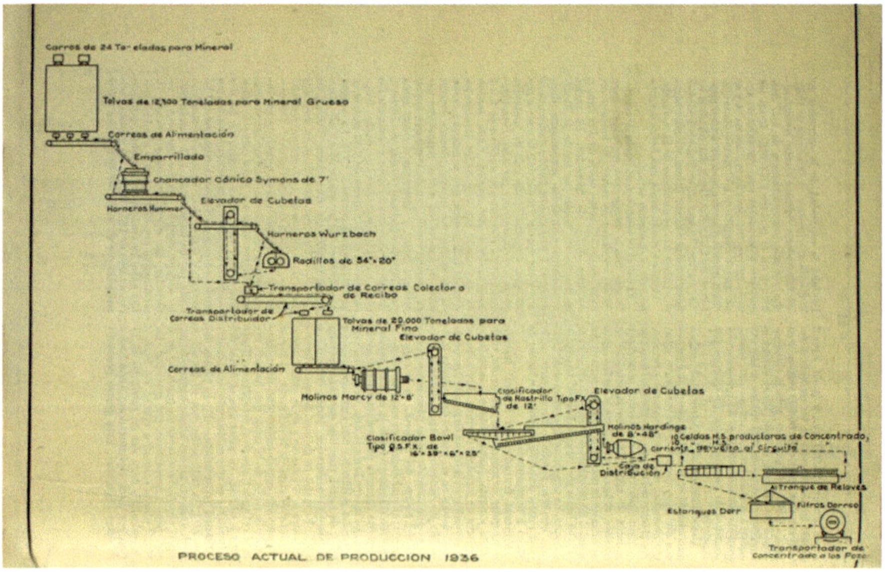

Fig. 1.2 Flowsheet of the Sewell concentrator; from Sutulov [5]

Harding ball mills. Marcy ball mill-spiral classifiers.

Fig. 1.3 Braden copper Company in Chile in the 1950s, grinding sections

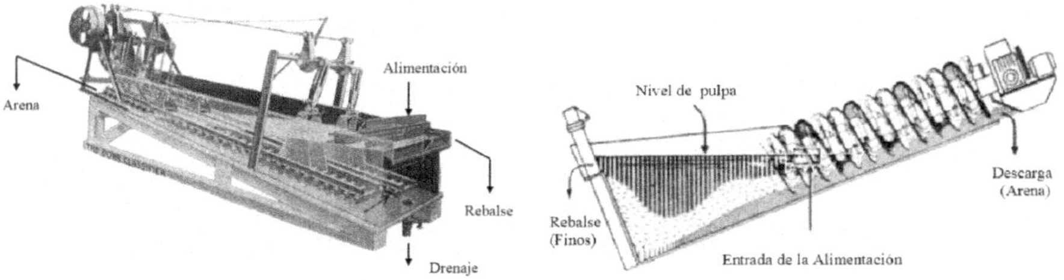

Fig. 1.4 Drawings of rake and spiral mechanical classifiers used at Sewell in 1936

1.1 Hydrocyclones

The major investment in infrastructure and enormous space needed to house these numerous circuits in concentrators forced companies to increase the size of the mills and thus reduce their numbers and introduce classifiers that occupied less floor space. By the end of the 1950s, Sewell had replaced rake and spiral mechanical classifiers with hydrocyclones. See Fig. 1.5.

The first recorded hydrocyclone patent was awarded to Bretney in the United States in 1891, and it was applied in 1914 in a phosphate mine. Nevertheless, Driessen was responsible for the major developments in the application of hydrocyclones for Dutch State Mines, which in 1948 applied for patent rights for the DSM hydrocyclone, a hydrocyclone without a vortex finder.

The Dorr-Oliver company acquired the DSM hydrocyclone and marketed it as the "Dorrclone." In 1954, the Kellogg American company Krebs Equipment Engineers, which later became the Krebs Engineers firm, presented new hydrocyclone designs that became the starting point for new ideas and versions developed by other manufacturers. Once several hydrocyclone models had been ordered, perhaps because of its construction, simplicity and ease of use, it became, along with centrifugal slurry pumps, the most popular device in mineral processing plants [2].

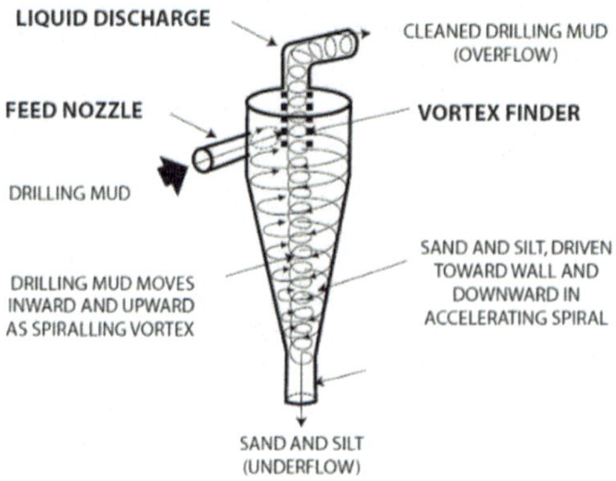

a) Conical hydrocyclone.

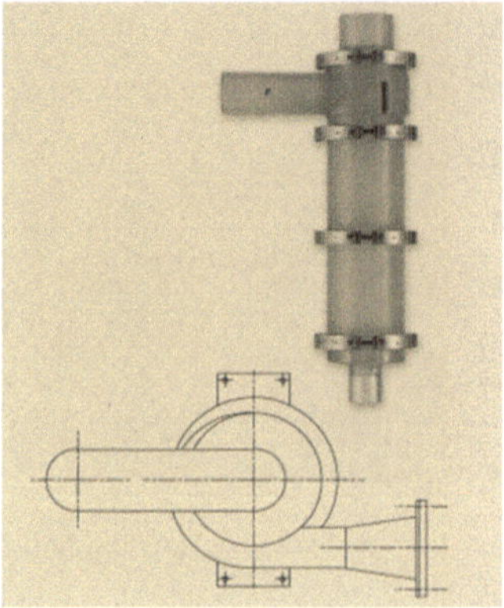

b) Flat-bottom hydrocyclone.

Fig. 1.5 Schematic view of a hydrocyclone

1.2 Hydrocyclone Applications

- **Classification**: Mill products contain particles of a wide range of sizes, some of which are larger than desired. Therefore, classification is necessary to remove particles smaller than the desired product and return the coarser particles to the mill for further size reduction. Classification in a hydrocyclone is accomplished through a balance of forces applied to the particles by centrifugal forces resulting from the tangential entrance of the pulp at a high speed, producing a swirling flow, with fine and lighter particles separated at the overflow, while coarse and heavy particles leave the underflow. See Fig. 1.5.
- **Desliming**: Hydrocyclones are currently the preferred method of desliming, the elimination of fine particles or clays from ground ores before a concentration process, where the useful product is the underflow of the hydrocyclone. These units are relatively inexpensive, have very high capacities and are easy to operate and maintain. See Fig. 1.6.
- **Densification**: In cases where the useful product is the denser fraction in the underflow, densification is used in some concentrators to increase thickening capacity. See Fig. 1.7.
- **Refining:** Removal of coarse particles in a suspension, such as in effluent cleaning of ultra-fine kaolin or bentonite minerals, where the overflow is the useful product. See Fig. 1.8.
- **Clarification:** Sometimes hydrocyclones can be installed as a double classification unit before thickeners or filters, as an adjunct to the main equipment. See Fig. 1.9.

It is interesting to note that each application has an optimum hydrocyclone geometry and circuit configuration; therefore, hydrocyclone manufacturers must have a

Fig. 1.6 Desliming hydrocyclones

Fig. 1.7 Densification hydrocyclone

Fig. 1.8 Refining hydrocyclone

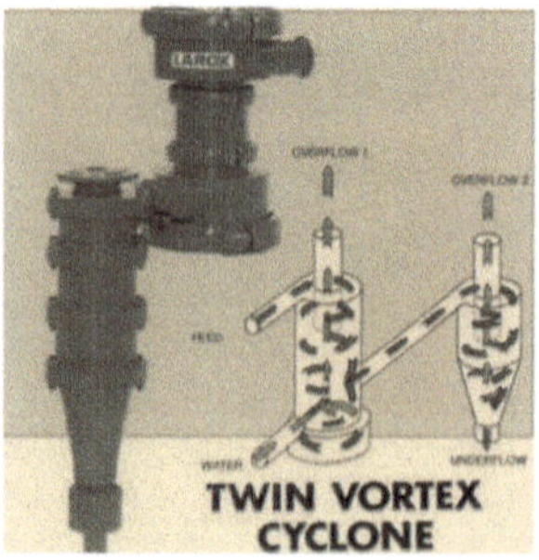

Larox Twin Vortex hydrocyclone. Heiskanen (1993).

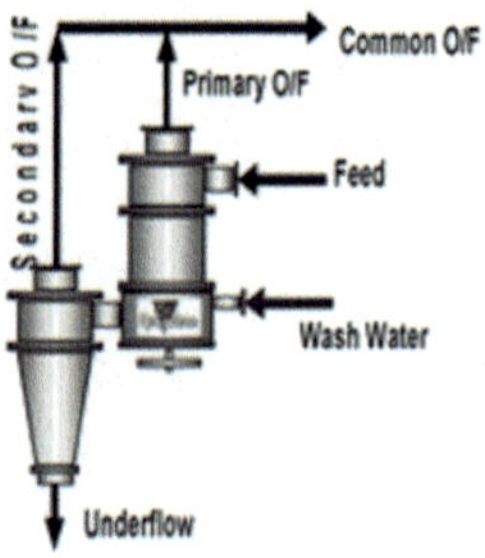

ReCyclone, Weir Minerals, South America. Castro et al. (2009).

Fig. 1.9 Double classification [3]

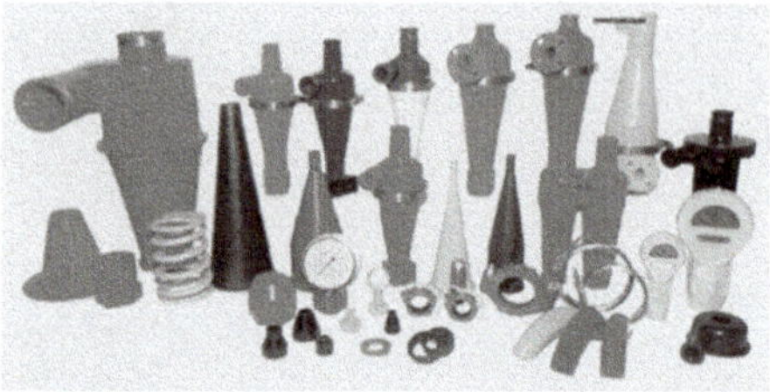

Fig. 1.10 O'Drill MCM company

wide range of varieties to offer. Figure 1.10 shows a range of hydrocyclone sizes and proportions. The diameter of a hydrocyclone must be proportional to the sizes of the particles to be classified; large diameter hydrocyclones are used in copper concentrators and small diameter units in the iron industry and with industrial minerals such as clays, where rigs come with larger numbers of smaller hydrocyclones.

See Fig. 1.8.

One of the problems encountered in connecting several hydrocyclones is that, as they are hydraulic devices, with linear feed distribution the pressure to every unit is different and, therefore, the operation of each hydrocyclone is not uniform. To solve this problem, rigs with radial feed distribution to all hydrocyclones were designed in order to deliver the same pressure to each of them. These rigs are standard today. See Fig. 1.11.

Sometimes the ore contains a significant amount of particles of the desired size or smaller, in which case the grinding classification circuit can be inverted; that is, the hydrocyclone comes before the mill, eliminating at the overflow particles that are already the desired size and returning oversized particles to the mill. See Fig. 1.10 for direct and inverse grinding-classification circuits (Fig. 1.12).

Capacity and cut size

Grinding circuits formed by rod-ball mills use hydrocyclones with a maximum diameter of 650 mm (26 in), producing final size separation between 150/300 microns.

Linear feed distribution.

Radial feed distribution.

Fig. 1.11 Feed distribution to hydrocyclones

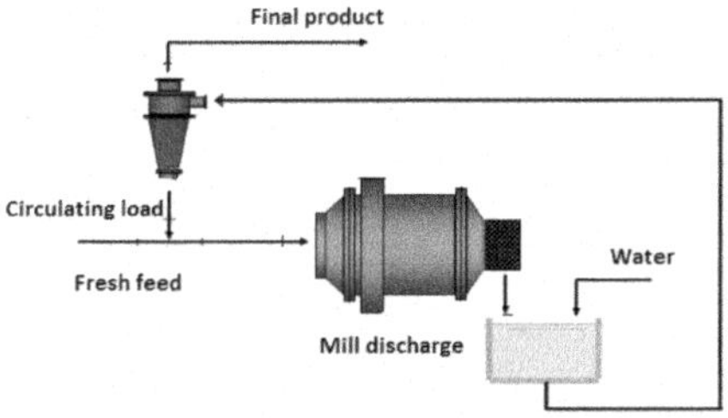

Direct circuit.

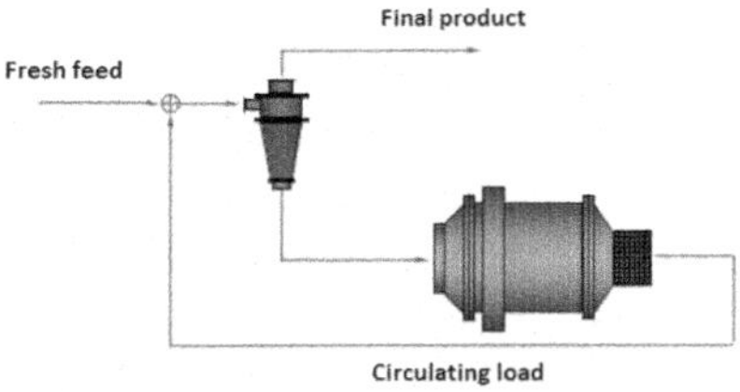

Inverse circuit.

Fig. 1.12 Alternative grinding-classification circuit types

In regrinding circuits, hydrocyclones produce sizes from 150 mm (6 in) to 375 mm (15 in), with particle size separation of 38/53 microns.

To increase profit in a mining operation, producers have two choices: enhancing recovery, which is a wise decision for low-grade deposits and limited reserves, or increasing treatment capacity for large deposits of high grade ores, because loss of recovery, in this case, is compensated for by increasing treatment and, ultimately, greater metal production. Increasing the grind size in a mining operation requires a minimal investment, as it is necessary only to adjust the hydrocyclone cut and increase the volume of the flotation plant, which entails a small expenditure. Increasing the cut size in a hydrocyclone is achieved by raising the concentration of solids in the feed, increasing tonnage and decreasing the water supply.

Flat-bottom hydrocyclones differ in design in that they have fully cylindrical bodies (see Fig. 1.5b). This configuration allows the cut size of cylindrical hydrocyclones to be extended to those achieved with larger-diameter conical cyclones and permits replacement of 650 mm conical hydrocyclones with 500 mm (20 in) flat-bottom hydrocyclones. Flat-bottom hydrocyclones have lower short circuits and less tendency to rope.

In conventional conical hydrocyclones, rope discharges produce fluctuations that cause inefficient operation; therefore, particle sizes must be increased in the overflow to reduce the quantity of particles in the underflow. For this type of hydrocyclone, the cylindrical part plays an important role. Varying the length of the cylinder by adding or removing cylindrical bodies allows the cutting size to be adjusted.

In many grinding circuits, 500 and 650 mm conical hydrocyclones have been replaced by 400 and 500 mm flat-bottom hydrocyclones, maintaining the cut size without losing milling capacity, or increasing milling capacity while keeping the cut size. Examples can be found in Codelco Chile's Chuquicamata and Andina mines and Anglo-American Company's Mantos Blancos mine.

Cut size in a hydrocyclone is proportional to its diameter; therefore, replacing several hydrocyclones with a smaller number of large diameter units prevents the obtainment of the desired particle size in the overflow. The minimum number of units recommended for circuits with large feed property variations is between four and six. Plants operating at high capacities with low-cost sterile materials or tailings may use coarser grinding to avoid the installation of new hydrocyclone rigs.

To reduce power and grinding medium consumption in copper ore processing, large semi-autogenous grinding mills (SAG) replaced primary rod mills in grinding classification circuits in the early 1980s. Crushing was reduced to a single step with feeds of up to 1500 mm (60 in) and products of < 200 mm (8 in) fed to the SAG mill. SAG mill discharges are classified into two fractions by vibrating screens. The oversize fraction, called pebbles, is returned to the SAG mill, while the fine fraction, together with the secondary ball mill discharge, is classified by a large-diameter hydrocyclone that delivers the fines to feed flotation. These gigantic grinding circuits can handle on the order of 6000 m^3/h, which must be pumped to the hydrocyclone batteries by large 1000 kW centrifugal pumps developed for this purpose.

References

1. Bouso JL (2001) Milestones at development of classification. In: Moly-Cop IX symposium on mineral processing, Puyehue, Chile
2. Bradley D (1965) The Hydrocyclone. Pergamon Press Inc., 122 East 55th St., New York 22, N.Y.
3. Castro O, López J, Switzer D (2009) Maximum classification efficiency through double classification using the ReCyclone. In: Malhotra D, Taylor P, Spiller E, Lever M (eds) Recent advances in mineral processing plant design. SME, 8307 Shaffer Parkway, Littleton, Colorado, USA 80127
4. Instituto de Ingenieros de Minas de Chile (1936) Monografía sobre la Braden Copper Company, Sewell, Chile. Soc Imp y Lit Universo, Santiago
5. Sutulov A (1975) Antecedentes Históricos de la producción de Cobre en Chile. Capítulo I en Zausschevich, A. and Sutulov, A., El Cobre Chileno, Corporación del Cobre, Santiago

Chapter 2
Classification Fundamentals

Abstract Classification in mineral processing is the operation of separating particle mixture components into two or more fractions according to size, with each resulting group more uniform in this property than the original mixture. There are two types of classifier: mechanical and hydraulic. During screening in the classical mechanical classifier, the material is subjected to the action of a series of screens through which small particles pass, while coarser particles greater than screen openings are retained. Like particle size, shape also has an important effect. The mechanism involved in separating particle sizes in a hydraulic classifier is sedimentation of the particles subjected to body forces such as gravity and centrifugal force. In this case, the factors affecting hydraulic classifications are the solid particle properties of size, density and shape and fluid temperature, density and viscosity.

In most cases the product of classification comprises two fractions, the coarser and finer fractions relative to the separation size. In a perfect operation the coarse fraction is called the "underflow," which contains all particles that are greater than the separation size, and the fine fraction is called the "overflow," which contains all finer particles. Even in very simple classification processes, the presence of random dispersion impedes prediction of the exact separation size. Other particle feed properties such as its mineralogical composition, particle size distribution and physicochemical conditions introduce additional complications. Finally, the fact that slurries are often highly concentrated further complicates the application of the sedimentation theory to this process, with the result that the underflow and overflow contain a considerable number of misplaced particles.

2.1 Concentration Measures

There are several ways of measuring fluid concentration; therefore, we must know how to convert one into another. Classification uses the concepts of concentration, fluid density, weight/volume fractions of solids and dilution.

F. Concha A. and J. L. Bouso A., *Fluid Mechanics Fundamentals of Hydrocyclones and its Applications in the Mining Industry*, Fluid Mechanics and Its Applications 126,
https://doi.org/10.1007/978-3-030-67913-2_2

If we designate m_s , m_f and M as the solid, fluid and total mass, respectively, ρ_s and ρ_f as the solid and fluid density, $\Delta\rho$ as their difference, w and φ as the weight and volume fractions of solids, ρ as the suspension density, also known as pulp density, and D as dilution, we can relate mass balances to the different variables. For example [3]:

$$\begin{aligned} \text{Pulp density } \rho &= \frac{\text{solid mass} + \text{sluid mass}}{\text{pulp volume}} \\ &= \rho_S\varphi + \rho_f(1-\varphi) \\ &= \rho_f + \Delta\rho\varphi \end{aligned} \tag{2.1}$$

$$\begin{aligned} \text{Solid fraction } w &= \frac{\text{solid mass}}{\text{total mass}} \\ &= \frac{\rho_s V_s}{\rho_s V_s + \rho_f V_f} = \frac{\rho_s V_s}{\rho_s V_s + \rho_f V_f} \\ &= \frac{\rho_s\varphi}{\rho_f + \Delta\rho\varphi} \end{aligned} \tag{2.2}$$

$$\begin{aligned} &\text{Solid volume fraction } \varphi = \text{solid volume/fluid volume} \\ &\varphi = \frac{m_s/\rho_s}{m_s/\rho_s + m_f/\rho_f} = \frac{\rho_f m_s}{\rho_f m_s + \rho_s m_f} = \frac{\rho_f w M_s}{\rho_f w M_s + \rho_s(1-w)M_s} \\ &\quad = \frac{\rho_f w}{\rho_f w + \rho_s(1 - w)} \end{aligned} \tag{2.3}$$

$$\begin{aligned} \text{Dilution } D &= \text{fluid mass/sold mass} \\ D &= \frac{\rho_f(1-\varphi)V_T}{\rho_S\varphi V_T} = \frac{\rho_f(1-\varphi)}{\rho_S\varphi} \end{aligned} \tag{2.4}$$

2.2 Variables that Affect Hydrocyclone Behavior

Among the variables that affect hydrocyclone behavior, we will address (1) design variables, (2) material parameters, (3) operating variables and (4) disturbances.

2.2.1 Design Variables

The following design variables characterize the gross behavior of the hydrocyclone: diameter d_c, the feed tube area A_F, the apex, or underflow, diameter d_u and the vortex, or overflow, diameter d_o (Fig. 2.1).

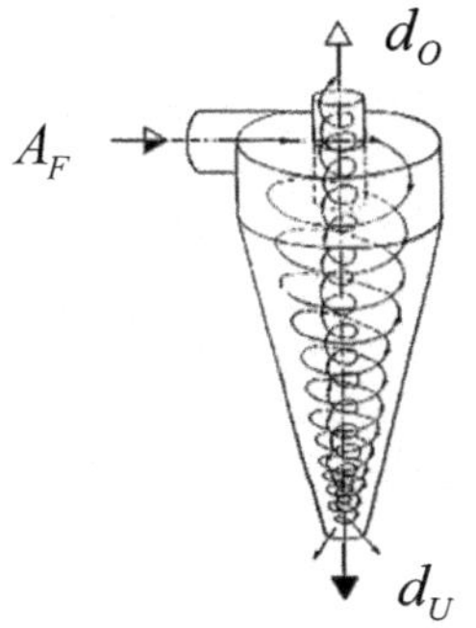

Fig. 2.1 Hydrocyclone

In general, hydrocyclones have a certain standard geometry. For example, hydrocyclones manufactured by Krebs [2] have geometries proportional to the hydrocyclone diameter:

Inlet area:	$A_F \propto d_c^2$.
Overflow diameter:	$d_O \propto 0.35 d_c$.
Underflow diameter:	$0.1 d_c < d_U < 0.35 d_c$.
Length of the cylindrical hydrocyclone part:	$L_c \propto d_c$.
Hydrocyclone cone angle:	$10° < \alpha < 20°$.

It is well established that the *reduced separation size* d_{50} of a hydrocyclone depends mainly on its diameter. The separation of small particles requires small hydrocyclones and the separation of larger particles larger hydrocyclones. For example, Arterburn [2] indicates that $d_{50} \approx d_c^{0.66}$, while Trawinski [33] gives $d_{50} \approx d_c^{0.50}$ and Pires and Massarani [27] suggest $d_{50} \approx d_c^{0.75}$.

These results show that, in selecting the hydrocyclone size, the material flow to be processed is not related to the hydrocyclone size, but rather to the number of hydrocyclones necessary to deliver the desired flow rate.

Data collected by Lynch [21] showed that hydrocyclones that were geometrically similar but with different diameters—from 4 inches (100 mm) to 15 inches (380 mm)—showed the same reduced classification function for a given feed material. Figure 2.5, in accord with [19], shows experimental data from two hydrocyclones, one a 12 inch (305 mm) unit and the other a 24 inch (610 mm) unit, operating under the same conditions on a copper ore,they present the same classification function and different d_{50} These observations lead to the conclusion that, for a given material, the reduced classification function represents the behavior of geometrically similar hydrocyclones.

It was recognized by Lynch [21], Bradley [5] and several other authors that the apex and vortex sizes also influence the separation size, which increases with the diameter of the vortex and the diameter of inlet tube area, and decreases as the apex diameter increases. For example, Lynch [21] indicates that both the vortex and the apex have an exponential relationship with the separation size:

$$d_{50} \propto \exp(d_O) \text{ and } d_{50} \propto \exp(-d_U).$$

Meanwhile, the pulp volume flow that a hydrocyclone is capable of treating is a function of the inlet tube area, and, since the feed area is proportional to $A \propto d_c^2$, the volume flow often is given as a function of the hydrocyclone diameter $Q_F \approx A_F^{0.5}$, so that $Q_F \approx d_c$, [33].

2.3 Material Parameter

In a classifier working at steady state, a mass balance can be expressed in terms of mass flows, total mass, solid mass and fluid mass, M, m_s and m_f, particle size x, the weight fraction $w(x)$ of particles of size x, $F(x)$ smaller than x and $R(x)$ greater than x. The sub-indices F, U and O represent the feed, underflow and overflow.

2.3.1 *Mass Balance*

Consider the following mass balance of pulp, solid, water and particle sizes of the feed *F*, underflow *U*, and overflow *O*, in t/h. ()* corresponds to pulp (Fig. 2.2).

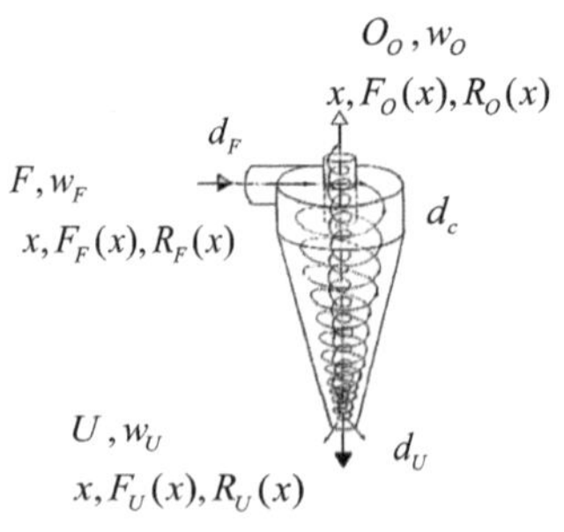

Fig. 2.2 Hydrocyclone

Pulp balance.

$$F^* = O^* + U^* \tag{2.5}$$

Solid balance:

$$w_O = \sqrt{\frac{F-U}{O}} \tag{2.6}$$

Water balance

$$\begin{aligned} F^*(1-w_F) &= O^*(1-w_O) + U^*(1-w_U) \\ F\frac{1-w_F}{w_F}(1-w_F) &= O\frac{(1-w_O)}{w_O} + U\frac{(1-w_U)}{w_U} \end{aligned} \tag{2.7}$$

Balance of particles of size *x:*

$$Ff_F(x) = Of_O(x) + Uf_U(x) \tag{2.8}$$

Balance of particles of size greater than x:

$$FR_F = OR_O(x) + UR_U(x) \tag{2.9}$$

Balance of particles of size smaller than x:

$$FF_F(x) = OF_O(x) + UF_U(x) \tag{2.10}$$

2.3.2 Selectivity Function

To characterize a hydrocyclone we use the *selectivity function* concept, $S(x)$, also known as the *Tromp curve*, which defines the mass of particles of size x in the feed, $F,\ f_F(x)$, that goes to the underflow, $U_U f_U(x)$:

$$S(x) = \frac{U_U f_U(x)}{Ff_F(x)} \tag{2.11}$$

Replacing (2.8) in (2.11) yields:

$$S(x) = \frac{(f_F(x) - f_O(x))f_U(x)}{(f_U(x) - f_O(x))f_F(x)} \tag{2.12}$$

To visualize the form of a selectivity function, use the mass balance in a hydrocyclone represented in Table 2.1.

Using Eq. (2.12) the *selectivity function* versus the average size is plotted, yielding Fig. 2.3.

It is interesting to note in Fig. 2.3 that below a certain particle size (40 μm in this case), finer particles go to the underflow, together with the oversize particles.

Table. 2.1 Size distribution around a hydrocyclone

Size x		Average size *x*	Feed		Undererflow		Overflow		Selectivity
Mesh #	μm	μm	F_F ton/h	$f_F(x)$	U_u ton/h	$f_u(x)$	O_o ton/h	$f_u(x)$	$S(x)$
28	600	505	0.000	0.00	0.000	0.00	0.00	0.00	1.000
35	425	357	2.122	3.49	2.122	8.60	0.00	0.00	1.000
48	300	252	1.678	2.76	1.678	6.80	0.00	0.00	1.000
65	212	178	3.028	4.99	2.812	11.40	0.22	0.60	0.929
100	150	126	5.904	9.72	3.923	15.90	1.98	5.49	0.664
150	106	89	10.266	16.91	4.860	19.70	5.41	14.99	0.473
200	75	63	12.414	20.44	3.627	14.70	8.79	24.37	0.292
270	53	45	8.752	14.41	2.270	9.20	6.48	17.98	0.259
-270	27	14	16.560	27.27	3.379	13.70	13.18	36.56	0.204
		Total	60.724	100.00	24.671	100.00	36.05	100.00	
			$w_F =$	48.60	$w_u =$	65.5			

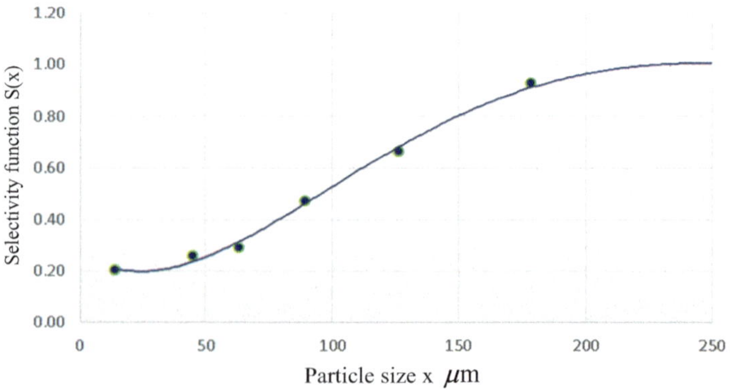

Fig. 2.3 Selectivity function

These misallocated particles are called the *short circuit* or *bypass of fines,* since in effect they are not separated by size; that is, they were not *classified*. Their amount is designed by $By(x)$. Note that in Fig. 2.3 the short circuit of fines is 20%.

2.3.3 Classification Function

For a given material fed to a hydrocyclone, the value of the short circuit depends on the operating condition, which we will study in Sect. 2.4; therefore, the selectivity function does not represent a real classification function characteristic of that material. We will define the selectivity function from which the short circuit has been subtracted as the *classification function.*

$$C(x) = \frac{Uf_U(x) - By(x)}{Ff_F(x) - By(x)} \tag{2.13}$$

Dividing by $Ff(x)$ yields

$$C(x) = \frac{Uf_U(x)/Ff_F(x) - By(x)/Ff_F(x))}{1 - By(x)/Ff_F(x)}$$

$$C(x) = \frac{S(x) - By(x)/Ff_F(x))}{1 - By(x)/Ff_F(x)} \tag{2.14}$$

Since the bypass is the part of the feed that has not been classified, it must have the size distribution of the feed, that is, $By(x) = a\, Ff_F(x)$. Substituting this value in Eq. (2.14) yields the *classification function.*

$$C(x) = \frac{S(x) - a}{1 - a} \tag{2.15}$$

From Table 2.1 the classification function $C(x)$ can be calculated. Table 2.2 shows the result.

The curves of Fig. 2.4 are obtained by plotting the selectivity and classification functions with values from Table 2.3.

Important parameters of the classification function are the *reduced separation sizes*—x/d_{50}, x/d_{75} and x/d_{25}—defined as the particle size values at which the classification functions are 50%, 75% and 25%; that is, 50%, 75% and 25% of the material goes to the underflow. Table 2.3 shows the values of the reduced separation size x/d_{50}.

Table 2.2 Balance including the classification function

Size x		Average size x	Feed		Undererflow		Overflow		Selectivity	Classification
Mesh #	µm	µm	F_F ton/h	$f_F(x)$	U_u ton/h	$f_u(x)$	O_o ton/h	$f_u(x)$	$\boldsymbol{S(x)}$	$\boldsymbol{C(x)}$
28	600	505	0.000	0.00	0.000	0.00	0.00	0.00	1.00	1.00
35	425	357	2.122	3.49	2.122	8.60	0.00	0.00	1.00	1.00
48	300	252	1.678	2.76	1.678	6.80	0.00	0.00	1.00	1.00
65	212	178	3.028	4.99	2.812	11.40	0.22	0.60	0.93	0.91
100	150	126	5.904	9.72	3.923	15.90	1.98	5.49	0.66	0.58
150	106	89	10.266	16.91	4.860	19.70	5.41	14.99	0.47	0.34
200	75	63	12.414	20.44	3.627	14.70	8.79	24.37	0.29	0.11
270	53	45	8.752	14.41	2.270	9.20	6.48	17.98	0.26	0.07
-270	27	14	16.560	27.27	3.379	13.70	13.18	36.56	0.20	0.00
		Total	60.724	100.00	24.671	100.00	36.05	100.00		
			$w_f =$	48.60	$w_u =$	65.5	a =	0.204	d50 =	113.83

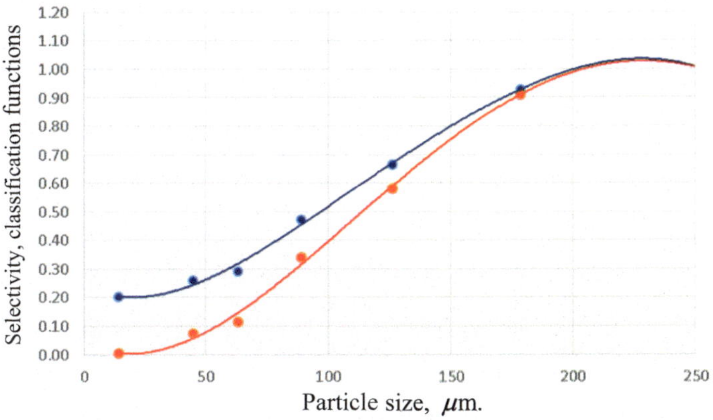

Fig. 2.4 Selectivity and classification functions. Blue is selectivity and red classification

2.3.4 Reduced Classification Function Equations

Defining the reduced classification function $C_R(x/x_{50})$ and plotting it with data from Table 2.3, showing x/d_{50}, x/d_{75} and x/d_{25}, gives Fig. 2.5.

$$d_{50} = 126 - \frac{126 - 89}{0.58 - 0.34} \times (0.58 - 0.5) = 114\,\mu\text{m}$$

$$d_{25} = 89.16 - \frac{89.16 - 63.05}{0.34 - 0.11} \times (0.34 - 0.25) = 78.8\,\mu\text{m}$$

$$d_{75} = 212.00 - \frac{178 - 126}{0.91 - 0.58} \times (0.91 - 0.75) = 153.0\,\mu\text{m}$$

There are several models to describe the reduced classification function. The simplest is the logistic equation of Finney [14] and Molerus (1977):

$$C_R(x) = \frac{1}{1 + (x/x_{50})^{\lambda}} \quad \text{with} \quad \lambda = \frac{2.1972}{\ln(S.I.)} \tag{2.16}$$

where $S.I. = d_{25}/d_{75}$ is the Selectivity Index. For example, in Table 2.3 and Fig. 2.5 we have: $SI = 78/153 = 0.52$ and $\lambda = -3.36$,

$$C(x) = \frac{1}{1 + (x/113.83)^{-3.36}}$$

Table 2.3 Balance selectivity, reduced classification and reduced size x/d_{50}

Size x			Feed		Underflow		Underflow		Selectivity	Clasfication	Reduced class
mesh #	μ m	Average μ m	t/h	f_A	t/h	f_U	t/h	f_O	s	c_i	x/d50
20	850	714	0.000	0.00	0.000	0.00	0.00	100.00	1.000	1.00	4.44
28	600	505	2.122	3.49	2.122	8.60	8.60	91.40	1.000	1.00	3.14
35	425	357	1.678	2.76	1.678	6.80	15.40	84.60	1.000	1.00	2.22
48	300	252	3.028	4.99	2.812	11. 40	26.80	73.20	0.929	0.91	1.57
65	212	178	5.904	9.72	3.923	15.90	42.70	57.30	0.664	0.58	1.11
100	150	126	10.266	16.91	4.860	19.70	62.40	37.60	0.473	0.34	0.78
150	106	89	12.414	20.44	3.627	14.70	77.10	22.90	0.292	0.11	0.55
200	75	63	8.752	14.41	2.270	9.20	86.30	13.70	0.259	0.07	0.39
270	53	45	16.560	27.27	3.379	13.70	100.00	0.00	0.204	0.00	0.17
-270	38	19	0.000	0.00	0.00	0.00	0.00	0.00	0.00	0.00	0.00
Total			60.724	100.00	24.671	100.00	419.31		a =	0.20	
			w_a =	48.6	w_d =	65.5			d50 =	113.83	

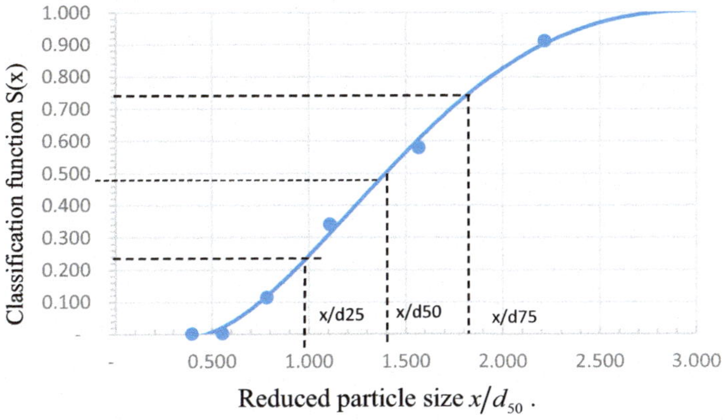

Fig. 2.5 Classification function

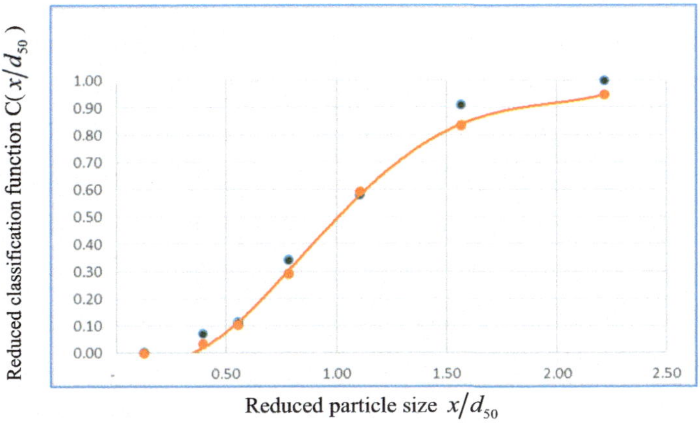

Fig. 2.6 Experimental data (blue) together with simulated reduced classification function (brown)

2.3.5 *Fish-Hook*

In some cases, very fine particles increase in the underflow beyond the degree that would be expected from classification. This effect, occurring mainly in mini-hydrocyclones, is known as the *fish-hook*. Figure 2.4 shows a selectivity function with a fish-hook [35] (Fig. 2.7).

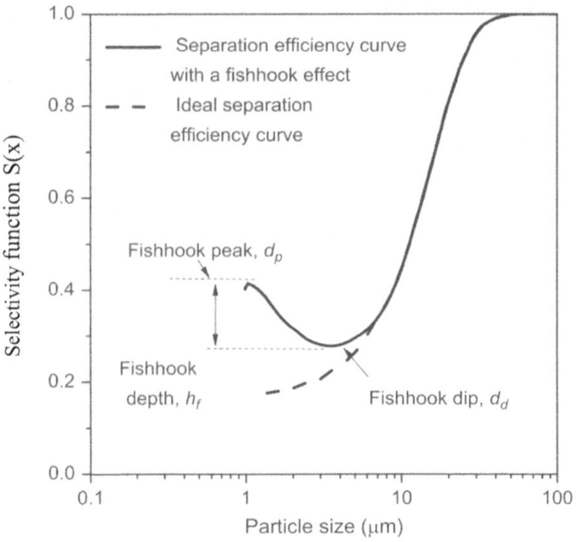

Fig. 2.7 Selectivity function with a fish-hook from [34, 35]

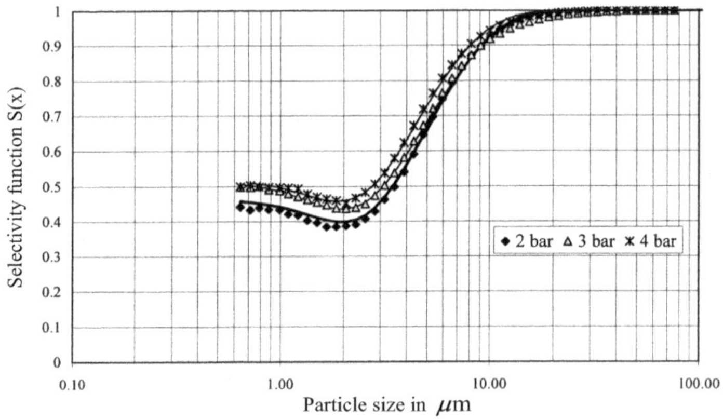

Fig. 2.8 Selectivity function for 2, 3 and 4 bar feed pressure. Solid lines indicate the model fit

According to Zhu and Liow [34, 35], several reasons for the existence of the fish-hook have been given:

1. Inaccuracies in the size analysis, including poor particle size resolution and inadequate resolution of particles sizes below 38 μm [13, 22].
2. Different densities of the particles in the feed, particularly with complex ores and dense media effects [13, 29].
3. Agglomeration of fines to coarse solids [13] and flocculation of the feed material [15].

4. Entrainment of fine material in different flow regimes present in the hydrocyclone [11–13, 18, 30].
5. Filtering effect of coarse particles on fines near the boundary layer [4, 13].
6. Shear-enhanced diffusion and inertial migration [16].
7. Hydrodynamic interaction between particles of different sizes, including changing the settling velocity and the capture of fines in the wakes of larger particles [20].

Discarding most of these possible causes for the appearance of a fish hook, Zhu and Liow [34, 35] assumed that the most probable reason for its formation is the entrainment of fine material in the wake of coarser particles in the different flow regimes present in the hydrocyclone. They did experiments to study the effects of inlet velocity, feed particle size distribution, particle shape and particle volume concentration and found that the principal factor was inlet velocity, which causes a fish hook to appear in the classification function. When inlet velocity is increased, smaller particles are captured in the wakes of the larger particles, an effect that increases as the fraction of large particles increases.

Franchon and Cilliers [12] developed a model for the selectivity function, including dispersion of fines and the presence of fish-hooks, as shown in Fig. 2.8.

Bourgeois and Majumder [6] investigated the possibility that the fish-hook effect could be error related. Using several data sets generated using a 100 mm diameter hydrocyclone operating under controlled dilute to dense regimes and following a statistical methodology, careful analysis of the partition functions yielded unambiguous evidence that the fish-hook effect is a real physical phenomenon.

It was Nesse et al. [24, 25] who demonstrated theoretically that the entrapment of small particles in the wake of larger particles is due to the turbulent diffusion of particles smaller than a threshold size, 6 μm in their work. They presented an equation and a figure showing the dimensionless turbulent diffusivity D_{tp}/v_t of the particle in terms of a dimensional size d_p/d^*

$$D_{tp} = v_t \frac{1}{1 + 10^4 \left(\frac{\varepsilon_t}{k_t}\right)^2 \frac{\rho_p^2 d^4}{\eta_\ell^2}}, \quad d^* = \left(\frac{k_t}{\varepsilon_t}\frac{\rho_\ell}{\rho_p}\right)^{0.5} \tag{2.17}$$

where D_{tp} is the coefficient of turbulent diffusion, v_ℓ is the kinematic viscosity of the fluid, k_t is the turbulent intensity, ε_t is the turbulent dissipation energy, ρ_p and ρ_ℓ are the particle and fluid density, respectively, d_p is the particle diameter and η_ℓ is the liquid dynamic viscosity. Figure 2.9 shows that particles smaller than size $0.3\,d_p/d^*$ follow the turbulent eddies transporting the corresponding particle size. Introducing turbulent diffusion to the balance equations in a 40 mm hydrocyclone, the authors showed that the classification functions present fish hooks at sizes smaller than 6 μm and different inlet volume concentrations, as shown in Fig. 2.10.

Chih et al. [9], using CFD with RSM and the Discrete Phase Model (DPM), showed the presence of fish-hooks at several inlet pressures through simulation and

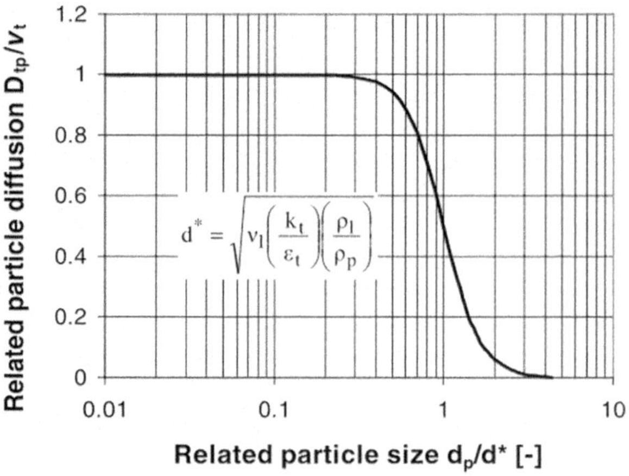

Fig. 2.9 Particle-size-dependent coefficient of turbulent diffusion (coffusivity) D_{tp} related to Eddy viscosity υ_t [24, 25]

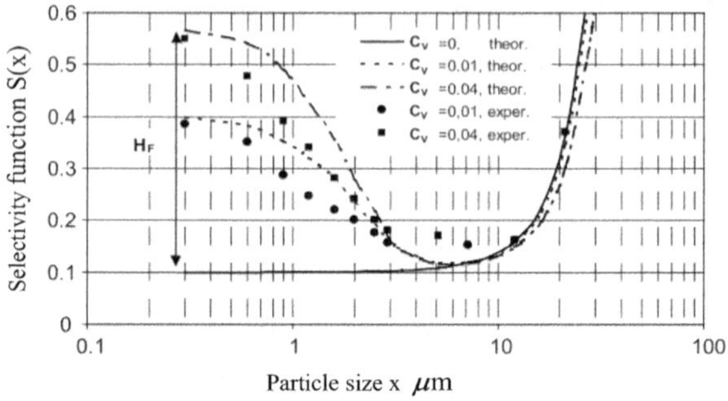

Fig. 2.10 Classification function with a fish-hook for ground quartz of less than 50 μm at different solids concentrations [24, 25]

experiments. In experiments the fish hook starts at 10 μm, while the simulation gives a value of 3 μm, with a greatly exaggerated minimum and maximum (Fig. 2.11).

Aminalroaya and Farzannegan [1] state that the fishhook is a more general effect in particle classification. They also observed a fishhook in screening studies. They suggest that the cause of this phenomenon is the attachment of very fine particles to coarse particles and the inability of screens to isolate these particles and thus transport them to the oversize flow stream. This observation verifies that it is the attachment of small particles to coarser particles that causes the fishhook in the classification function, whatever its classification mechanism.

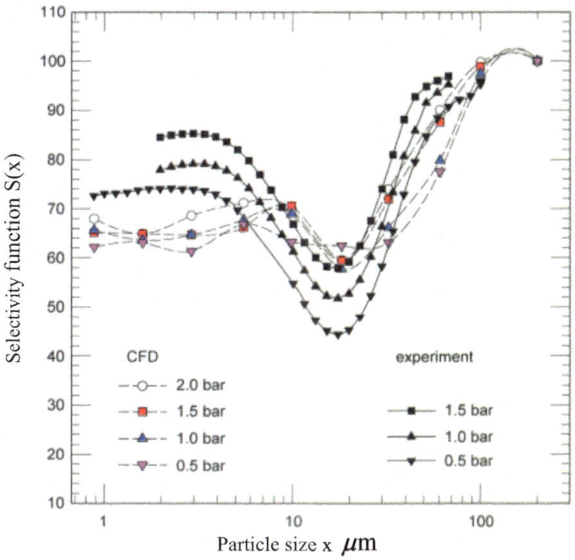

Fig. 2.11 Selectivity functions for inlet pressure values in a 30 mm hydrocyclone with a very high short circuit [9]

2.4 Operating Variables

Among the variables that affect the performance of hydrocyclone input and output variables are disturbances and control variables.

2.4.1 Input Variables

The main input variables of a hydrocyclone are feed volume flow Q_F, concentration φ_F viscosity η_F, pressure Δp, particle size x and size distribution $f_F(x)$ (see Fig. 2.6). A suspension of solid particles has non-Newtonian behavior; that is, the relationship between the shear rate and the shear stress is not linear. This fact is important in a hydrocyclone, especially when the feed is highly concentrated.

Inlet flow rate and pressure

Several researchers such as [5, 21] and others indicated that an increase in the feed rate, with all other variables held constant, produces a decrease in the separation size (d_{50}). In fact, this effect is due to the inlet pressure, since the feed volume flow is proportional to the pressure $Q_F \propto \Delta p$. It has been shown that d_{50} decreases with increasing Δp. For example, Trawinski [23] and Pires and Massarani [27] suggest $d_{50} \propto \Delta p^{-0.52}$, Arterburn [2] and (2002) give $d_{50} \propto \Delta p^{-0.28}$ and Olson and Turner [26] estimate $d_{50} = 3.27\Delta p^{-0.28}$, with Δp expressed in kPa. If the feed flow

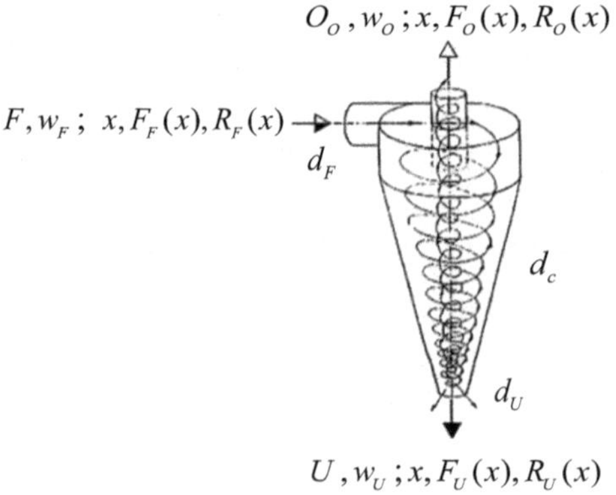

Fig. 2.12 Operating variables in a hydrocyclone

increases, to maintain the particle size in the overflow at a given constant pressure, it is necessary to decrease the flow rate in the hydrocyclones by increasing their number proportionally (Fig. 2.12).

Solid density and fluid density

The solid and fluid dnsity difference $\Delta\rho = \rho_s - \rho_f$, where ρ_s and ρ_f are the densities of the solid and the fluid, respectively, has a remarkable effect on the hydrocyclone cut size. As this difference increases, the centrifugal force on the solid particles increases, sending materials with smaller particle sizes to the underflow. For a given material [2] suggests that d_{50} decreases as $d_{50} \approx \Delta\rho^{-0.5}$.

Concentration

The concentration of the feed is considered the most important variable in a grinding classification circuit. Researchers agree that the cut size (d_{50}) in the hydrocyclone increases with increasing feed concentration. For example, [2] propose that $d_{50} \propto (1 - 1.9\varphi)^{-1.43}$, while [27] indicate a far lower influence at high concentrations, $d_{50} \propto \exp(4\varphi)$, and [21] and [28] also give exponential relationships.

According to these findings and, as the normal feed concentration of hydrocyclones in a closed circuit grinding ranges between $0.1 < \varphi < 0.25$, d_{50} can vary by about 180% with the suspension concentration. It is precisely this property that is used to control grinding-classification circuits.

The high-concentration effect of the pulp in a comminution circuit is related to the viscosity of the suspension [8, 17, 23, 31, 32].

Rheology

Bustamante et al. [7] indicate that the effect of viscosity on classification in a hydrocyclone is only part of a larger and complex dynamic effect inside it, which is present in the treatment of fine china clay minerals. They did experiments with kaolin with sizes less than 44 microns in a 100 mm diameter hydrocyclone. Using elements of rheology, they developed a phenomenological model of the effect of rheology on the cut size (d_{50}):

$$d_{50} = \frac{1}{\tau_t}\left[\frac{\eta}{\lambda(\varphi(\Delta\rho + 1)) + 1}\tau_r\right]^{0.5} \tag{2.18}$$

where τ_t and τ_r are the shear stresses in radial and tangential directions, respectively, η is the shear viscosity of the suspension, φ is the suspension volume fraction, $\Delta\rho$ is the solid liquid density difference and λ a particle shape factor.

2.4.2 *Output Variables*

Among the output variables, the particle size in the overflow $f_O(x)$ and the underflow solid volume fraction are of particular interest. The particle size distribution $f_O(x)$ depends on the classification curve $C(x)$, d_{50} and the short circuit "*a*". There is an interrelationship between $f_O(x)$ and φ_U, as the water content in the underflow influences the short circuit "*a*".

2.5 Balance Adjustment

Determination of the particle size distribution in the various stages of a mineral processing plant is a complicated procedure due to errors committed in sampling and the particle size analysis. Therefore, mass balances in grinding mills and hydrocyclones never provide exact values. To attain the best estimate of these properties, mass balance adjustments are necessary. In this section, we will present one of the simplest methods of mass balance adjustment [10]. Consider a mass balance in a hydrocyclone (Fig. 2.13).

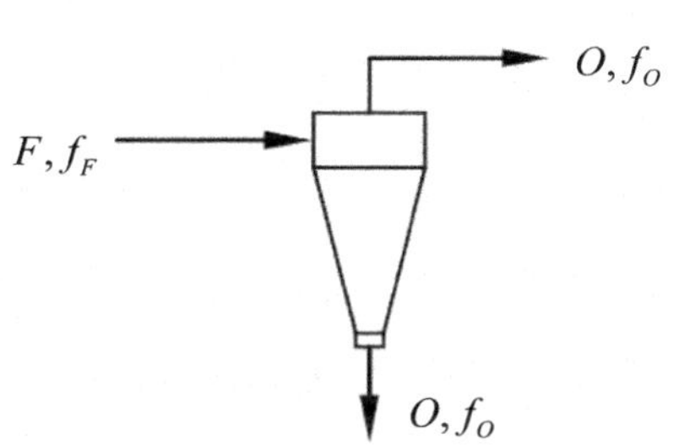

Fig. 2.13 Hydrocyclone showing feed, overflow and underflow parameters

Consider a mass balance of the component x of the size distribution $f(x)$ around the hydrocyclone. Writing f_Z for $f_Z(x)$, the mass balance yields, with $Z = F,\ U$ and O yields

$$\begin{aligned} F f_F &= U f_U + O f_O \\ F &= U + O \\ \frac{U}{O} &= \frac{f_F - f_O}{f_U - f_F} \end{aligned} \tag{2.19}$$

Defining the circulating ratio as $C = U/O$,

$$C = \frac{f_F - f_O}{f_U - f_F} \tag{2.20}$$

where $F,\ U$ and O are the solid flow rate in the feed, underflow and overflow. The circulation ratio is obviously the same for each particle size. When applying this mass balance to a sample taken from a concentrator, there are always sampling and size analysis errors that do not allow the balance to be closed correctly. The problem is that errors are usually different for the different particle sizes. This discrepancy is evidenced by different values of C for the different sizes. Therefore, we need a procedure to correct the values of the frequency function $f_Z(x)$ for each particle size so that all values of C are the same for all particle sizes. This procedure is called *balance adjustment* and consists of minimizing the sum of the squares of the closure errors.

If all the data were correct, Eq. (2.20) could be written in the form

$$(f_F(x_i) - f_O(x_i)) - C(f_U(x_i) - f_F(x_i)) = 0;\quad \text{for}\, x_1 < x_i < x_n$$

But consider the error e_i for each particle size in the sample,

$$(f_F(x_i) - f_O(x_i)) - C(f_U(x_i) - f_F(x_i)) = e_x;\quad \text{for}\, x_1 < x_i < x_n \tag{2.21}$$

and the sum squared errors E, defined by

$$\begin{aligned} E &= \sum_i (e_i)^2 = \sum_i ((f_F(x_i) - f_O(x_i) - C(f_U(x_i) - f_F(x_i)))^2 \\ &= \sum_i ((f_F(x_i) - f_O(x_i))^2 + C^2(f_U(x_i) - f_A(x_i))^2 \\ &\quad -2C(f_F(x_i) - f_O(x_i))(f_U(x_i) - f_F(x_i))) \end{aligned} \tag{2.22}$$

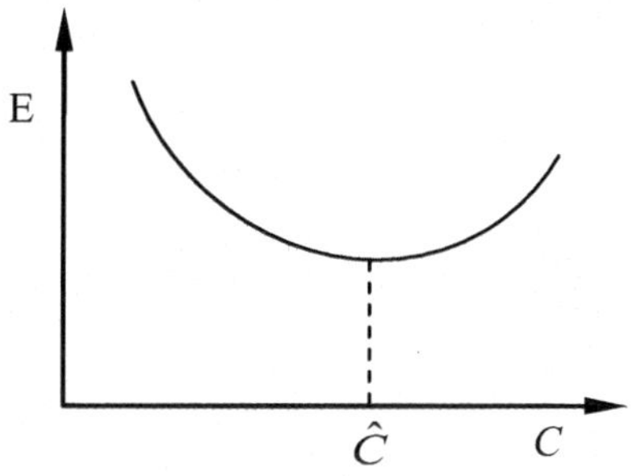

Fig. 2.14 Error E as a function of C

The purpose of the balance adjustment is to find the value $\hat{C}$ of C that minimizes the error E, which will occur for $dE/dC = 0$ (see Fig. 2.14).

To find the minimum of the function E, differentiate E with respect to C and equate it to 0:

$$\begin{aligned}\frac{dE}{dC} &= \frac{d}{dC}\Big((f_F(x) - f_O(x))^2 + \hat{C}^2 \sum\nolimits_i (f_U(x) - f_F)^2 \\ &\quad -2C \sum\nolimits_i (f_F(x) - f_O(x))(f_U(x) - f_F(x))\Big) = 0 \\ &= 2\hat{C} \sum\nolimits_i (f_U(x_i) - f_F(x_i))^2 \\ &\quad - 2\sum\nolimits_i (f_F(x_i) - f_O(x_i))(f_U(x_i) - f_F(x_i)) = 0\end{aligned} \tag{2.23}$$

Therefore

$$\hat{C} = \frac{\sum_i (f_F(x_i) - f_O(x_i))(f_U(x_i) - f_F(x_i))}{\sum_i (f_U(x_i) - f_F(x_i))^2} \tag{2.24}$$

where $\hat{C}$ is the best estimate of the circulation ratio. From (2.21) the error in the particle size x_i is:

$$r_i = (f_F(x_i) - f_O(x_i)) - \hat{C}(f_U(x_i) - f_F(x_i)); \text{ for } i = 1, \ldots, n \tag{2.25}$$

If we add an adjustment Δ_Z to each particle size to get a better estimate, we can write:

$$\begin{aligned}r_i &= (f_F(x_i) - \Delta_F) - (f_O(x_i) - \Delta_O) \\ &\quad - \hat{C}((f_U(x_i) - \Delta_U) - (f_F(x_i) - \Delta_F)) = 0; \text{ for } 1 < z < n\end{aligned} \tag{2.26}$$

Subtracting (2.25) and (2.26) gives the constraint for the particle size in the feed, underflow and overflow:

$$r_i = (\Delta_A - \Delta_R) - \hat{C}(\Delta_D - \Delta_A)\,; \ \text{for}\, i = 1, \ldots, n \tag{2.27}$$

The problem is to find the best values of Δ_Z, so we define the function S as the sum of the squared errors of Δ_Z

$$S_i = \sum\nolimits_i \left(\Delta_F^2 + \Delta_U^2 + \Delta_O^2\right), \ \text{for}\, i = 1, \ldots, n \tag{2.28}$$

To minimize S_i, subjected to the constraint (2.25), we use the method of Lagrangian multipliers, defining the Lagrangian function L:

$$L = S_i + 2\sum_{k=1}^{n} \lambda_k r_i\,, \ \text{for}\, i = 1, \ldots, n \tag{2.29}$$

Replacing (2.27) and (2.28) yields

$$\begin{aligned} L &= \sum\nolimits_i \left(\Delta_F^2 + \Delta_U^2 + \Delta_O^2\right) \\ &\quad + 2\sum_{i=1}^{n} \lambda_i \Big(r_i - (\Delta_F - \Delta_O) + \hat{C}(\Delta_U - \Delta_F)\Big); \quad \text{for}\, i = 1, \ldots .n \\ &= \sum\nolimits_i \left(\Delta_F^2 + \Delta_U^2 + \Delta_O^2\right) \\ &\quad + 2\sum_{i=1}^{n} \lambda_i \Big(r_i - \Delta_F\Big(1 + \hat{C}\Big) + \Delta_O + \hat{C}\Delta_D\Big); \quad \text{for}\ i = 1, \ldots .n \end{aligned} \tag{2.30}$$

Differentiating L with respect to each Δ_F, Δ_U, Δ_O and λ_i gives

$$\frac{\partial L}{\partial \Delta_F} = 2\Delta_F - 2\lambda_i\Big(1 + \hat{C}\Big) = 0; \ \Rightarrow \ \Delta_F = \lambda_i\Big(1 + \hat{C}\Big), \ \text{for}\ i = 1, \ldots, n \tag{2.31}$$

$$\frac{\partial L}{\partial \Delta_U} = 2\Delta_U + 2\lambda_i\hat{C} = 0; \quad \Rightarrow \Delta_D = -\lambda_i\hat{C} \ \ \text{for}\ i = 1, \ldots, n$$

$$\frac{\partial L}{\partial \Delta_R} = 2\Delta_R + 2\lambda_i = 0; \quad \Rightarrow \Delta_R = -\lambda_i \ \ \text{for}\, i = 1, \ldots, n \tag{2.32}$$

$$\frac{\partial L}{\partial \lambda_i} = 2\left(r_i - \Delta_F\left(1+\hat{C}\right) + \Delta_O + \hat{C}\Delta_U\right) = 0\,,\ \text{for}\, i = 1, \ldots, n \tag{2.33}$$

and solving for r_i

$$r_i = \Delta_F\left(1+\hat{C}\right) - \Delta_O - \hat{C}\Delta_U\,, \qquad \text{for}\, i = 1, \ldots, n \tag{2.34}$$

Substituting the values of $\Delta_{(\cdot)}$ from (2.31) to (2.33) in r_i yields

$$\begin{aligned} r_i &= \lambda_i\left(1+\hat{C}\right)^2 + \lambda_i + \lambda_i\hat{C}^2, \quad \text{for}\, i = 1, \ldots, n \\ &= \lambda_i\left(\left(1+\hat{C}\right)^2 + \left(1+\hat{C}^2\right)\right), \quad \text{for}\, i = 1, \ldots, n \end{aligned} \tag{2.35}$$

Now, solving for λ_i

$$\lambda_i = \frac{r_i}{\left(1+\hat{C}\right)^2 + \left(1+\hat{C}^2\right)}, \qquad \text{for}\, i = 1, \ldots, n \tag{2.36}$$

and replacing (2.36) in expressions (2.31) to (2.33) we obtain the best values for the particle size corrections:

$$\Delta_F = \frac{2\left(1+\hat{C}\right)}{2\left(1+\hat{C}\right)^2 + \left(1+\hat{C}^2\right)} r_i, \quad \text{for}\, i = 1, \ldots, n \tag{2.37}$$

$$\Delta_U = -\frac{\hat{C}}{2\left(1+\hat{C}\right)^2 + \left(1+\hat{C}^2\right)} r_i, \quad \text{for}\, i = 1, \ldots, n \tag{2.38}$$

$$\Delta_O = -\frac{1}{2\left(1+\hat{C}\right)^2 + \left(1+\hat{C}^2\right)} r_i, \quad \text{for}\, i = 1, \ldots, n \tag{2.39}$$

Example The performance of a grinding-classification circuit in a copper concentrator is to be evaluated. Samples of the feed, underflow and overflow of a hydrocyclone were screened, yielding the particle size analysis in the following table. Since errors in sampling and in particle sizing were expected, the circulating ratios C and $\hat{C}$ were calculated.

Size	Feed	Underflow	Overflow	C	C	
Mesh	f_A	f_D	f_R	$C = \frac{D}{R} = \frac{f_A - f_R}{f_D - f_A}$	$\sum_i (f_A - f_R)(f_D - f_A)$	$\sum_i (f_D - f_A)^2$
8	0.32	0.34	0.00	16.00	0.01	0.00
10	1.09	1.20	0.00	9.91	0.12	0.01
14	3.00	3.11	0.00	27.27	0.33	0.01
20	3.93	5.57	0.00	2.40	6.45	2.69
28	5.74	7.55	0.00	3.17	10.39	3.28
35	9.31	12.51	0.00	2.91	29.79	10.24
48	14.46	18.39	0.93	3.44	53.17	15.44
65	19.09	21.66	6.22	5.01	33.08	6.60
100	12.08	11.93	12.85	5.13	0.12	0.02
150	8.01	6.39	13.54	3.41	8.96	2.62
200	4.94	3.19	10.20	3.01	9.21	3.06
270	3.00	1.95	8.90	5.62	6.20	1.10
400	2.67	1.37	7.89	4.02	6.79	1.69
Bottom	12.36	4.84	39.30	3.58	202.59	56.55
Total	100	100	100	6.78	367.18	103.33

$$\hat{C} = \frac{\sum_i (f_A - f_R)(f_D - f_A)}{\sum_i (f_D - f_A)^2} = 3.55 \tag{2.40}$$

From the table we see that C varies for all particle sizes and that the average would be 6.78. Calculating the corrected frequency functions, the adjusted value is $\hat{C} = 3.55$. See the following table.

r_i	Feed	Underflow	Overflow	Feed	Underflow	Overflow	$\hat{C}$
$r_i = (f_A - f_R) - \hat{C}(f_D - f_A)$	Δ_F	Δ_U	Δ_O	f_F	f_u	f_o	$r_i = (f_A - f_R) - \hat{C}(f_D - f_A)$
0.25	0.0329885	−0.0257437	−0.0072448	0.3	0.4	0.0	3.55
0.70	0.0926484	−0.0723013	−0.0203471	1.0	1.3	0.0	3.55
2.61	0.3457621	−0.2698270	−0.0759350	2.7	3.4	0.1	3.55
−1.90	−0.2514656	0.1962396	0.0552260	4.2	5.4	−0.1	3.55
−0.69	−0.0916564	0.0715271	0.0201293	5.8	7.5	0.0	3.55
−2.06	−0.2731052	0.2131268	0.0599784	9.6	12.3	−0.1	3.55
−0.43	−0.0576242	0.0449690	0.0126552	14.5	18.3	0.9	3.55
3.74	0.4953315	−0.3865486	−0.1087830	18.6	22.0	6.3	3.55
−0.24	−0.0314061	0.0245088	0.0068973	12.1	11.9	12.8	3.55
0.23	0.0300153	−0.0234235	−0.0065919	8.0	6.4	13.5	3.55
0.96	0.1270123	−0.0991183	−0.0278940	4.8	3.3	10.2	3.55
−2.17	−0.2874281	0.2243041	0.0631239	3.3	1.7	8.8	3.55
−0.60	−0.0795902	0.0621109	0.0174793	2.7	1.3	7.9	3.55
−0.22	−0.0289539	0.0225951	0.0063588	12.4	4.8	39.3	3.55
0.17	0.0225284	−0.0175808	−0.0049476	100.0	100.0	100.0	3.55

References

1. Aminalroaya A, Farzanegan A (2015) The "fish-hook" phenomenon and using separation sharpness parameter in investigation of vibration screen performwance. J Mater Environ Sci 6(7)
2. Arterburn RA (1982) Design and installation of comminution circuits, Chapter 32 (Mular A, Jergensen G, eds). SME-AIME, New York, NY, pp 592–607
3. Austin L, Concha F (1994) Diseño y Simulación de Circuitos de Molienda Clasificación. CYTED, Cettem. Ltd.
4. Austin LG, Klimpel RR (1981) An improved method for analyzing classifier data. Powder Technol 29:277–281
5. Bradley D (1965) The hydrocyclone. International Series of Monographs in Chemical Engineering, vol 4 (Dankwerts PV, ed). Pergamon Press, London
6. Bourgeois F, Majumder AK (2013) Is the fish-hook effect in hydrocyclones a real phenomenon? Powder Technol 237:367–375
7. Bustamante O, Gomez HA, Chica LM (2010) Rheology slurry effects over the classification in small diameter hydrocyclones. In: XXV international mineral processing congress
8. Castro O (1990) An investigation of pulp rheology effects and their application to the dimensionless type hydrocyclone models. M. Eng.Sc. Thesis, JKMRC, University of Queensland
9. Chih YH, Wu S-J, Wu R-M (2011) Particles separation and tracks in a hydrocyclone. Tamkang J Sci Eng 14(1):65–70
10. Concha F (2011) Mass balance adjustment, Classification Presentation, Grinding-Classification course. Department of Metallurgical Engineering, University of Concepción
11. Del Villar R, Fich JA (1992) Modelling the cyclone performance with a size dependent entrinment factor. Miner Eng 5:661–669
12. Frachon M, Cilliers J (1999) A general model for hydrocyclone partition curves. Chem Eng J 73:53–59
13. Finch JA (1983) Modelling a fishhook in hydrocyclone selectivity curves. Powder Technol 36:127–129
14. Finney DJ (1964) Statistical method in biological assay. 2. Aufl. Charles Griffin & Cc. Ltd., London
15. Gerhard C (2001) Untrersuchungen zum trennverhalten von hydrozyclones niedriger trennkorngrösse. Theis,Universität Erlangen-Nürenberg
16. Hwang JK, Hsueh W, Nagase Y (2008) Mechanism of particle separation in small hydrocyclone. Dry Technology 26:1002–1010
17. Kawatra, SK (1996) On-line measurement of viscosity and determination of flow types for mineral suspensions. Int J Miner Process 47:3–4
18. Kelsall D, Holmes J (1960) Improvement in classification efficiency in hydraulic cyclones by water injection. In: Proceedings of the fifth mineral processing, congress IMM, London, pp 159–170
19. Klimpel RR (1982) The influence of dispersant on the sizing performance of a 24 in hydrocyclone. Powder Technol 31:255–262
20. Kraipech W, Chen W, Parma FJ, Dyakowski T (2002) Modelling the fish-hook effect of the flow within hydrocyclones. Int J Miner Process 66:50
21. Lynch AJ (1977) mineral crushing and grinding circuits, Chapter 5 and 6. Elsevier Scientific Publishing Co., New York, NY, pp 87–126
22. Nageswararao K (2000) A critical analysis of the fish-hook effect in hydrocyclone classifiers. Chem Eng J 80:251–256
23. Napier-Munn TJ (1990) The effect of dense medium viscosity on separation efficiency. Coal Preparation, 8:3–4
24. Neesse Th, Schneider M, Dueck J, Golyk V, Buntenbach S, Tiefel H (2004) Hydrocyclone operation at the transition point rope/spray discharge. Miner Eng 17:733–737
25. Neesse Th, Schneider M, Golyk V, Tiefel H (2004) Measuring the operating state of the hydrocyclone. Miner Eng 17:697–703

26. Olson TJ, Turner PA (2002) Hydrocyclone selection for plant design. In: Mular A, Halbe DN, Barratt DJ (eds) Mineral processing plant design, practice and control proceedings, vol 1. SME, p 884
27. Pires RP, Massarani G (1980) Revista Brasilera de Tecnología 11:289–299
28. Plitt L (1976) A mathematical model of the hydrocyclone classifier. CIMM Bull 69
29. Rouse BD, Clayton JS, Brooks GF (1978) Confirmation of modelling techniques for small diameter hydrocyclones. In: Wood P (ed) Proceedings of the third international conference on hydrocyclones, Elsevier, BHRA, Cranfield, pp 7–19
30. Roldan-Villasana EJ, Williams RA, Dyakowski T (1993) The origin of the fish-hook effect in hydrocyclones separators. Powder Technol 77:243–250
31. Shi X, BMiller JN (2002) Effects of food gums on viscosities of starch suspensions during pasting. Carbohydr Polym 51:7–8
32. Tavares LM, Souza LLG, Lima JRB, Possa MV (2002) Modeling classification in small-diameter hydrocyclones under variable rheological conditions. Miner Eng 15(8):613–622
33. Trawinski H (1969) Filtration and separation, pp 359–373
34. Zhu G, Liow JL (2014) Experimental study of particle separation and the fish-hook effect in a mini-hydrocyclone. Chem Eng Sci 111:95
35. Zhu G, Liow J (2014) Experimental study of particle separation and the fish-hook effect in a mini-hydrocyclone. Chem Eng Sci 111:94–105

Chapter 3
Sedimentation of Particulate System

Abstract Sedimentation as the mechanism of the hydrocyclone classification is the subject this Chapter. The settling velocity of a sphere is analyzed in Stokes, Euler and Prandtl's regimes leading to the development of a unified drag coefficient and an explicit equation for the settling velocity. The sedimentation of a suspensions of spheres and particles of arbitrary shape follows.

Sedimentation is the settlement of a particle, or a suspension of particles, in a fluid by the effect of an external force, which can be gravity, a centrifugal force or any other body force. For many years engineers and researchers in the field of particle technology have been looking for a simple equation to relate the sedimentation rate of particulate suspensions in a fluid to their size, shape and concentration as parameters [1, 2, 4, 5, 11, 12–14]. These works first solve the sedimentation of a particle in a fluid and then introduce corrections due to interactions between particles, through which the sedimentation of a suspension is dramatically diminished. The starting point is Navier–Stokes equations [7, 9].

$$
\begin{aligned}
&\nabla \cdot \mathbf{v} = 0 \\
&\rho\left(\frac{\partial \mathbf{v}}{\partial t} + \underbrace{\nabla \mathbf{v} \cdot \mathbf{v}}_{\text{Convective forces}}\right) = -\nabla p + \underbrace{\mu \nabla^2 \mathbf{v}}_{\text{Diffusive force}} + \rho \mathbf{g}
\end{aligned}
\tag{3.1}
$$

where $\mathbf{v}$ is the fluid velocity, p and $\mathbf{g}$ are the pressure and gravity force and ρ and μ are the fluid density and viscosity.

Depending on the relative values of the *convective* and *diffusive* forces, measured by the Reynolds number ($\text{Re} = \text{convective}/\text{diffusive}$), we can distinguish three types of flows: Stokes flow when $\text{Re} \to 0$, Euler's flow when $\text{Re} \to \infty$ and Prandtl's flow in the intermediate cases. The main problem is to determine the hydrodynamic force of the fluid acting on particles moving in that fluid.

F. Concha A. and J. L. Bouso A., *Fluid Mechanics Fundamentals of Hydrocyclones and its Applications in the Mining Industry*, Fluid Mechanics and Its Applications 126,
https://doi.org/10.1007/978-3-030-67913-2_3

3.1 Hydrodynamic Force on a Sphere in Stokes Flow

$$\begin{aligned} &\nabla \cdot \mathbf{v} = 0 \\ &\rho \frac{\partial \mathbf{v}}{\partial t} = -\nabla p + \mu \nabla^2 \mathbf{v} + \rho \mathbf{g} \end{aligned} \tag{3.2}$$

Due to the linearity of the differential equation in Stokes flow, the velocity, pressure and hydrodynamic force at any point of space in a steady flow are linear functions of the relative solid–fluid velocity. For the flow past a sphere [11].

$$\begin{aligned} &v_r(R) = 0 \\ &v_\theta(R) = 6\pi R(u - u_0) \end{aligned} \tag{3.3}$$

$$p(r) = \mu P(u - u_0) \tag{3.4}$$

The hydrostatic force, buoyancy, is

$$F_b = -\rho \frac{4}{3} \pi R^3 g$$

For the hydrodynamic force $F_D(r)$, the linear function depends on the size and shape of the particle and on the fluid viscosity. Solving the boundary value problem and neglecting the Basset term of added mass yields the hydrodynamic force.

$$F_D(r) = -6\pi \mu R(u(r) - u_0) \tag{3.5}$$

where, μ and u_0 are the fluid viscosity and velocity and u and R are the spherical particle velocity and diameter. In dimensionless form the hydrodynamic force, known as drag coefficient C_D, is

$$\mathrm{C}_D = \frac{\mathrm{F}_D}{\left(1/2\rho_f \mathrm{u}^2\right)\left(\pi R^2\right)} \tag{3.6}$$

Substituting (3.5) in (3.6), the drag coefficient of the sphere in Stokes flow is obtained:

$$\mathrm{C}_D = \frac{24}{\mathrm{Re}} \tag{3.7}$$

Sedimentation Dynamics

To calculate the settling velocity of a spherical particle under the effect of gravity in a quiescent fluid, we perform a balance of forces:

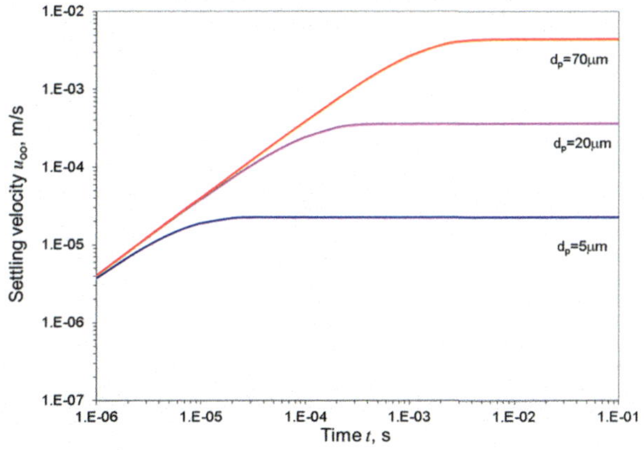

Fig. 3.1 Transient settling velocity of spheres in a quiescent fluid

$$\underbrace{\rho_p V_p \dot{u}(t)}_{\text{Acceleration}} = \underbrace{\Delta\rho V_p g}_{\text{Net weight}} - \underbrace{6\pi\mu R u(t)}_{\text{Drag force}} \underbrace{-(1/2)\rho_p V_p \dot{u}(t)}_{\text{Induced mass force}} \tag{3.8}$$

The induced mass force will be explained later in this text.

$$\dot{u}(\text{t}) + \frac{2}{3}\frac{18\mu}{\rho_p d^2}\text{u(t)} - \frac{2\Delta\rho}{3\rho_p}g = 0 \tag{3.9}$$

The solution of which is

$$u(t) = \frac{1}{18}\frac{\Delta\rho d^2 g}{\mu}\left(1 - \exp\left(-\frac{2}{3}\frac{\mu}{18\rho_p d^2}t\right)\right) \tag{3.10}$$

Example The transient settling velocity of quartz spherical particles with a density of 2.65 g/cm^3 and diameters of 5 μm, 20 μm and 70 μm, in water with a density of 1 g/cm^3 and viscosity of 1 centipoise, is calculated with Eq. (3.10), giving the result shown in Fig. 3.1.

The terminal settling velocities were: 2.25×10^{-5} m/s for the 5 μm sphere; 3.6010^{-4} m/s for the 20 μm sphere and 4.41×10^{-3} m/s for the 70 μm sphere. These terminal velocities correspond to an infinite time in Eq. (3.10), where the exponential term is zero. The corresponding Reynolds numbers are Re = 1.12×10^{-4}; 7.140×10^{-3} and 3.08×10^{-1} respectively.

3.2 Hydrodynamic Force on a Sphere in Euler's Flow

$$\nabla \cdot \mathbf{v} = 0$$
$$\rho \frac{\partial \mathbf{v}}{\partial t} + \nabla \mathbf{v} \cdot \mathbf{v} = -\nabla p + \rho \mathbf{g} \tag{3.11}$$

In Euler's equation, the velocity of the sphere at any point in space is also a linear function of the relative solid fluid velocity, but in this case the hydrodynamic force and the pressure are not. Since no viscosity exists, there is no friction, and the hydrodynamic force depends on particle acceleration. Meanwhile, the convective term in Euler's equation $\nabla \mathbf{v} \cdot \mathbf{v} = 1/2\ v^2$ is non-linear, so the pressure is also non-linear. The results are:

Radial velocity

$$v_r(R) = 0 \tag{3.12}$$

Tangential velocity

$$v_\theta(R) = \left(\frac{3}{2} \sin\theta\right) u_0 \tag{3.13}$$

Pressure distribution

$$p(\theta) - p_0 = \frac{1}{2} \rho_f \left(v_\theta^2 - u_0^2\right) \tag{3.14}$$

Hydrostatic force and pressure coefficient

$$F_h = \rho g V_p, \quad \text{and} \quad \mathrm{C_p} = 1 - \frac{9}{2} \sin^2\theta \tag{3.15}$$

Hydrodynamic force for unsteady motion

$$F_h = -\rho(b + \dot{v}_\theta) V_p F_h \tag{3.16}$$

Figure 3.2 represents Eqs. (3.13) and (3.15) graphically.

On the surface of the sphere, the tangential velocity becomes zero at $\theta = 0$, the front stagnation point. Always remaining on the surface, θ increases until reaching a maximum of $v_\theta = (3/2)u_0$, then decreases again to zero at $\theta = \pi$, the rear stagnation point.

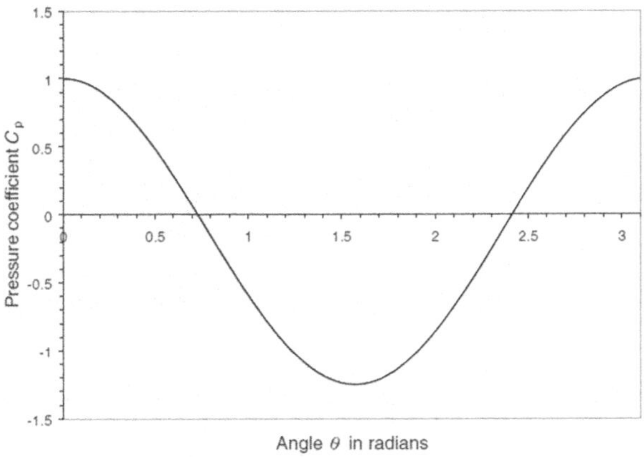

Fig. 3.2 Pressure coefficient in terms of the distance over the sphere and dimensionless tangential velocity on the surface of a sphere in Euler's flow

3.3 Prandtl's Flow

For intermediate values of the Reynolds number, inertial and viscous forces in the fluid are on the same order of magnitude. In this case, the flow may be divided into two parts, an *external inviscid flow* far from the particle and an *internal flow* near the particle, where the viscosity plays an important role. Concha [7, 9]. This picture forms the basis of the *boundary layer theory,* Goldstein [11], Schichting [21], Meksyn [17], Rosenhead [20].

In the external inviscid flow, Euler's equations are applicable, and the velocity and pressure distribution may be obtained from Eqs. (3.13) and (3.14). As the Reynolds number increases from the front stagnation point, the *boundary layer* over the sphere also increases with the angle. Due to the action of viscosity, at the surface of the sphere the tangential velocity is zero and at the outer limit of the boundary layer the tangential velocity reaches 99% of the value in Euler's flow.

For Reynolds numbers exceeding 10,000 the energy dissipation produced by the viscous flow within the boundary layer retards the flow and, at a certain point, the flow reverses its direction and forces the fluid particles outward and away from the solid particle. This phenomena is called *boundary layer separation*, which occurs at an *angle of separation* of $\theta_s = 85.2°$, diminishing to $\theta_s = 84°$ at Re $\approx$ 150,000 and remaining at this value beyond [15] (Fig. 3.3).

The pressure, which is highest at the front of the sphere, diminishes in the boundary layer, preventing its recovery behind the separation point, remaining at a constant value of $C_P = -0.4$, called *base pressure.* [11, 18], resulting in an asymmetrical pressure distribution with a higher pressure at the front than at the rear of the sphere while outside the boundary layer the tangential fluid velocity increases, reaching a Reynolds number of 1.5×10^5. Beyond this value, the boundary layer separates at an angle of $\theta_s = 84°$. The dimensionless thickness of the boundary layer was given by McDonald and Fish [16].

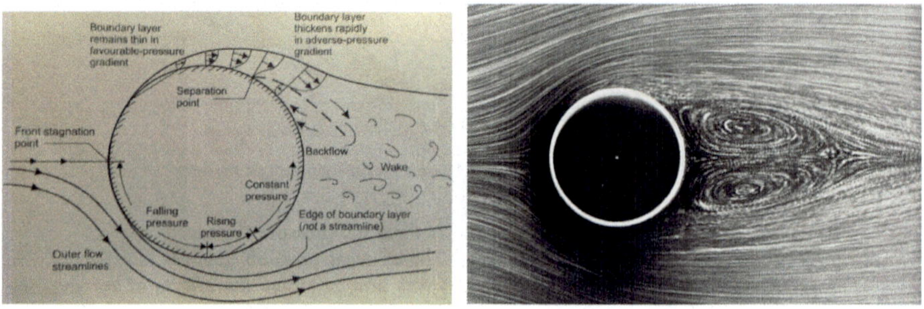

Fig. 3.3 The boundary layer region of viscous flow near a spherical particle

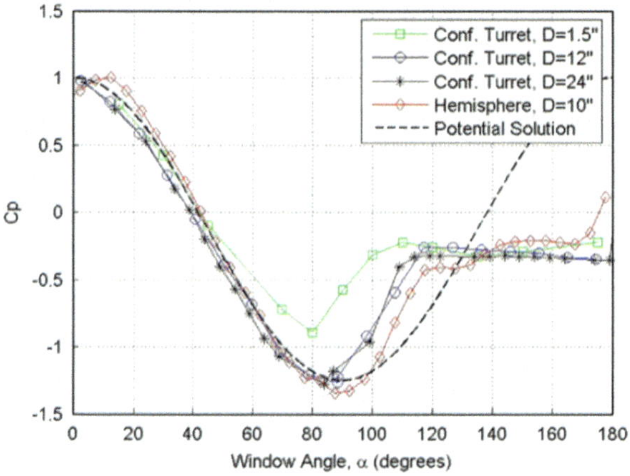

Fig. 3.4 Pressure coefficient as a function of the distance from the front stagnation point over the surface of a sphere in boundary layer flow [12]

$$\frac{\delta}{R} = \frac{9.06}{\mathrm{Re}^{1/2}} \tag{3.17}$$

where δ is the thickness of the boundary layer and R radius of the sphere (Fig. 3.4).

3.4 Sedimentation Velocity of a Sphere in Prandtl's Flow $1 < \mathrm{Re} < 1.5 \times 10^{-5}$

Using the information obtained from Prandtl's flow, we developed the hydrodynamic force and drag coefficient on the sphere with $\theta_s = 85.2$ Concha and Almendra [4].

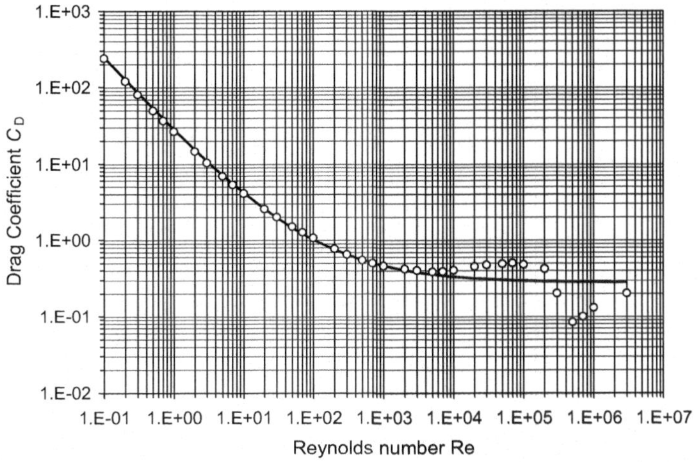

Fig. 3.5 Drag coefficient versus Reynolds number for a sphere. The continuous line is a simulation with Eq. (3.19) and the circles correspond to standard data from Lapple and Shepherd [14]

$$F_D = 2\pi a^2 \int_0^{\pi} p(\theta) \sin\theta \cos\theta d\theta$$

Calculating this integral for the front of the sphere from $\theta = 0$ to the angle of separation $\theta = \theta_s$,

$$F_D = 2\pi a^2 \left(\int_0^{\theta_s} p(\theta) \sin\theta \cos\theta d\theta + \int_{\theta_s}^{\pi} p(\theta) \sin\theta d(\sin\theta) \right) \tag{3.18}$$

The drag coefficient $C_D = F_D / \pi a^2$ results in (Fig. 3.5)

$$C_D = 0.284 \left(1 + \frac{9.06}{\mathrm{Re}_\infty^{1/2}} \right)^2 \quad \mathrm{Re} = \frac{d u \rho_f}{\mu_f} \tag{3.19}$$

The drag coefficient settling velocity of a sphere in a quiescent fluid can be obtained from Eq. (3.19) with good accuracy up to Re = 3000 once the drag coefficient is known for the corresponding Reynolds number in (3.19). Unfortunately, the drag coefficient for a falling sphere and the Reynolds number are both functions of the sphere size and velocity, and therefore for many years the only way to obtain the settling velocity was with an iterative method. Concha and Almendra [4] combined the two dimensionless numbers C_D and Re, yielding two different dimensionless numbers that were each dependent on only one variable, size or velocity,

$$C_D \mathrm{Re}_\infty^2 = \left(\frac{4}{3}\frac{\Delta\rho\rho_f g}{\mu_f^2}\right) d^3 \quad \frac{\mathrm{Re}_\infty}{C_D} = \left(\frac{3}{4}\frac{\rho_f^2}{\Delta\rho\mu_f g}\right) u_\infty^3 \tag{3.20}$$

Define the dimensional parameters

$$P = \left(\frac{3}{4}\frac{\mu_f^2}{\Delta\rho\rho_f g}\right)^{1/3} \quad Q = \left(\frac{4}{3}\frac{\Delta\rho\mu_f g}{\rho_f^2}\right)^{1/3} \tag{3.21}$$

Observe that P is a sole function of size and Q a sole function of velocity, so that Eq. (3.20) become a dimensionless diameter d^* and a dimensionless velocity u^*,

$$C_D \mathrm{Re}^2 = \left(\frac{d}{P}\right)^3 = d^{*3} \quad \frac{\mathrm{Re}}{C_D} = \left(\frac{u}{Q}\right)^3 \equiv u^{*3} \tag{3.22}$$

Using these equations, the following explicit terminal settling equation for a sphere was obtained by Concha and Almendra [4], Concha [8], Concha [7, 9].

$$u^* = \frac{20.52}{d^*}\left(\left(1 + 0.0921 d^{*3/2}\right)^{1/2} - 1\right)^2 \tag{3.23}$$

and the inverse

$$d^* = 0.07 u^{*2}\left(1 + \left(1 + 68.49 u^{*-3/2}\right)^{1/2}\right)^2 \tag{3.24}$$

Figure 3.6 shows the comparison of Eq. (3.23) with experimental data Concha and Almendra [4].

Example The settling velocity of a sphere with a density of 2.50 g/cm^3 and diameter of 1000 μm, in water with a density of 1 g/cm^3 and viscosity 1 cp is calculated in the following table.

Data	
d_p (microns)	1000
d_p (cm)	0.1
ρ_s (g/cm^3)	2.5
ρ_f (g/cm^3)	5
μ_f (poises)	0.01

Results	
P (cm)	0.00370766

(continued)

(continued)

Results	
Q (cm/s)	1.87007597
d* (g/cm3)	26.9711626
U00* (g/cm3)	5.11045596
u (cm/s)	9.56
Re	238.92

3.5 Sedimentation of a Suspension of Spheres

When a suspension of spheres of equal size settle in a viscous fluid the neighboring spheres hinder their settling. This effect is more significant with increasing particle concentration. Concha and Almendra [5, 11] developed an equation for the settling velocities and the drag coefficient that takes this effect into consideration (Fig. 3.7).

$$u^* = f_p(\varphi) f_q(\varphi)\left(\left(1 + 0.0921 \times f_p^{-2/3}(\varphi) \times d^{*3/2}\right)^{1/2} - 1\right)^2 \tag{3.25}$$

where

$$f_p(\varphi) = (1-\varphi)^{-2.033}, \quad f_q(\varphi) = (1-\varphi)^{-0.167} \tag{3.26}$$

$$C_D = 0.28(1-\varphi)^{-2.01}\left(1 + \frac{9.08(1-\varphi)^{-1.83}}{\mathrm{Re}^{1/2}}\right)^2 \tag{3.27}$$

Example The calculation of the drag coefficient, the settling velocity and Reynolds number of particles of $d_p = 300\mu\text{m}$ and solid and fluid density $\rho_p = 2.65\ \text{g/cm}^3$ and $\rho_f = 1.00\ \text{g/cm}^3$, water viscosity $\mu = 0.01$ cp and a suspension of 20% solids (volume fraction of solids).$\varphi = \rho_f \times w/\rho_s(100 - w) + \rho_f w)$ is given below.

Data	Results
d_p (cm) = 0.03	P (cm) = 0.004
ρ_s (g/cm^3) = 2.65	Q (cm/s) = 1.930
ρ_f (g/cm^3) = 1.00	$\varphi = 0.086$
μ_f (poises) = 0.01	$f_p = 1.201$
w (%) = 20.00	$f_q = 1.015$
	d* = 8.353

(continued)

(continued)

Data	Results
	U* (φ) = 1.226
	u (cm/s) = 2.367
	Re = 7.101

3.6 Sedimentation of Particles of Arbitrary Shape

Concha and Christiansen [6] described the drag coefficient and the settling velocity of particles of arbitrary shape by the following equations [6, 7, 9].

$$C_D(\psi, \lambda, \varphi) = \tilde{C}_0(\psi, \lambda, \varphi)\left(1 + \frac{\delta_0(\psi, \lambda, \varphi)}{\boldsymbol{Re}^{1/2}}\right)^2$$

$$u_p^*(\psi, \lambda, \varphi) = \frac{1}{4}\frac{\tilde{\delta}_0^2(\psi, \lambda, \varphi)}{d_e^*}\left(\left(1 + \frac{4}{\tilde{C}_0^{1/2}(\psi, \lambda, \varphi)\tilde{\delta}_0^2(\psi, \lambda, \varphi)} d_e^{*3/2}\right)^{1/2} - 1\right)^2$$

where

$$\tilde{C}_0(\psi, \lambda, \varphi) = C_0 f_A(\psi) f_C(\lambda) f_p(\varphi) f_q^{-2}(\varphi)$$

$$\tilde{\delta}_0(\psi, \lambda, \varphi) = \delta_0 f_B(\psi) f_D(\lambda) f_F(\varphi) f_p^{1/2}(\varphi) f_q^{1/2}(\varphi)$$

In the suspension ψ, λ and φ are the experimental sphericity of the particles, the density ratio of solid and fluid and the volume fraction of solid. The experimental sphericity may be obtained by performing sedimentation, or fluidization experiments, calculating the drag coefficient for the particle using the volume equivalent diameter d_e and obtaining the sphericity defined as the surface of a sphere having the same volume (drag) as the particle. The volume equivalent sphere is $d_e = (6V/\pi)^{1/3}$, where V is the volume of the particle.

Similarly as in the case of isometric particles, they assumed that the functions $\tilde{C}_0$ and $\tilde{\delta}_0$ may be written in the form:

A unified correlation for the drag coefficient and the sedimentation velocity of irregular particles can be obtained defining a modified drag coefficient, modified Reynolds number, modified equivalent diameter and modifies settling velocities (Figs. 3.8 and 3.9).

Where C_{DM}, Re_M, d_{eM} and u_{pM} are given by:

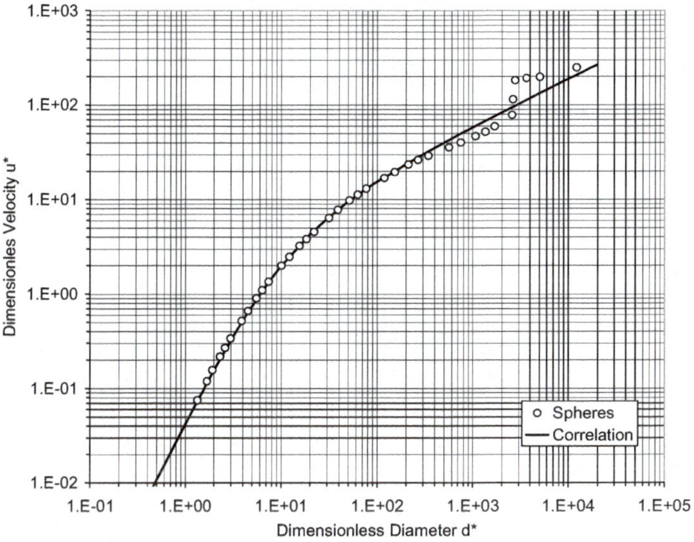

Fig. 3.6 Dimensionless velocity versus dimensionless diameter for the sedimentation of spheres according to Eq. (3.23) of Concha and Almendra's equation [4], Concha [7, 9]. Circles are data from Lapple and Shepherd [14]

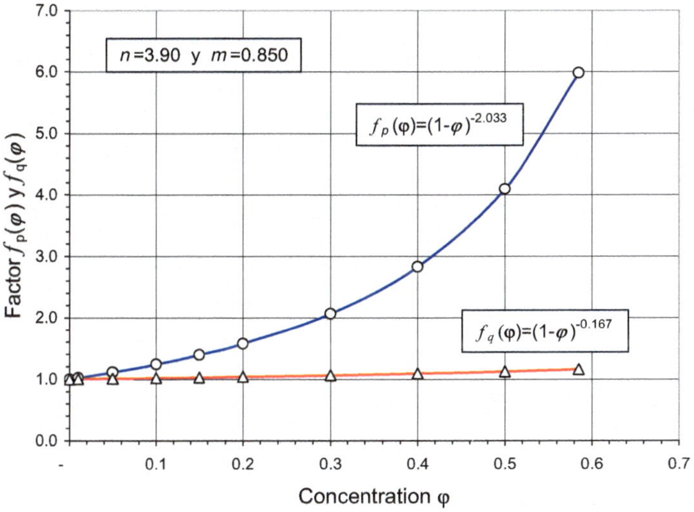

Fig. 3.7 Effect of concentration parameters from data of [19]

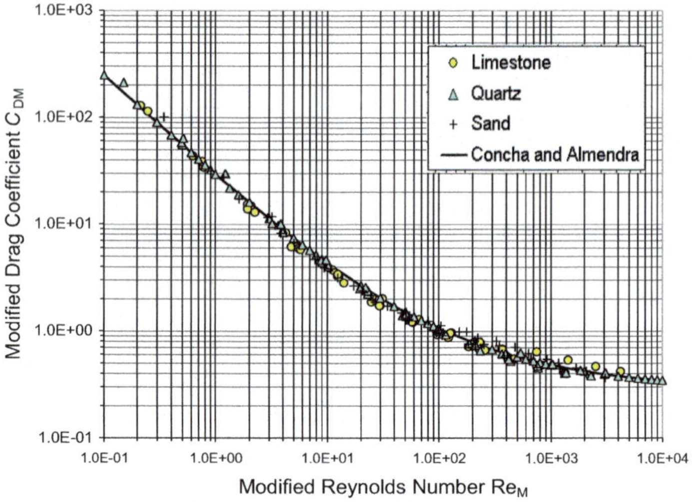

Fig. 3.8 Unified drag coefficient versus Reynolds number for quartz, limestone and sand particles (the same data of Fig. 4.20 [6])

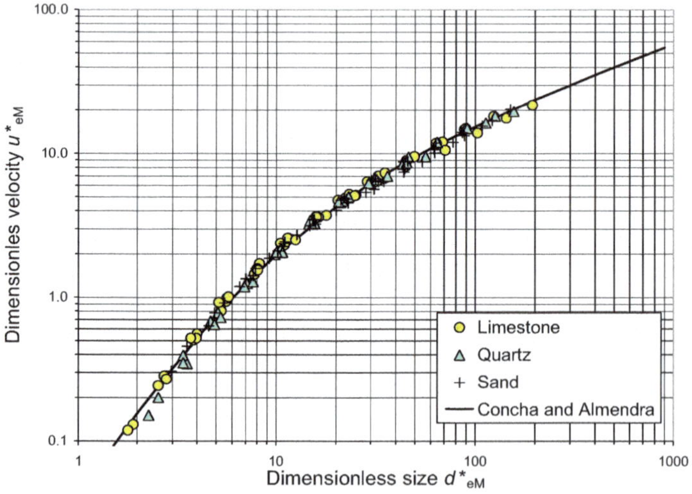

Fig. 3.9 Unified sedimentation velocity versus size for limestone, quartz and sand particles (the same data of Fig. 4.20 [6])

$$C_{DM} = \frac{C_D(\psi, \lambda)}{f_A(\psi) f_C(\lambda) f_p(\varphi)} \quad \boldsymbol{Re}_M = \left(\frac{\boldsymbol{Re}}{f_B^2(\psi) f_D^2(\lambda) f_p^2(\varphi)} \right)$$

$$d^*_{eM} = \frac{d^*_e(\psi,\lambda)}{f_p(\varphi)\times\left(f_A(\psi)^{1/2}\times f_B^2(\psi)\times f_C(\lambda)^{1/2}\times f_D^2(\lambda)\right)^{2/3}}$$

$$u^*_{pM} = \frac{u^*_p(\psi,\lambda)}{f_B(\psi)f_D(\lambda)f_q(\varphi)}$$

with

$$f_A(\psi) = \frac{5.31 - 4.88\psi}{0.62}$$

$$f_B(\psi) = \left\{\frac{5.31 - 4.48\psi}{0.62}\times 0.843\log\frac{\psi}{0.065}\right\}^{-1/2}$$

$$f_C(\lambda) = \lambda^{1/18} \quad f_D(\lambda) = \lambda^{-1/36}$$

$$f_p(\varphi) = (1-\varphi)^{-2.033}, \quad f_q(\varphi) = (1-\varphi)^{-0.167}$$

References

1. Barnea E, Mednick RL (1975) Correlation for minimum fluidization velocity. Trans Inst Chem Eng 53:278–281
2. Batchelor GK (1967) An introduction for fluid dynamics. Cambridge University Press, Cambridge, p 262
3. Boundary Layers, Case studies, Engineering, November, 15 2016.
4. Concha F, Almendra ER (1979) Settling velocities of particulate systems, 1. Settling velocities of individual spherical particles. Int J Mineral Process 5:349–367
5. Concha F, Almendra ER (1979) Settling velocities of particulate systems, 2. Settling velocities of suspensions of spherical particles. Int J Mineral Process 6:31–41
6. Concha F, Christiansen A (1986) Settling velocities of particulate systems, 5. Settling velocities of suspensions of particles of arbitrary shape. Int J Mineral Process 18:309–322
7. Concha F (2014) Solid-liquid separation in the mining industry. Springer, Dordrecht
8. Concha F (2001) Manual de Espesamiento y Filtración, Cettem, Edmundo Larenas 285 Concepción, Chile
9. Concha F (2014) Solid-fluid separation in the mining industry, op.ci.t, p 51
10. Fage A (1937) Experiments on a sphere at critical Reynolds Number. Rep Mem Aero Res Council London, N° 1766:108
11. Goldstein S (ed) (1965) Modern development in fluid dynamics, vol I and II. Dover, New York, 702p
12. Happel J, Brenner H (1965) Low Reynolds hydrodynamics. Prentice-Hall Inc., Englewood Cliffs, p 220
13. Heywood H (1962) Uniform and non-uniform motion of particles in fluids. In: Proceeding of the symposium on the interaction between fluid and particles, Institute of Chemical Engineering, London, pp 1–8
14. Lapple CE, Shepherd CB (1940) Calculation of particle trajectories. Ind Eng Chem 32:605

15. Lee K, Barrow H (1968) Transport process in flow around a sphere with reference to the transfer of mass. Int. Jl. Heat and mass Transfer 11:1020
16. McDonald H, Fish RW (1973) Practical calculations of transitional boundary layers. Int J Heat Mass Transf 16(9):1729–1744
17. Meksyn D (1961) New methods in laminar boundary-layer theory. Pergamon Press, London
18. Lighthill MJ (1963) Boundary layer theory. In: Rosenhead (ed) Laminar boundary layers. Oxford University, p 108
19. Richardson JF, Zaki WN (1979) Sedimentation and fluidization. Part 1. Trans Inst Chem Eng 32:35–53
20. Rosenhead L (ed) (1963) Laminar boundary layers. Clarendon Press, Oxford, p 687
21. Schlichting H (1968) Boundary layer theory. McGraw-Hill, New York, p 21

Chapter 4
Flow of Fluids in Circular Pipelines

Abstract This chapter deals with the flow of fluids in circular pipelines as the basis to understand the feed, underflow and overflow piping necessary and the prevention of sedimentation in the tubes. Explicit equations of the settling velocities of suspensions in pipes for Newtonian in turbulent flow are presented and a general explicit equation for the velocity for turbulent flow of a Newtonian fluid in pipe flow is developed. Several critical transport velocities to avoid sedimentation are presented and the settling velocity of suspensions of spheres in non-Newtonian fluid is developed.

The rectilinear flow of any fluid in a tube is described by the following field variables: fluid density $\rho(r, t)$, velocity $v(r, t)$, pressure $p(r.t)$ and viscous or shear stresses $\tau(r, t)$. For developed flows these field variables must satisfy the continuity and momentum equations [3] (Fig. 4.1)

$$\begin{aligned} v_z &= v_z(r) \\ \tau(r) &= -\frac{1}{2}\frac{\Delta p}{L}r \end{aligned} \tag{4.1}$$

4.1 Newtonian Fluid in Laminar and Turbulent Flow

If the fluid is *Newtonian*, the stresses are a linear function of the shear rate $\dot{\gamma} \equiv \partial v_z/\partial r$

$$\tau = \mu\dot{\gamma} \tag{4.2}$$

and in that case, from

$$\mu\frac{\partial v_z}{\partial r} = -\frac{1}{2}\frac{\Delta p}{L}r \tag{4.3}$$

F. Concha A. and J. L. Bouso A., *Fluid Mechanics Fundamentals of Hydrocyclones and its Applications in the Mining Industry*, Fluid Mechanics and Its Applications 126,
https://doi.org/10.1007/978-3-030-67913-2_4

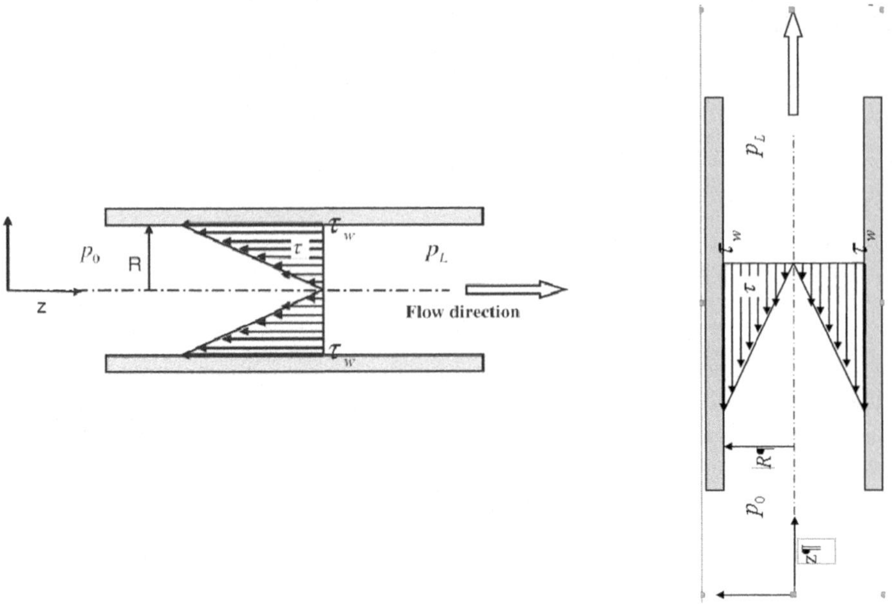

Fig. 4.1 Description of fluid flow showing the shear stresses

The velocity $v(r)$, average $\overline{v}_z$, maximum velocities v_m, volume flow rate Q and the wall shear rate at $\dot{\gamma}_w$ are obtained from the integration of (4.3).

$$v_z(r) = \frac{1}{4}\frac{\Delta p R^2}{\mu L}\left(1 - \left(\frac{r}{R}\right)^2\right), \quad \overline{v}_z = \frac{1}{8}\frac{\Delta p R^2}{\mu L},$$
$$v_m = \frac{1}{4}\frac{\Delta p R^2}{\mu L}, \quad Q = \frac{1}{8}\frac{\Delta p \pi R^4}{\mu L},$$
$$\dot{\gamma}_w = \frac{1}{2}\frac{\Delta p R}{\mu L} \equiv \frac{8\overline{v}_z}{D} \equiv \frac{32}{\pi}\frac{Q}{D^3}, \quad \tau_w = \frac{1}{2}\frac{\Delta p R}{\mu L} \tag{4.4}$$

where R and D are the radius and diameter of the pipe respectively. (Fig. 4.2).

Mechanical Energy Balance

Bernoulli's mechanical energy equation indicates that head balance, pressure head and velocity head are constant for two points 1 and 2 in an open flow.

$$\underbrace{\frac{\Delta p}{\rho g}}_{pressurehead} + \underbrace{\frac{1}{2g}\left(v_{z1}^2 - v_{z2}^2\right)}_{velocityhead} + \underbrace{z_{z1} - z_{z2}}_{head} = 0 \tag{4.5}$$

where $\Delta p = p_1 - p_2 > 0$.

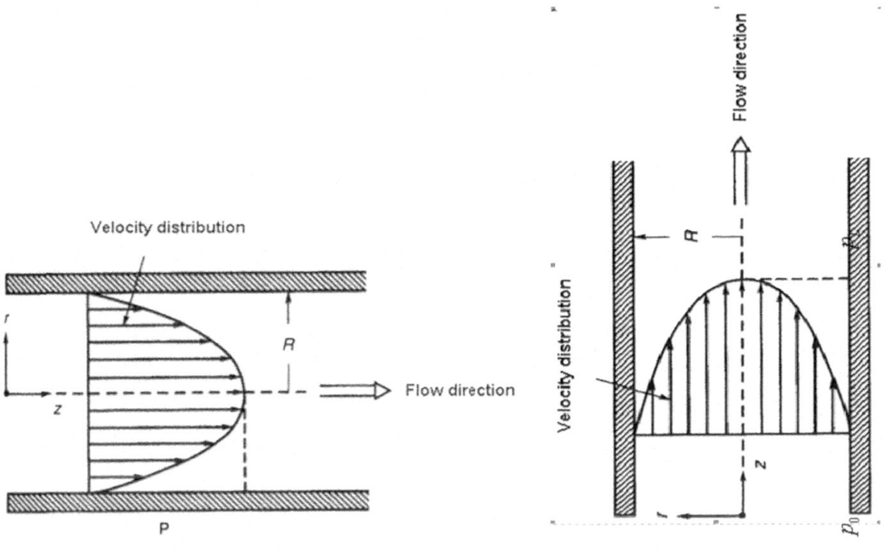

Fig. 4.2 Velocity distribution of the flow of a Newtonian fluid in a circular tube

To apply this equation to conduits, two additional terms should be added, one positive term for the additional total head H_T to drive the fluid through the pipe and one negative term for the total head losses $\sum h_L$ on the pipe walls and fittings.

$$\underbrace{\frac{\Delta p}{\rho g}}_{pressure\ head} + \underbrace{\frac{1}{2g}\left(v_{z1}^2 - v_{z2}^2\right)}_{velocity\ head} + \underbrace{z_{z1} - z_{z2}}_{head} = \underbrace{H_T}_{Total\ head} - \underbrace{\sum h_L}_{Friction\ head} \tag{4.6}$$

Then the total head H_T, for a steady flow in a conduit where $z_{z1} = z_{z2}$, can be written in the form

$$H_T = \underbrace{z_{z1} - z_{z2}}_{head} + \underbrace{\frac{\Delta p}{\rho g}}_{pressure\ head} + \underbrace{\frac{1}{2g}\left(v_{z1}^2 - v_{z2}^2\right)}_{velocity\ head} + + \underbrace{\sum h_L}_{friction\ head} \tag{4.7}$$

and in graphical form (Fig. 4.3).

Consider the following hydraulic system (Fig. 4.4).

Here the system head is 40–5 = 35 m and total head loss is:

$$\sum h_L = h_{friction} + h_{fittings} + h_{entrance} + h_{outlet} \tag{4.8}$$

where the fittings include valves and elbows, the outlet of tank 1 and the inlet of tank 2, which are given in the table in the terms X in the friction factor $f_{fittings} = X \times \overline{v}_z^2/2g$, the number of velocity heads and the friction on the pipe $h_{friction}$ (Table. 4.1).

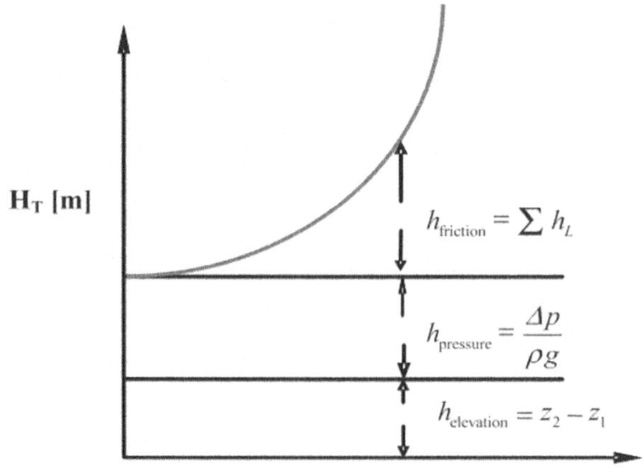

Fig. 4.3 Total head in terms of elevation, pressure head and friction head

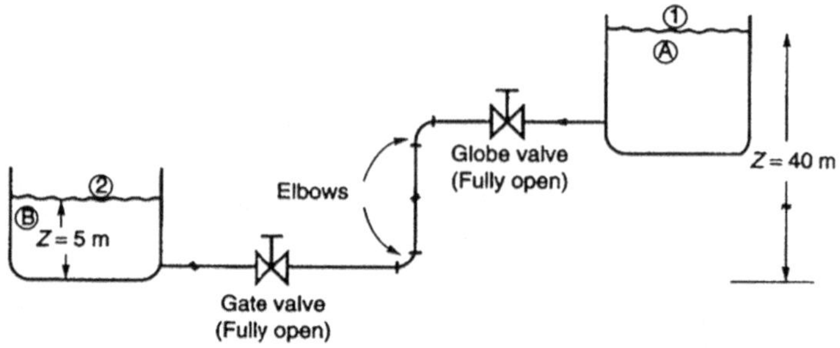

Fig. 4.4 Scheme of a hydraulic system

Laminar Flows

The *friction coefficient* (*Fanning friction factor* f) is defined as the ratio of the friction at the wall τ_w to the kinetic energy $(1/2)\rho\overline{v}_z^2$ of the flow in pipelines.

$$f = \frac{\tau_w}{(1/2)\rho\overline{v}_z^2} \tag{4.9}$$

Substituting $\tau_w = \tau(R)$ from Eq. (4.1) and $\overline{v}_z$ from (4.4) yields

$$f = \frac{\Delta p}{L}\frac{R}{\rho\overline{v}_z^2}, \quad f = \frac{16}{\mathrm{Re}} \tag{4.10}$$

In hydraulics, the custom is to use the *Darcy-Weisbach friction Factor* $\lambda = 4f$, thus

$$\lambda = \frac{64}{\mathrm{Re}} \tag{4.11}$$

Table. 4.1 Head loss in fittings

Fittings	X
45° elbow	0.3
90° elbow	0.7
90° square elbow	1.2
Exit from leg of T-piece	1.2
Entry into leg of T-piece	1.8
Unions and couplings	small
Globe valve (fully open)	1.2–6.0
Gate valve (fully open)	0.15
Gate valve (3/4 open)	1.0
Globe valve (1/2 open)	4.0
Globe valve (1/4 open)	16
Sudden expansion	$\left(1 - (D_1/D_2)^2\right)^2$
Discharge into a large tank	1
Sudden contraction	$X = 0.7867(D_2/D_1)^6 - 1.3322(D_2/D_1)^4 + 0.1816(D_2/D_1)^2 + 0.363$
Outlet of a large tank	0.5

In the laminar regime, with Re $<$ 2100, the friction factor is independent of the roughness of the tube and diminishes when the Reynolds number increases, as shown by Eqs. (4.10) and Fig. 4.6.

Turbulent Flows

Depending on the *Reynolds number*, defined as the ratio of inertial to viscous forces,Re $= \rho v_z D/\mu$, and the surface asperities of the pipes, the flow of a Newtonian fluid may occur in *laminar* or *turbulent regimes.*

As the Reynolds number increases, the inertia forces, which are proportional to the velocity squared, begin to dominate. The momentum exchange between randomly moving masses of fluid throughout the stream gives rise to turbulent motion, which causes a significant increase in the resistance compared with the case of laminar flow. Pipes can have smooth or rough surfaces. A rough surface is characterized by a relative roughness ϵ, defined by

$$\epsilon = \frac{\varepsilon}{D} \tag{4.12}$$

where ε is the is the average height of the asperities and D the tube diameter. Even flows over rough surfaces appear to be smooth in the case of laminar flows, because the roughness of the pipe walls does not affect the flow resistance. Under these conditions, Re $<$ 1, the *friction coefficient* or *friction factor* is a function of the Reynolds number alone. For low velocities in rough tubes, the friction factor decreases with an increase in Reynolds number, but increases at higher velocities. This effect is explained by the fact that at low flows the fluid viscous sublayer height δ in the tube wall is larger than the roughness protuberances ($\delta > \varepsilon$) and the fluid moves smoothly past the irregularities. At higher velocities the sublayer becomes thinner than the roughness protuberances ($\delta < \varepsilon$), which increases the friction factor and pressure drop. The tubes may be considered smooth if the heights of the asperities are smaller than the thickness of the laminar sublayer. See Fig. 4.5.

Experience shows that the friction factor depends on both the Reynolds number and tube roughness, represented by a *roughness Reynolds number* defined by $\mathrm{Re}_{\in} = \in \sqrt{f/2}\mathrm{Re}$. The Moody diagram gives the *Darcy-Weisbach friction factor* λ for commercial pipes, with the Fanning friction factor $f = \lambda/4$.

According to Colebrook [6], the equation for the turbulent flow in a pipe in terms of the Reynolds number, friction factor and dimensionless pipe roughness can be written in the form

$$\frac{1}{\sqrt{f}} = -4\log\left(\frac{1.25}{\mathrm{Re}\sqrt{f}} + \frac{\varepsilon^*}{3.7}\right) \tag{4.13}$$

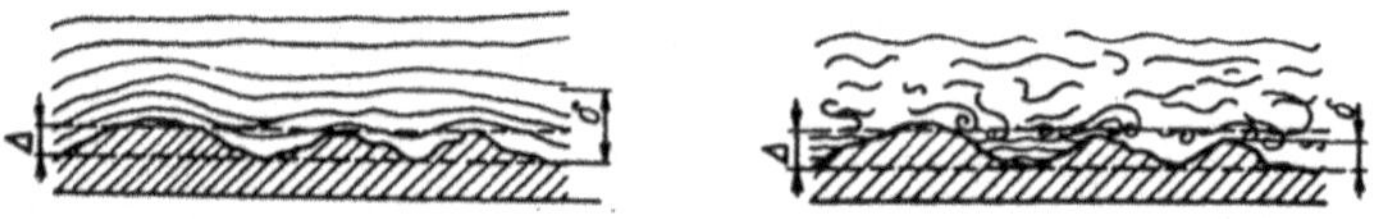

Fig. 4.5 Flow past rough tube walls for different viscous-sublayer-to-roughness-asperity ratios

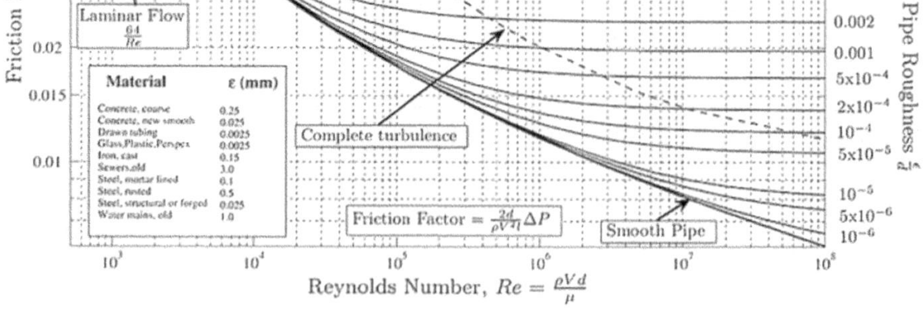

Fig. 4.6 Moody diagram

This equation permits the calculation of the friction factor without using the Moody Diagram. Unfortunately, the friction factor appears on both sides of the equation, such that f must be calculated iteratively.

General Explicit Equation for the Velocity of a Newtonian Fluid in Pipe Flow
Concha [4] developed an explicit equation for the velocity in a pipe for turbulent flows, without having to iterate in terms of the friction factor, when the diameter and roughness of the pipe, pressure gradient and fluid properties are known.

Observe that Colebrook's Eq. (4.14) involves dimensionless numbers only; therefore, we offer the definition of the following new dimensionless numbers:

$$\frac{1}{\sqrt{f}} = -4\log\left(\frac{1.25}{\mathrm{Re}\sqrt{f}} + \frac{\varepsilon^*}{3.7}\right) \tag{4.14}$$

$$\mathrm{Re}^2 f = \left(\frac{\rho^2 D^2 \overline{v}_z^2}{\mu^2}\right) \times \left(\frac{\Delta p}{L}\frac{D}{2\rho\overline{v}_z^2}\right) \tag{4.15}$$

$$\frac{\mathrm{Re}}{f} = \left(\frac{\rho D\overline{v}}{\mu}\right) \times \left(\frac{L 2\rho\overline{v}_z^2}{\Delta p D}\right) \tag{4.16}$$

Since Eq. (4.15) is a function of the pipe diameter only and Eq. (4.16) a function of the fluid velocity only, we can write

$$\begin{aligned} \mathrm{Re}^2 f &= \left(\frac{D}{\mathbb{D}}\right)^3 \equiv D^*, \quad \text{where } \mathbb{D} = \left(\frac{\rho}{2\mu^2}\frac{\Delta p}{L}\right)^{1/3} \\ \frac{\mathrm{Re}}{f} &= \left(\frac{\overline{v}_z}{\mathbb{V}}\right)^3 = \overline{v}_z^*, \quad \text{where } \mathbb{V} = \left(\frac{2\rho^2}{\mu}\frac{L}{\Delta p}\right)^{1/3} \end{aligned} \tag{4.17}$$

where

$$\mathbb{D} = \left(\left(\frac{\rho}{2\mu^2}\right)\left(\frac{\Delta p}{L}\right)\right)^{-1/3} \text{ and } \mathbb{V} = \left(\left(\frac{2\rho^2}{\mu}\right)\left(\frac{\Delta p}{L}\right)^{-1}\right)^{-1/3} \tag{4.18}$$

From (4.17),

$$\mathrm{Re} = D^*\overline{v}_z^* \quad \text{and} \quad f = \frac{D^*}{\overline{v}_z^{*2}} \tag{4.19}$$

Using the expression (4.19), we can write Colebrook's Eq. (4.14) in the form

$$\overline{v}_z^* = -4D^{*1/2}\log\left(\frac{1.25}{D^{*3/2}} + \frac{\varepsilon^*}{3.7}\right) \tag{4.20}$$

This is an explicit equation for the dimensionless average velocity in a pipe in terms of the pressure gradient, pipe properties—diameter and roughness—and fluid properties—density and viscosity. The friction factor is implicit in the equation.

4.2 Non-Newtonian Fluid Flows in Circular Pipelines

The transport of water in pipes allows the calculation of the flow speed when the driving force, pressure gradient or gravity and the fluid properties are known.

Water is a Newtonian fluid in which there is a linear relationship between shear stress and shear velocity,

$$\tau = \mu\gamma \tag{4.21}$$

where τ and $\dot{\gamma}$ are the shear stress and shear rate, respectively, and μ is the constant viscosity. (Fig. 4.7).

The case of concentrate and tailings transport, for example in the copper industry, can also be described by the law of fluid mechanics, although in this case the relationship between stress and shear velocity is non-linear. Such fluids are called non-Newtonian fluids. For instance, laboratory experiments on 63% solid copper flotation tailings gave the following rheogram, which shows the shear stress-shear rate relationship.

Figure 4.8 clearly shows that the relationship between the viscosity and shear rate is non-linear and can be written in the form

$$\tau = \eta(\dot{\gamma})\dot{\gamma} \tag{4.22}$$

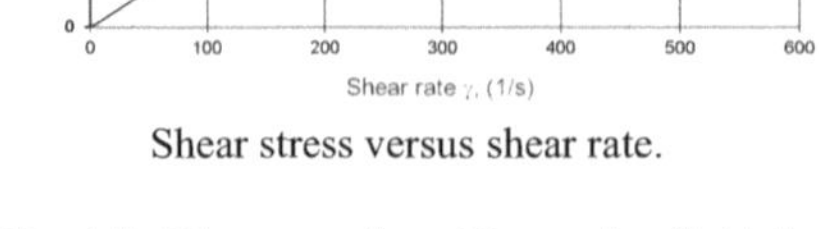

Shear stress versus shear rate.

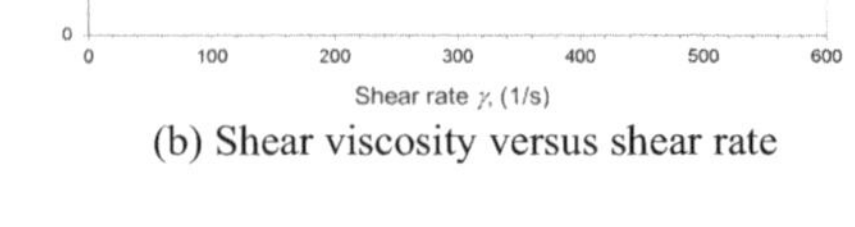

(b) Shear viscosity versus shear rate

Fig. 4.7 Rheogram for a Newtonian fluid, in this case, water

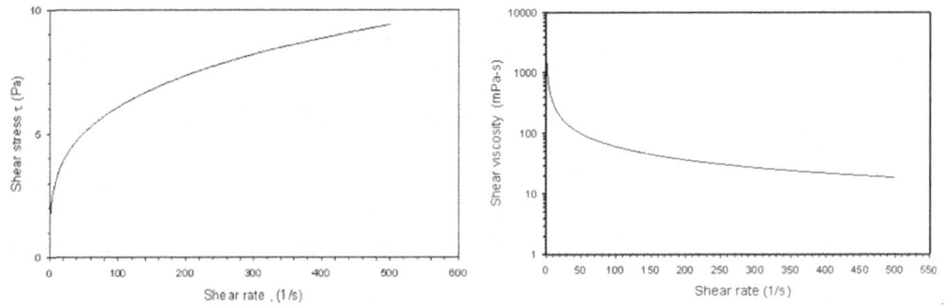

Shear stress versus shear rate. (b) Shear viscosity versus shear rate

Fig. 4.8 Rheogram for a 63% solid copper flotation tailing at pH 10.1

where τ and η are the fluid shear rate and viscosity respectively.

Modeling Non-Newtonian fluids

Two of the most used Newtonian fluid models are Bingham and power-law fluids. The first is used for high speed flows and the second for slow flows.

Bingham-Plastic Models

If we look at the Rheogram in Fig. 4.8, we see that at high shear rate, $\dot{\gamma} > 200\ \text{s}^{-1}$ in this case, the relationship between shear stress and shear rate is linear and can be described by the tangent to the curve at high shear rate, a constant viscosity K. See Fig. 4.9.

If we use the blue line in Fig. 4.9 as the constant viscosity ($K = 0.0069$), we must define a second constant for the shear stress at shear rate zero ($\dot{\gamma} = 0,\ \ \tau_y = 6.0179$ Pa), which is called *yield stress* $\tau = \tau_y + K\dot{\gamma}$.

The effect of the suspension concentration on the yield stress can be expressed by Eq. (4.23), where φ_m is the concentration of the packed bed of particles and a, b and c are empirical parameters.

Fig. 4.9 Bingham model for the copper flotation tailings of Fig. 4.8

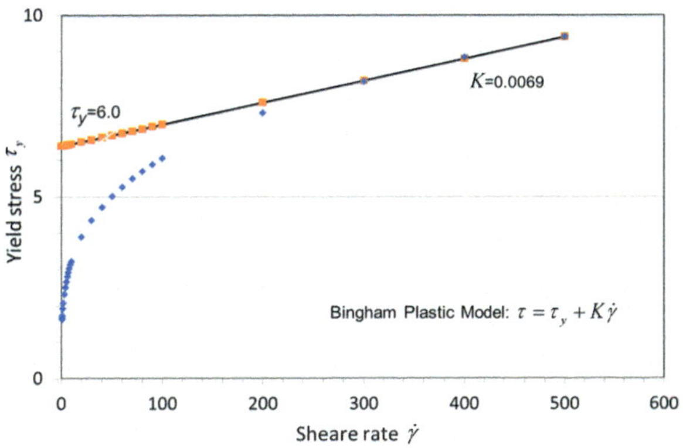

$$\frac{\tau_y}{\tau_m} = \frac{b(\varphi/\varphi_m)^c}{(1-\varphi/\varphi_m)^a} \tag{4.23}$$

Power-Law Model

If we are interested in slow flows, for example with < 200 s^{-1}, the rheogram in Fig. 4.9 shows that the Bingham-plastic model does not give the correct answer. In these cases, the power-law model is better and gives a correct value for the nonlinear viscosity in the entire range of shear rates.

From Fig. 4.8b we see that a possible model for the shear viscosity is $\eta(\gamma) = m\dot{\gamma}^{\kappa}$, where κ can be obtained from a plot of the log of the viscosity η versus the log of the shear rate $\dot{\gamma}$ (see Fig. 4.10). The intercept of the straight line with value $\dot{\gamma} = 1$ gives the constant coefficient of position $m = 51872$ and the slope gives $\kappa = 0.1933$. With this parameter the shear rate viscosity would be $\eta = \dot{\gamma}^{\kappa}\dot{\gamma}$, and, defining $n = \kappa + 1$, we can write the constitutive equation and the respective plot as

$$\tau = m\dot{\gamma}^n \tag{4.24}$$

For a given shear rate, the effect of the concentration on the shear viscosity η can be seen in Eq. (4.25) and Fig. 4.10.

$$\eta = \left(1 + \frac{\varphi}{0.5}\right)^{-z} \tag{4.25}$$

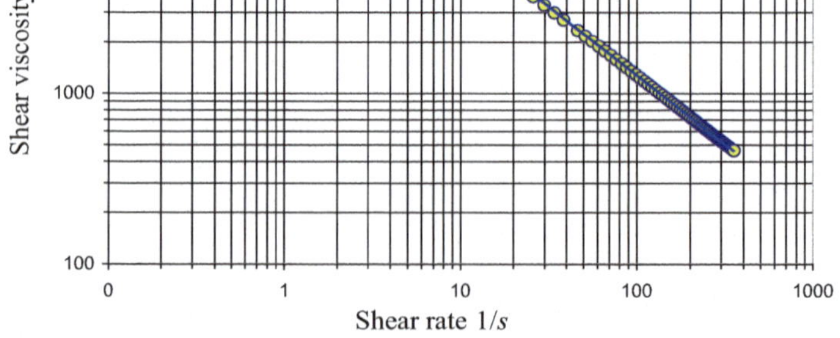

Fig. 4.10 Viscosity versus shear rate for the tailings of Fig. 12.11

Settling Velocities in Non-Newtonian Fluids

From the work of Concha [3], the drag coefficient $C_D(n)$ and settling velocity of suspension of spherical particles $u^*(n)$ in a power-law model type of Non-Newtonian fluid may be expressed as

$$C_D(n) = 0.284X(n)\left(1 + \frac{9.06}{\mathrm{Re}(n)^{1/2}}\right)^2 \tag{4.26}$$

$$u^*(n) = \frac{1}{4}\frac{(9.06)^2}{d^*(n)}\left(\left(1 + \frac{4}{(0.284X(n))^{1/2}(9.06)^2}d^{*3/2}\right)^{1/2} - 1\right)^2 \tag{4.27}$$

where the $\mathrm{Re}(n)$ and the constitutive equation for the Non-Newtonian fluid are given by Concha [3] and τ and $\dot{\gamma}$ are the shear stress and shear rate, respectively:

$$\mathrm{Re}(n) = \frac{\rho u^{2-n} d^n}{m}; \quad \tau(\dot{\gamma}) = m\dot{\gamma}^n$$

and $X(n)$ is a parameter proposed by Chabra [1]:

$$X(n) = -1.1492 \times n^2 + 0.8734 \times n + 1.2778 \tag{4.28}$$

The definitions of the dimensionless diameter and velocity are given by

$$d^*(n) = \frac{d}{P_n} \quad \text{and} \quad u^*(n)\frac{u}{Q_n} \tag{4.29}$$

where P_n and Q_n are given by

$$P_n = \left(\frac{3}{4}\frac{m^{\frac{2}{2-n}}}{\Delta\rho g \rho_f^{\frac{n}{2-n}}}\right)^{\frac{2-n}{2+n}} \quad \text{and} \quad Q_n = \left(\left(\frac{4}{3}\right)^n \frac{m\Delta\rho^n g^n}{\rho_f^{n+1}}\right)^{\frac{1}{2+n}} \tag{4.30}$$

The drag coefficient C_D(n) and the dimensionless settling velocity u^*(n) are given by Eqs. (4.26) and (4.27). They can be expressed graphically in terms of the plastic Reynolds number and the dimensionless particle diameter. (Figs. 4.11 and 4.12).

Transport Equation for Non-Newtonian Fluids
Bingham fluids

We saw at the beginning of this chapter that all fluids must obey the following relationships for the shear rate in the bulk of the fluid and at the wall:

$$\tau(r) = -\frac{1}{2}\frac{\Delta p}{L}r \quad \tau(r) = -\frac{1}{2}\frac{\Delta p}{L}r \quad \tau_w = \frac{1}{2}\frac{\Delta pR}{\mu L} \tag{4.31}$$

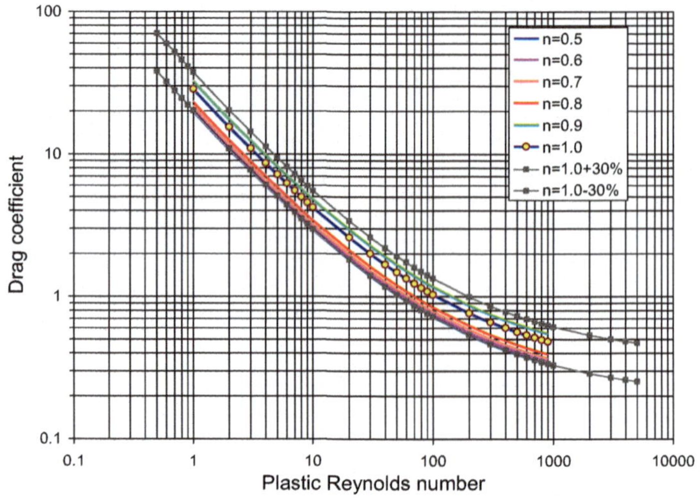

Fig. 4.11 $C_D(n)$ from Eq. (4.26)

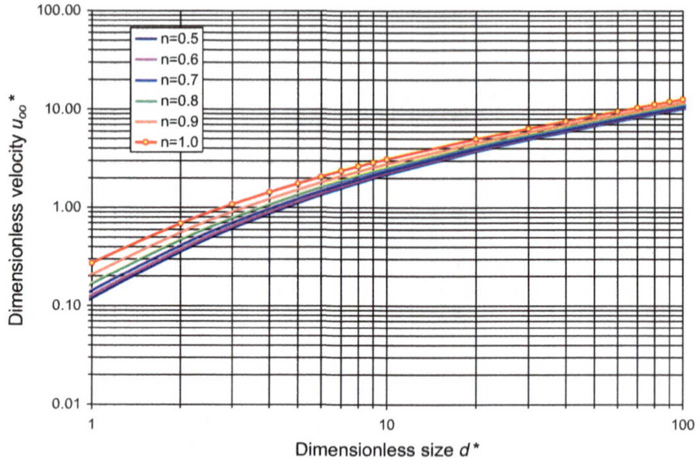

Fig. 4.12 $u^*(n)$ and $d^*(n)$ from Eq. (4.27)

For Bingham-type non-Newtonian fluids the constitutive equation for the shear stress is $\tau = \tau_y + K\dot{\gamma}$. If R_y denotes the tube radius at which the shear stress is τ_y, from (4.31) we can write

$$\tau_y = \tau_w \frac{R_y}{R} \tag{4.32}$$

Flow Fields

Since the constitutive equation for a Bingham fluid is the sum of two terms, the velocity flow field may be separated for $r \geq \tau_y$ [3].

For $\tau(r) \geq \tau_y$

$$\frac{\partial v_z(r)}{\partial r} = \frac{1}{2}\frac{\Delta p}{\eta L}\left(R_y - r\right) \tag{4.33}$$

Integrating this equation yields

$$v_z(r) = -\frac{1}{2}\frac{\Delta p R^2}{\eta L}\left(\frac{R_y}{R}\left(1 - \frac{r}{R}\right) - \frac{1}{2}\left(1 - \left(\frac{r}{R}\right)^2\right)\right) \tag{4.34}$$

and for $\tau(r) \leq \tau_y$

$$\frac{\partial v_z(r)}{\partial r} = \frac{1}{2}\frac{\Delta p}{\eta L} r \tag{4.35}$$

Integrating gives

$$v_z(r) = -\frac{1}{2}\frac{\Delta p R^2}{\eta L}\left(1 - \left(\frac{R_y}{R}\right)^2\right) \tag{4.36}$$

Volume flow rate

For a known velocity, the volume flow rate is $Q = \pi/4 D^2 \overline{v}_z(r)$ and can be obtained by substituting (4.34) and (4.36), resulting in

$$Q = \frac{\pi}{2}\frac{\Delta p R^4}{\eta L}\left(\frac{3}{8}\left(\frac{R_y}{R}\right)^4 + \frac{2}{3}\left(\frac{R_y}{R}\right)^3 - \frac{1}{4}\left(\frac{R_y}{R}\right)^2 + \frac{1}{2}\frac{R_y}{R} - \frac{1}{4}\right) \tag{4.37}$$

Average velocity

$$\overline{v}_z = \frac{1}{2}\frac{\Delta p R^2}{\eta L}\left(\frac{3}{8}\left(\frac{R_y}{R}\right)^4 + \frac{2}{3}\left(\frac{R_y}{R}\right)^3 - \frac{1}{4}\left(\frac{R_y}{R}\right)^2 + \frac{1}{2}\frac{R_y}{R} - \frac{1}{4}\right) \tag{4.38}$$

Maximum velocity

$$v_{\mathrm{m}} = \frac{1}{4}\frac{\Delta p R^2}{\eta L}\left(1 - \frac{R_y}{R}\right)^2 \tag{4.39}$$

Power-law fluids

For power-law fluids, Eq. (4.31) is also valid and in this case the constitutive equation for the shear stress is [3]

$$\tau = m\left(\frac{\partial \overline{v}_z}{\partial r}\right)^n \tag{4.40}$$

Flow field

From (4.31) and (4.40), the velocity field results in

$$v_z(r) = \frac{nR}{m+1}\left(\frac{\Delta p R}{2mL}\right)^{1/n}\left(1-\left(\frac{r}{R}\right)^{(n+1)/n}\right) \tag{4.41}$$

Volume flowrate

$$Q = \pi R^2 \frac{n}{n+1}\left(\frac{\Delta p R^{n+1}}{2mL}\right)^{1/n} \tag{4.42}$$

Average velocity

$$\overline{v}_z = \frac{n}{n+1}\left(\frac{\Delta p R^{n+1}}{2mL}\right)^{1/n} \tag{4.43}$$

Maximum velocity

$$v_m = \frac{nR}{n+1}\left(\frac{\Delta p R}{2mL}\right) \tag{4.44}$$

Shear stress and shear rate at the wall

$$\tau_w = m\frac{3n+1}{4n}\left(\frac{8\overline{v}_z}{D}\right)^n \quad \text{and} \tag{4.45}$$

Pressure drop

$$\Delta p = \frac{2mL}{R^{n+1}}\left(\frac{3n+1}{n}\frac{Q}{\pi R^2}\right) \tag{4.46}$$

Reynolds number [10]

$$\mathrm{Re}_M = \frac{\rho \overline{v}_z^{2-1} D^n}{8^{n-1} m \left(\frac{3n+1}{4n}\right)^n} \tag{4.47}$$

4.3 Critical Transport Velocity

Suspension flow patterns in horizontal piping depend on the transport velocity. At low speeds the particles will form a bed at the base of the pipe and will not be transported by the fluid. As the velocity increases, the particles located on the surface of the sediment will begin to move. At a higher speed, the sediment will move like a cloud in a jumping motion and some particles will be suspended by the fluid and transported at its speed. At even higher speeds, all particles will be suspended, and the fluid and particles will behave like a homogeneous mixture.

In Fig. 4.13, the velocities v_{M1} to v_{M4} correspond to different concentration ratios of the suspended material in the tube to the total concentration.

Mixture velocity v_{Mi}	Flow pattern	Ratio of volume fraction to total volume fraction
v_{M1}	Homogeneous suspension	1.0
v_{M2}	Asymmetric suspension	0.7–1.0
v_{M3}	Moving bed with asymmetric suspension	0.2–0.7
v_{M4}	Stationary bed with some particles in suspension	0 Metzner 0.2

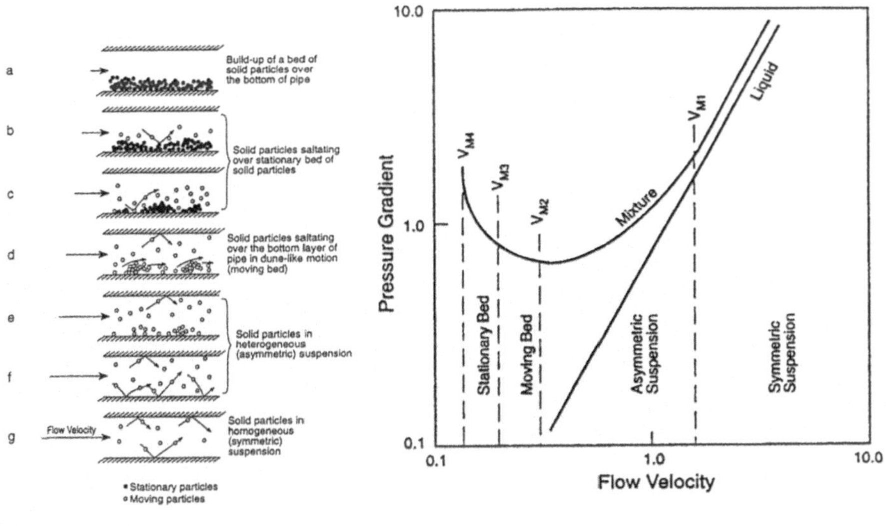

Fig. 4.13 Evolution of the suspension of particles forming a bed in a tube

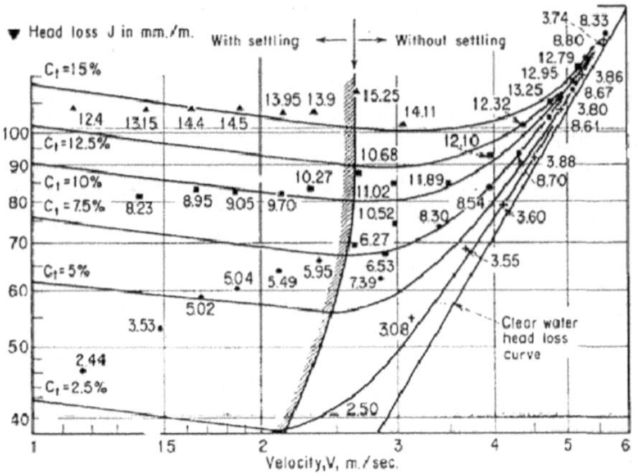

Fig. 4.14 Head loss versus average transport velocity for 0.44 mm fine sand at several concentrations [5]

Condolios and Chapus [5] presented the following experimental results. (Fig. 4.14).

When pressure is applied to a pipe containing sediment, the fluid moves, and as the particles begin to suspend, the pressure decreases. Between the beginning of the movement of the particle bed and complete suspension there is a range of speeds at which the pressure of the fluid is minimum and beyond which it increases again. This speed is called the *critical transport velocity.*

Numerous correlations have been suggested to quantify the critical velocity of particle suspensions of sizes from 50 μm to 5 mm and pipes from 50 mm (2 in) to 300 mm (12 in). Chien [2] reviewed the work on transport velocity of Wilson; Durand; Spells; Newitt et al.; Cairn et al.; Govier; Schulz; Sinclair; Condolios and Chapus; Yufin and Lopasin; Zandi and Govatos; Babcock, Shook; Bain and Bonninton; Charles; Thomas; Oroskar and Turian; and Gillies and Shook.

Based on extensive experience with suspensions from the copper and iron mining industries, Rayo [11] proposed a modification of the Durand expression of critical velocity (1953) for a vast number of particle sizes and size distributions and tube diameters, where F_L is a parameter:

(a) For fine solids with a narrow particle distribution in small pipes,

$$v_L = 1.1 \times F_L(\varphi, d_{50}) \times \left(2gD\Delta\rho/\rho_f\right)^{0.5}; \quad \text{for} \quad \varphi < 0.20 \qquad (4.48)$$

(b) For large solid particles and wide particle size distribution in small pipes,

$$v_L = F_L(\varphi, d_{50}) \times \left(2gD\Delta\rho/\rho_f\right)^{0.5} \times \left(\frac{d_{80}}{d_0}\right)^{0.1}; \quad \text{for} \quad \varphi < 0.20 \qquad (4.49)$$

(c) For fine solids and narrow particle distribution in large pipes,

$$v_L = 1.25 \times F_L(\varphi, d_{50}) \times \left(2gD\,\Delta\rho/\rho_f\right)^{0.25} \qquad (4.50)$$

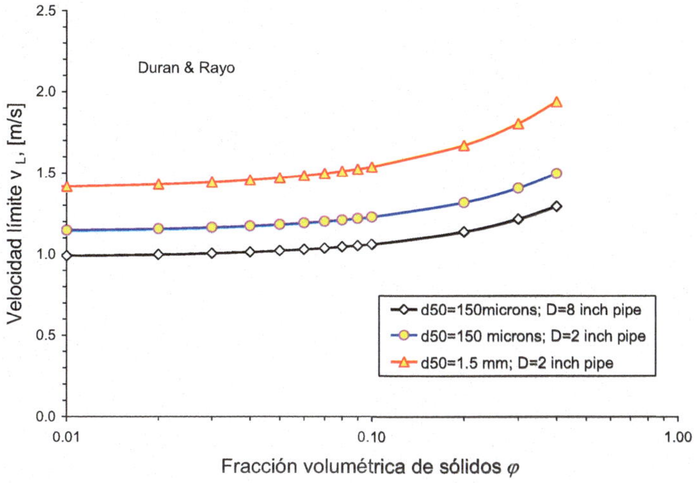

Fig. 4.15 Critical transport velocity according to Durand and Rayo (1996)

(d) F_L parameter:

$$F_L = (0.2154\varphi + 0.1584) \times \ln(d_{50}) + (1.112\varphi + 1.19);$$
$$\text{for } 5 < d_{50}(\mu\text{m}) < 500 \quad (4.51)$$

Example of Durand and Rayo's equations (Fig. 4.15).

In 1986, Jergensen and Samuell [8] developed a simple empirical equation with the pipe diameter, particle size, volumetric fraction of solids and densities of the solid and liquid as variables.

$$v_L = (0.2154\varphi + 0.1584) \times \ln(d_{50}) + (1.11\varphi + 1.19)\ ;\ \text{for } 5 < d_{50} < 500\ ;\ \text{in } \mu\text{m} \quad (4.52)$$

References

1. Chabra RP (2007) Bubbles, Drops, and Particles in Non-Newtonian Fluids, CRC, 204 B/W Illustrations, p 771
2. Chien AF (1994) Critical velocity of sand-fluid mixtures in horizontal pipe flow. ASME, FED-189, pp 231–246
3. Concha F (2014) Solid-liquid separation in the mining industry, fluids and its applications, vol 105. Springer, Heidelberg
4. Concha F (2008) Settling velocities of particulate systems 15: velocities in turbulent Newtonian Flows. Int J Mineral Process 88:89–93
5. Condolios E, Chapus EE (1963) Designing solids handling pipelines. Chem Eng 70:131–138

6. Colebrook CF, White CM (1937) Experiments with fluid friction factor in roughened pipes. In: Proceedings of the Royal Society of London. Series A: Mathematical and Physical Sciences, 161:367–381
7. Durand R, Condolios E (1952) Hydraulic transport of coal and solids material in pipes. In: Proceedings of Colloquium on Hydraulic Transportation, London
8. Jergensen GV, Slurry SEJ (1986) Transportation in concentration circuits. In: Mular AL, Anderson MA (eds) Transportation of tailings design and installation of concentration and dewatering circuit. Society of Mining Engineering Inc.
9. Mc. Elvain RE, Cave I (1973) Transportation of tailings. In: Aplin CL, Argall GO (eds) Tailings disposal today. Miller Freeman, San Francisco
10. Metzner AB, Reed, JC (1955) Flow of non-Newtonian fluids, correlation of the laminar, transition, and turbulent-flow regions. AIChE Journal 1
11. Rayo J (1977) Transporte Hidráulico de Sólidos por Tuberías. Revista Minerales 137

Chapter 5
Flow Pattern in Hydrocyclones

Abstract In this chapter, models of hydrocyclones will be ordered from a fluid dynamics perspective and an attempt will be made to present the velocity and pressure distribution inside the device. With experimental values of the velocity distributions, mass and linear momentum, equations were solved for the simplest inviscid approach combining a free vortex with a sink, rotational inviscid models, models based on the exact solution of the Reynolds equation for highly idealized conditions and solutions using different mathematical techniques such as asymptotic expansions, similarity solutions and boundary layer flows.

Several hydrocyclone models have been presented since the second half of the twentieth century. In this chapter, they will be ordered from a fluid dynamics perspective [36] and an attempt will be made to present the velocity and pressure distribution inside the device. With experimental values of the velocity distributions, mass and linear momentum, equations were solved for: the simplest inviscid approach combining a free vortex with a sink, rotational inviscid models, models based on the exact solution of the Reynolds equation for highly idealized conditions and solutions using different mathematical techniques such as asymptotic expansions, similarity solutions and boundary layer flows.

5.1 Introduction

Since the flow enters at the periphery and top of the hydrocyclone and leaves through the top and the bottom of the vessel at its center, there is a three-dimensional fluid motion, called *vortex motion*, with tangential, axial and radial velocities. The vortex motion produces a low-pressure zone in the center of the device that sucks air from outside, forming an air core. The shape of this air column has been described as cylindrical and conical but based on simple observation, these descriptions do not fit its actual shape.

While in normal operation a hydrocyclone has axial and tangential velocities on the order of several meters per second; the radial velocity is one order of magnitude

F. Concha A. and J. L. Bouso A., *Fluid Mechanics Fundamentals of Hydrocyclones and its Applications in the Mining Industry*, Fluid Mechanics and Its Applications 126,
https://doi.org/10.1007/978-3-030-67913-2_5

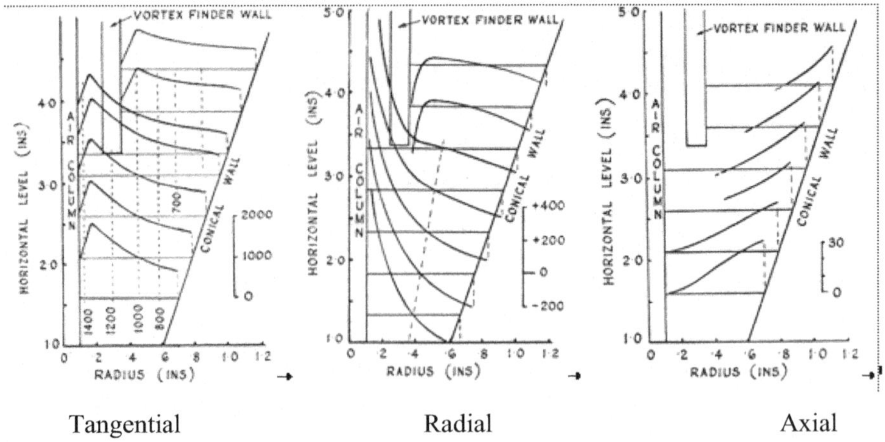

Fig. 5.1 Tangential, axial and radial velocity components according to [70]

smaller. This very fact is the reason that the radial velocity distribution in the device is not very well known. In the presence of highly turbulent tangential and axial velocities, the fluctuations of these velocities are on a similar order of magnitude than the radial velocity, making it difficult to measure.

Flow patterns in hydrocyclones have been studied by many authors: [5, 6, 7, 29, 42, 61, 64, 63, 65, 67, 71, 70, 89, 72, 98, 99, 117, 35].

The first and most cited velocity measurement was presented by Kelsall [71, 70], who used small cyclones with diluted feeds containing fine aluminum particles. With proper illumination, the motion of the particles was followed with a microscope with rotating objectives. The following figure shows the tangential, axial and radial velocity distributions obtained by Kelsall (Fig. 5.1).

Kelsall [70] describes his findings as follows: "Beneath the bottom of the vortex finder, envelopes of constant tangential velocity are cylinders coaxial with the cyclone, and, at any horizontal level, starting from near the conical wall, this component of velocity increases with decrease in radius according to the relationship $vr^n = K$, with $0 < n < 1$, where K is a constant." This holds in the region near the center of the cyclone, where the velocity reaches a maximum value at a radius somewhat smaller than that of the inside of the vortex finder. As the radius decreases further the tangential velocity decreases such that $v \propto r$.

The two most significant features of the vertical velocity distribution are the approximately conical envelope-defining regions of zero vertical velocity. In regions between the conical wall and the zero-velocity envelope, all liquid flows downwards toward the apex and between the envelope and the central axis, or wall of the air core column if present, all liquid flows upwards toward the vortex finder.

The limited information about the inward radial velocity component tends to indicate that, at any horizontal level, near the conical wall the following expression is valid: $v_r = v_z \tan\theta$, where θ is the half angle of the cone.

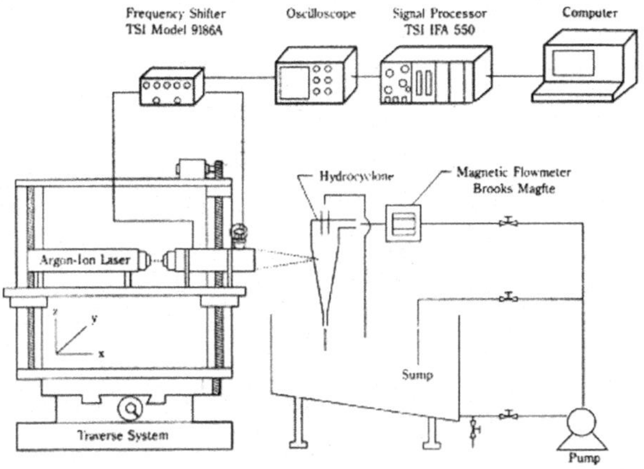

Fig. 5.2 Laser Doppler speedometer system and hydrocyclone

It was not until twenty years after Kelsall's measurements that other measurements were made. During this time, the development of laser Doppler velocimetry (LDV) significantly improved velocity measurements. Very accurate measurements of the tangential and axial velocity distributions were made by Hsieh and Rajamani ([63, 65]). Figure 5.2 shows a schematic view of the laser Doppler velocimeter system and hydrocyclone used by Hsieh and Rajamani [99].

The glass hydrocyclone had an inside diameter of 75 mm with an inlet diameter of 23 mm, vortex-finder of 23 mm, apex of 11 mm and total height of 244 mm. Over 200 hundred measurements of the two velocity components, tangential and axial, at each location, 1 mm radial direction and 10 mm axial direction, were taken. The fluid medium was water with fine size silicon carbide powder to enhance the data collection ability of the receiving optics. Due to the magnitude of the fluctuating component of the velocities (turbulence), which for the tangential velocity are on the order of 20–50 cm/s, as seen in Fig. 5.3, measurement of the radial velocity was not possible, since it was on the same order of magnitude as the turbulence. Therefore, the radial velocity component was calculated from the continuity equation.

Figure 5.3 shows the axial velocity measurements in the vortex-finder region. The asymmetry of the axial velocity predominates near the region of the inflow and gradually becomes almost symmetric in the lower part of the cone. Further, there is more than one reversal in the flow direction in the region between the vortex-finder and the hydrocyclone wall.

Figure 5.4 shows the distribution for the three components measured and calculated by Hsieh and Rajamani [63] (Fig. 5.5).

Quian et al. [98] also used LDV to measure the velocity distribution in conventional and water-sealed hydrocyclones, both with an 82 mm diameter body, 12 mm internal-diameter vortex finder, 6 mm internal-diameter apex and a feed pressure of 0.9 kg/cm^2.

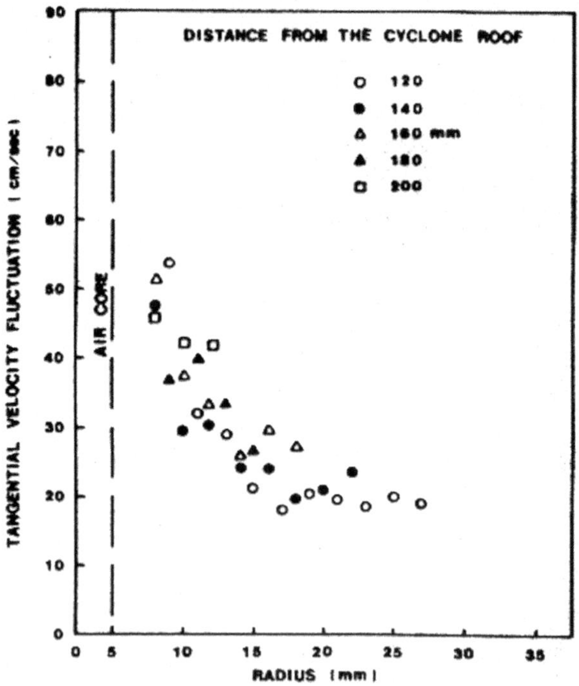

Fig. 5.3 Turbulent fluctuation of the tangential component, according to Hsieh and Rajamani [99]

In the cylindrical section the tangential and radial velocities are very low, while the axial velocity is in the range of 0.5 m/s. In the conical section the value of the tangential velocity varied from 2 m/s close to the wall to approximately 4 m/s at its maximum value. The authors confirmed Kelsall's expression $v_\theta r^n = K$ and gave the values of the parameter $n = 0.432$ and the constant $K = 7.95$. They noted an 8 mm diameter air core rotating at 10,300 rpm. For the axial velocity they obtained values of around 1.5 m/s, which they approximated with the expression $v_z = \ln\bigl(r/(a + br)\bigr)$, where a and b are constant depending on the distance from the vortex finder, as seen in Table 5.1. Finally, the radial velocity increased from close to zero near the hydrocyclone wall to around −4 m/s near the air core. We must note that this result is the opposite of Kelsall's and different from that of Hsieh and Rajamani (Fig. 5.6).

Chine and Concha [29] used a two-beam, 300 mW laser Doppler velocimeter to determine the tangential and axial velocity fields for the water flow in a 102 mm modular hydrocyclone. The body of the equipment could be changed to transform it into a conical or flat-bottom hydrocyclone. During the tests the pressure drop and the diameter of apex and vortex were varied and the axial and tangential velocities and their turbulence intensity were measured. The results show that the inlet pressure affects only the magnitude of the velocities but does not change the flow pattern. The tangential velocity is similar in both hydrocylone types, while the axial velocity is different. In both hydrocyclones the axial velocity is a function of the radial position

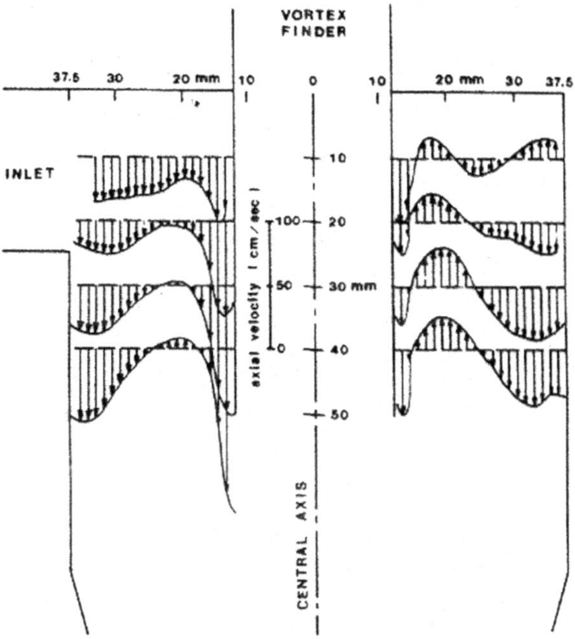

Fig. 5.4 Axial velocity profile in the inlet region, according to Hsieh and Rajamani [99]

but, while it is a linear function of the vertical coordinate in the cylindrical hydrocyclone, this is not the case for the conical vessel. Figures 5.7 and 5.8 show results for the tangential velocity in cylindro-conical and flat-bottom hydrocyclones.

The measurement of the velocity distribution in a hydrocyclone is the first step in predicting the flow pattern from a fluid mechanics perspective. The simplest approach is to use inviscid isochoric flows. The three-dimensional nature of the flow produces a secondary anisotropic flow near the entry region. Since the classification action of the equipment is performed below the level of the bottom of the vortex finder tube, an axial-symmetric simplification of the flow is possible without losing the main features of the vortex motion. The flow enters with a combination of tangential and radial velocities, the ratio of which is dependent on the geometry of the actual entry. To model the feeding of the hydrocyclone in an axial-symmetrical model, a feeding ring substitutes for the actual shape of the feeding tube. This imaginary device has a certain height that, in combination with the radial entry velocity, yields the volumetric flow into the device. The fast inlet velocity, on the order of several meters per second, produces a highly turbulent flow in the device. Turbulence modeling is crucial to obtaining reasonable results on flow patterns.

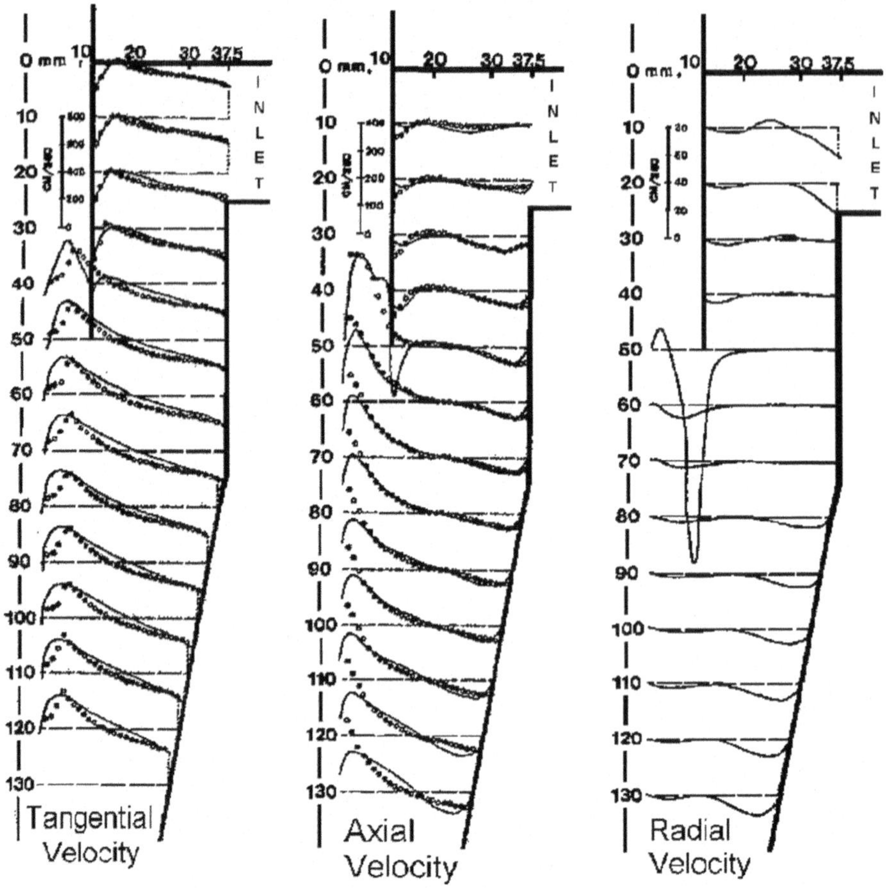

Fig. 5.5 Tangential, axial and radial velocity components according to Hsieh and Rajamani [61]

Table 5.1 Axial velocity parameter values

Distance from the vortex finder in mm	A	B
45	−6.906	1.429
75	−7.579	1.506
105	−6.924	1.670
135	−5.142	1.745
165	−3.482	1.733

5.2 Fluid Dynamics Equations

The Navier Stokes equations have been for many years the starting point to solve fluid dynamic problems involving simple fluids such as liquids and gases at moderate

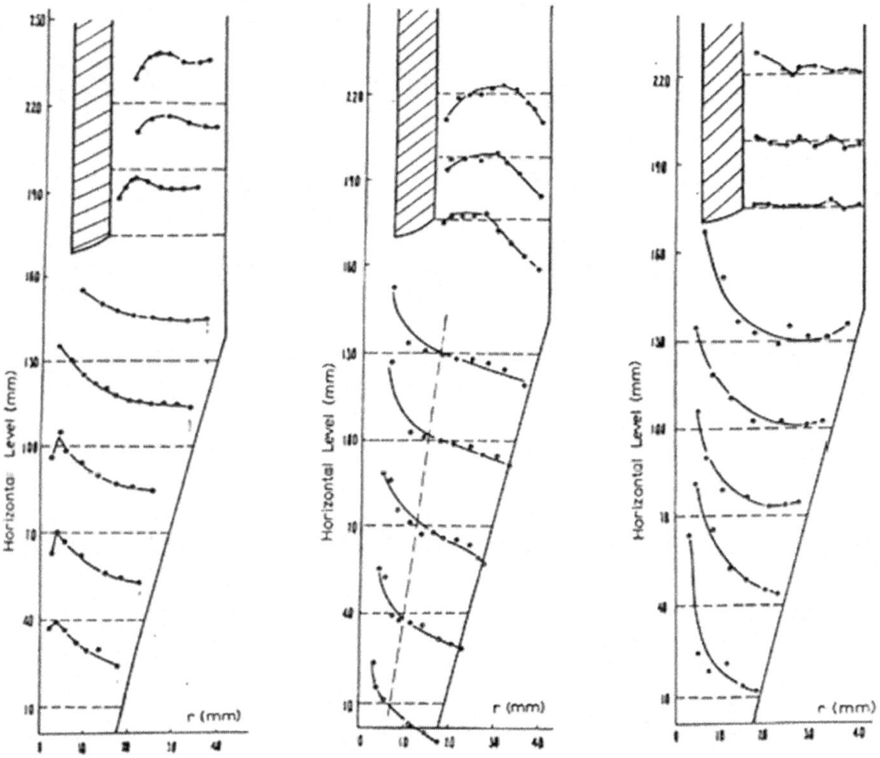

Fig. 5.6 Tangential, axial and radial velocity components according to Quian et al. [98]

temperature and pressure. For laminar incompressible flows they may be represented by the equations:

$$\frac{\partial \rho \boldsymbol{v}}{\partial t} + \nabla \cdot \rho \boldsymbol{v}\boldsymbol{v} = \nabla \cdot \boldsymbol{T} + \boldsymbol{b}$$
$$\nabla \cdot \boldsymbol{v} = 0 \tag{5.1}$$

where ρ and $\boldsymbol{v}$ are the density and velocity of the fluid and $\boldsymbol{T}$ and $\boldsymbol{b}$ are the tensor and external force field, respectively. To study specific problems, it is necessary to set a boundary value problem stipulating boundary conditions.

The turbulent flow of a Newtonian fluid is modeled with the instantaneous and time-averaged velocity, stress and external force fields of the Navier–Stokes equations, $\boldsymbol{v} = \overline{\boldsymbol{v}} + \boldsymbol{v}'$, $\boldsymbol{T} = \overline{\boldsymbol{T}} + \boldsymbol{T}'$ and $\boldsymbol{b} = \overline{\boldsymbol{b}} + \boldsymbol{b}'$. Reynolds Equations are obtained by integrating the resulting equations for a short time:

$$\frac{\partial \rho \overline{\boldsymbol{v}}}{\partial t} + \nabla \cdot \rho \overline{\boldsymbol{v}}\overline{\boldsymbol{v}} = \nabla \cdot \left(\overline{\boldsymbol{T}} - \rho \overline{\boldsymbol{v}'\boldsymbol{v}'}\right) + \overline{\boldsymbol{b}}$$

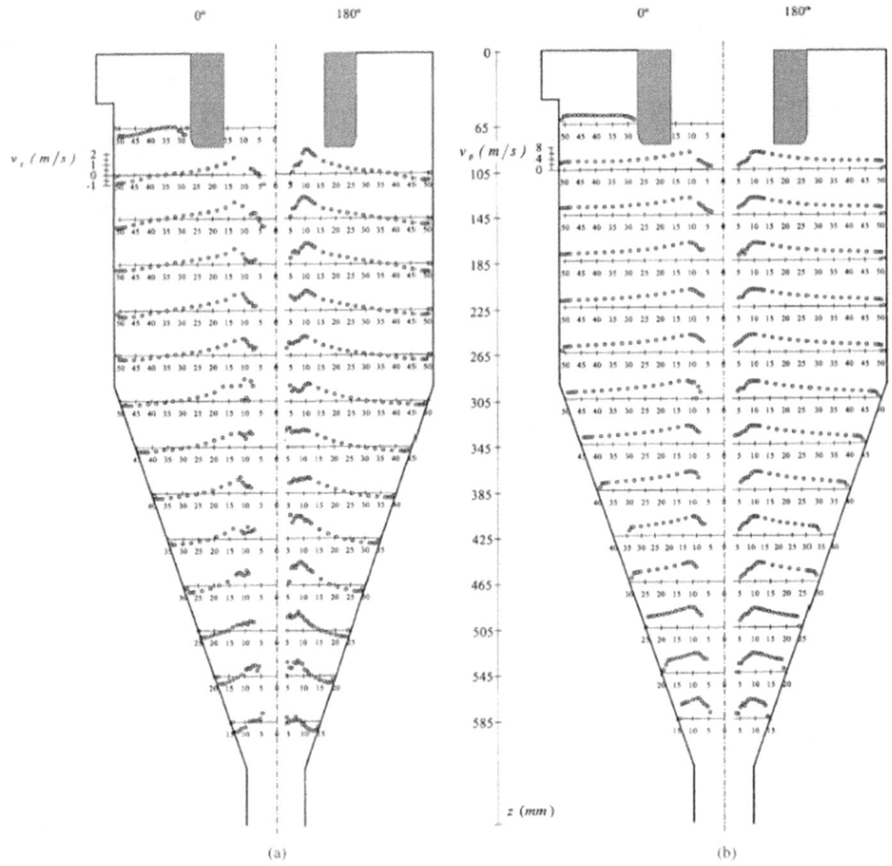

Fig. 5.7 **a** Axial and **b** tangential velocity profiles for cylindro-conical hydrocyclones

$$\nabla \cdot \overline{\boldsymbol{v}} = 0 \tag{5.2}$$

where $\overline{\boldsymbol{v}}$, $\overline{\boldsymbol{T}}$ and $\overline{\boldsymbol{b}}$ are the mean velocity, mean stress tensor and mean body force fields, while $\boldsymbol{v}'$ is the fluctuating velocity field, ρ is the constant fluid density and $\boldsymbol{T}_t = -\rho\overline{\boldsymbol{v}'\boldsymbol{v}'}$ are the Reynolds stresses. In most cases, researchers have used Reynolds equations to model hydrocyclones.

Unfortunately, the Navier Stokes equations or Reynolds equations, depending on the case, are non-linear and, therefore, difficult to solve. There are several approaches—exact solutions, extreme solutions, similarity and asymptotic models—to solving these boundary-value problems:

(1) ***Exact solutions***. The first and most common approach to solving these equations is to find exact solutions for highly idealized situations for which the non-linear terms vanish due to the geometry or symmetry of the flow. These

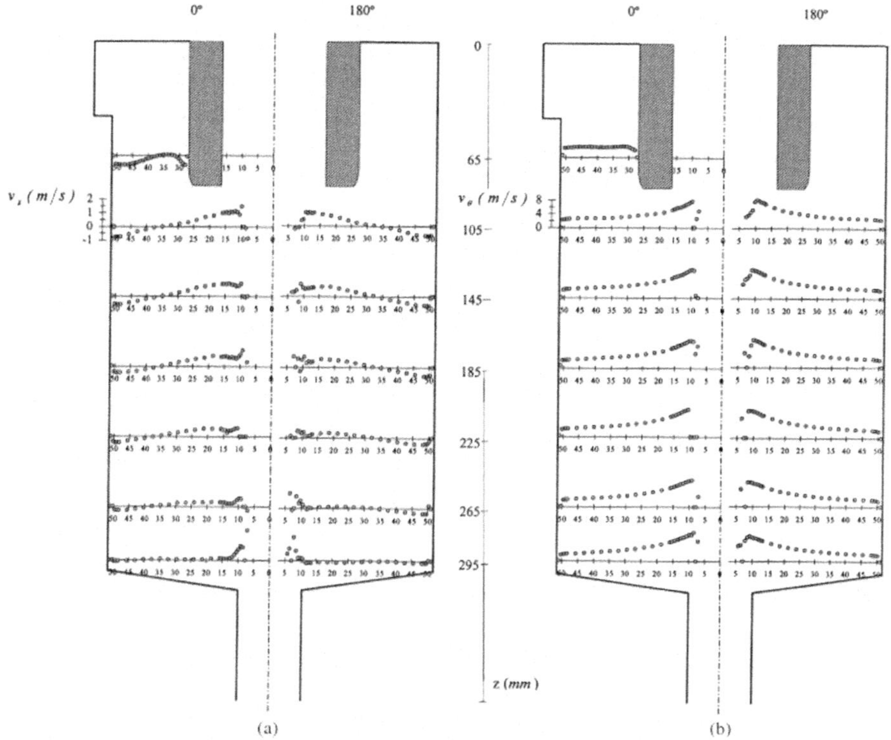

Fig. 5.8 **a** Axial and **b** tangential velocity profiles for flat-bottom hydrocyclones

solutions are rarely valid for real configurations, but often give an indication of the flow and pressure distribution to be expected in the real problem.

(2) ***Extreme solutions***. The difficulties encountered in obtaining general solutions to the flow equations have shown that it is advantageous to divide the flow into categories depending on the relative value of the several terms of the equations and to neglect the terms of lower orders of magnitude. To perform this task the Navier Stokes equations are written in dimensionless form:

$$\begin{gathered} \frac{1}{St}\frac{\partial \boldsymbol{v}^*}{\partial t^*} + \nabla^* \boldsymbol{v}^* \cdot \boldsymbol{v}^* = -\frac{1}{Ru}\nabla^* p^* + \frac{1}{\mathrm{Re}}\nabla^{*2}\boldsymbol{v}^* + \frac{1}{Fr}\boldsymbol{b}^* \\ \nabla^* \boldsymbol{v}^* = 0 \end{gathered} \tag{5.3}$$

with

$$St = \frac{t_0 u_0}{L}, \quad Ru = \frac{\rho u_0^2}{p_0}, \quad Re = \frac{\rho u_0 L}{\mu}, \quad Fr = \frac{u_0^2}{Lg}$$

where the stress tensor $\boldsymbol{T}$ was expressed as the sum of a pressure p and an extra stress modeled as a Newtonian fluid, L, t_0, u_o and p_0 are characteristic parameters of the flow, *St, Ru, Re* and *Fr* are the Strouhal, Ruark, Reynolds and Froude dimensionless numbers and the asterisk indicates dimensionless field variables. The Reynolds number is the ratio of the convective to the diffusive forces in the flow; therefore, it is a good parameter to divide the flow into three categories:

(a) ***Euler's isochoric flows***

Euler's isochoric flows, or inviscid flows, are obtained if the convective forces are much greater than the diffusive, or viscous, forces. In this case the Reynolds number is infinite, making the second term on the right side of (5.55) vanish:

$$\rho\frac{\partial v}{\partial t} + \nabla \boldsymbol{v} \cdot \boldsymbol{v} = -\nabla p + \boldsymbol{b}$$
$$\nabla \cdot \boldsymbol{v} = 0 \tag{5.4}$$

(b) ***Stokes flow***

Stokes flow is obtained if the convective forces are much smaller than the viscous forces. In this case the Reynolds number is very small (Re $\rightarrow$ 0), making the second term on the left side of (5.55) negligible:

$$\rho\frac{\partial v}{\partial t} = -\nabla p + \mu\nabla^2 \boldsymbol{v} + \boldsymbol{b}$$
$$\nabla \cdot \boldsymbol{v} = 0 \tag{5.5}$$

(c) ***Prandtl's flow***

Prandtl's flow, or boundary layer flow, is obtained when the Reynolds number is high but convective forces are on the same order of magnitude as the diffusive forces.

5.3 Model Based on the Exact Solutions of the Reynolds Equation

The first and most common approach to solving Navier Stokes equations is to find exact solutions for highly idealized situations for which the non-linear terms vanish due to the geometry or symmetry of the flow. These solutions are rarely valid for real configurations, but often give an indication of the flow and pressure distribution to be expected in the real problem.

In general, the turbulent velocity distribution in a hydrocyclone is three dimensional, as we have already discussed, but the flow can be approximated, except for the region close to the entrance, as an axial-symmetric motion. Kelsall [71] was the first author to determine the essential features of the flow pattern in a hydrocyclone. With a fair degree of approximation, his work shows that the radial and tangential velocities depend only on the r coordinate. Then, neglecting the body forces, Reynolds equations in cylindrical coordinates, with the turbulence represented by a constant eddy viscosity ν_ε much greater than the kinematical shear viscosity ν_t, can be written in the form:

$$\frac{\partial}{\partial r}(r v_r) + \frac{\partial}{\partial z}(r v_z) = 0 \tag{5.6}$$

$$v_r \frac{\partial v_r}{\partial r} - \frac{v_\theta^2}{r} = -\frac{1}{\rho}\frac{\partial p}{\partial r} + \nu_\varepsilon \frac{\partial}{\partial r}\left(\frac{1}{r}\frac{\partial}{\partial r}(r v_r)\right) \tag{5.7}$$

$$\frac{v_r}{r}\frac{\partial}{\partial r}(r v_\theta) = \nu_\varepsilon \frac{\partial}{\partial r}\left(\frac{1}{r}\frac{\partial}{\partial r}(r v_\theta)\right) \tag{5.8}$$

$$v_r \frac{\partial v_z}{\partial r} + v_z \frac{\partial v_z}{\partial z} = \nu_e \left(\frac{1}{r}\frac{\partial}{\partial r}\left(r\frac{\partial v_z}{\partial r}\right) + \frac{\partial^2 v_z}{\partial z^2}\right) \tag{5.9}$$

The approach followed by various authors to solve this set of equations depends on their assumptions regarding the form and dependence of the equations. Their guideline was the experimental velocity distribution obtained by authors such as Kellsal [71], Knowles et al. [72], Hsieh and Rajamani [63, 98]. Concha [35] organized these models based on the stated assumptions.

(a) **Models with $v_r = v_r(r), v_\theta = v_\theta(r)$ and $v_z = 0$**

All early authors ([101, 102, 111, 114] considered a cylindrical hydrocyclone with the feed entering through the entire cylindrical mantle at $r = R$ and having a radial velocity v_R and a tangential velocity $v_{\theta R}$ at the entrance (see Figs. 5.9 and 5.10).

Thus, the flow rate into the hydrocyclone is given by $Q = 2\pi R L v_R$ and the circulation at $r = R$ is $\Gamma(R) = 2\pi R v_{\theta R}$, where L is the height of the cylinder. Rietema and Krajenbrink [101] analyzed two cases, a hydrocyclone with and without an air core. The boundary conditions for a hydrocyclone without and with an air core, respectively, were:

$$\text{Without air core:}\quad v_\theta(R) = v_{\theta R},\quad v_r(R) = v_R,\quad v_r(0) = 0 \tag{5.10}$$

$$\text{With an air core:}\quad v_\theta(R) = v_{\theta R},\quad v_r(R) = v_R,\quad v_r(R_a = R_o) = 0 \tag{5.11}$$

where R_a is the radius of the air core and R_o is the radius of the overflow opening.

Since the hydrocyclone feed entered through the entire cylindrical mantle they assumed no axial velocity. This is equivalent to assuming that a linear sink evacuates the fluid at the cylinder axis. Thus, from the continuity Eq. (5.6) and boundary condition (5.11) for $r = R$, the radial velocity becomes:

$$v_r = R v_R / r \tag{5.12}$$

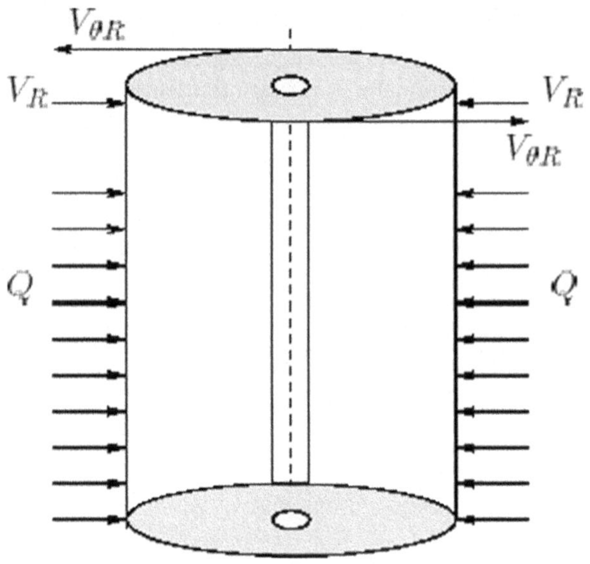

Fig. 5.9 Physical model of a cylindrical hydrocyclone

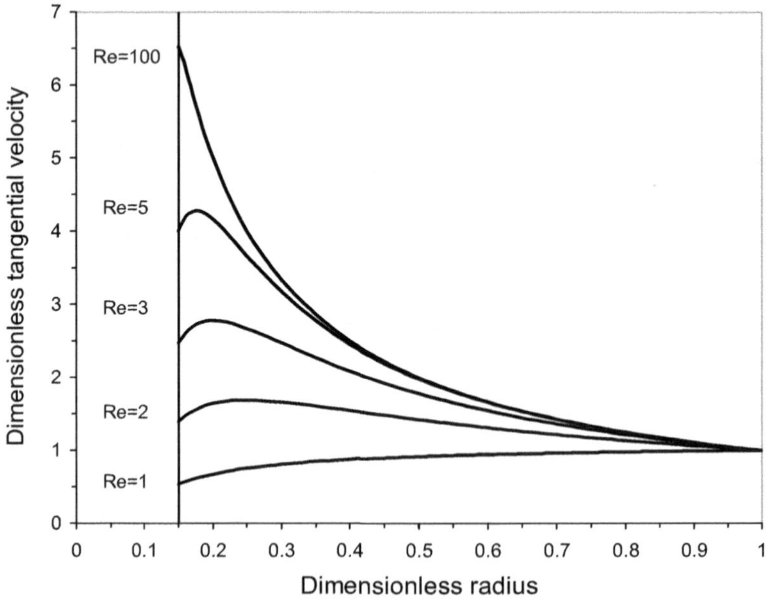

Fig. 5.10 Tangential velocity profile according to Rietema and Krajenbrink [101] for a cylindrical hydrocyclone with a dimensionless air core the size equal to the overflow $R_a^* = R_o^* = 0.15$

Introducing Eqs. (5.11) and (5.12) into Eq. (5.8), they solved for $v_\theta(r)$ with the radial Reynolds number, defined by $\mathrm{Re} = R v_R / \nu_e$, as a parameter. Their solutions are given in Eqs. (5.13) and (5.14) and Figs. 5.11 and 5.12. For a cylindrical hydrocyclone with an air core the size of the overflow R_o:

$$v_\theta^* = \begin{cases} \dfrac{\mathrm{Re} R_o^{*2-\mathrm{Re}} - 2r^{*2-\mathrm{Re}}}{\mathrm{Re} R_o^{*2-\mathrm{Re}} - 2} & \text{for } \mathrm{Re} > 0 \neq 2 \\ \dfrac{1 - 2\ln R_o^*/r^*}{1 - 2\ln R_o^*} & \text{for } \quad \mathrm{Re} = 2 \\ 1 & \text{for } \quad \mathrm{Re} = \infty \end{cases} \tag{5.13}$$

For a cylindrical hydrocyclone without an air core:

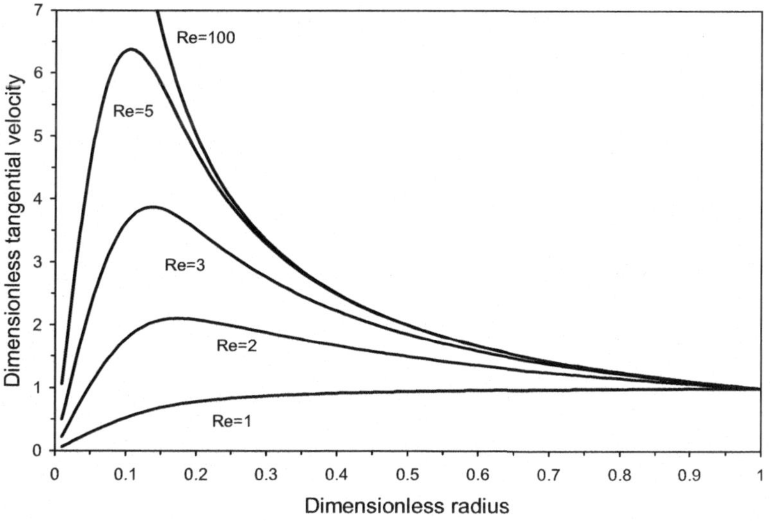

Fig. 5.11 Tangential velocity profile according to Rietema and Krajenbrink [101] for a hydrocyclone without air core and a dimensionless overflow opening of $R_0^* = 0.15$

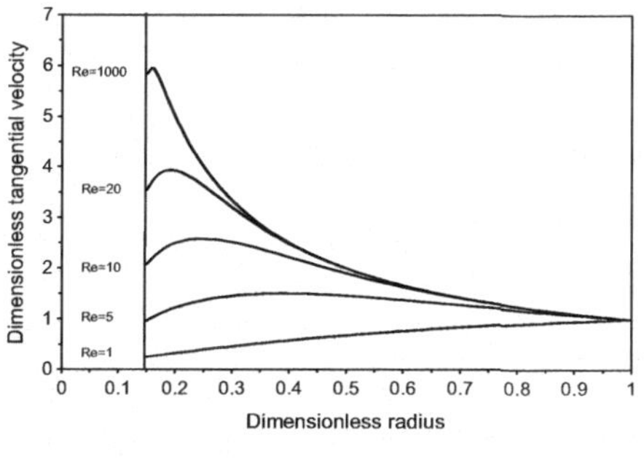

Fig. 5.12 Tangential velocity profile according to Rietema with $R_a^* = 0.15$ [102]

$$v_\theta^* = \begin{cases} \frac{R_o^{*2-\text{Re}}(2-\text{Re}-2\exp(-\text{Re}/2))+\text{Re}\exp(\text{Re}/2)r^{*2-\text{Re}}}{R_o^{*2-\text{Re}}(2-\text{Re}-2\exp(-\text{Re}/2))+\text{Re}\exp(\text{Re}/2)} & \text{for Re} > 0 \neq 2,\ 1 > r^* > R_o^* \\ \frac{1+2\ln(R_o^*/r^*)-\exp(1)}{1+2\ln R_o^*-\exp(1)} & \text{for} \quad \text{Re} = 2, \quad 1 > r^* > R_o^* \\ \frac{R_o^{*2-\text{Re}}(2-\text{Re})+\left(1-\exp\left(-r^*/R_0^*\right)^2(\text{Re}/2)\right)}{R_o^{*2-\text{Re}}(2-\text{Re}-2\exp(-\text{Re}/2))+\text{Re}\exp(\text{Re}/2)} & \text{for Re} > 0 \neq 2,\ R_o^* > r^* > 0 \\ \frac{\exp\left(1-(r^*/R_o^*)\right)-\exp(1)}{1+2\ln R_o^*-\exp(1)} & \text{for} \quad \text{Re} = 2, \quad R_o^* > r^* > 0 \\ 1 & \text{for} \quad \text{Re} = \infty, \quad 1 > r^* > 0 \end{cases} \tag{5.14}$$

Comparing their result with experimental information, Rietema and Krajenbrink concluded that the correct radial Reynolds number for this problem is close to $\text{Re} \approx 3$. In other words, the turbulent eddy viscosity should be a constant on the order of $\nu_e \approx 3 \times 10^{-4}$ m/s. In a further analysis, these authors indicated that, if wall effect were considered, the tangential velocity would depend on the z coordinate.

In 1961, Rietema published a study assuming that the radial velocity was constant ($v_r = v_R$) in the entire flow field. He claimed that this assumption was based on Kelsall's measurements. (as a matter of fact, Kelsall's experiments show that the radial velocity is proportional to the radial distance from the axis). With $v_r = v_R$, Rietema solved Eq. (5.7) and pointed out that these results were valid for a conical hydrocyclone with a Reynolds number of Re $\approx$ 10 (turbulent kinematical eddy viscosity of $\nu_e = 10^{-4}\,\text{m}^2/\text{s}$. His solution, for a dimensionless air core of $R_a^* = 0.15$, is given in Eqs. (5.15) to (5.17) and Fig. 5.13.

$$v_\theta^* = C_2(\text{Re}) - C_1\left(\text{Re}, R_a^*\right)\exp\left(-\text{Re}r^*\right)\left(\frac{r^*}{\text{Re}} + \frac{1}{\text{Re}^2}\right) \tag{5.15}$$

with

$$C_1\left(\text{Re}, R_a^*\right) = \frac{\text{Re}^2}{-(1+\text{Re})\exp(-\text{Re}) + \left(\left(R_a^{*2}/2\right) + R_a^*\text{Re} + 1\right)\exp\left(-\text{Re}R_a^*\right)} \tag{5.16}$$

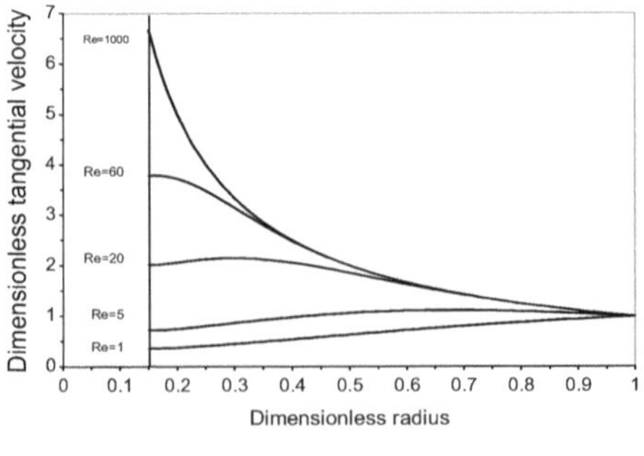

Fig. 5.13 Tangential velocity profile according to Rietema with $R_a^* = 0.15$ [103]

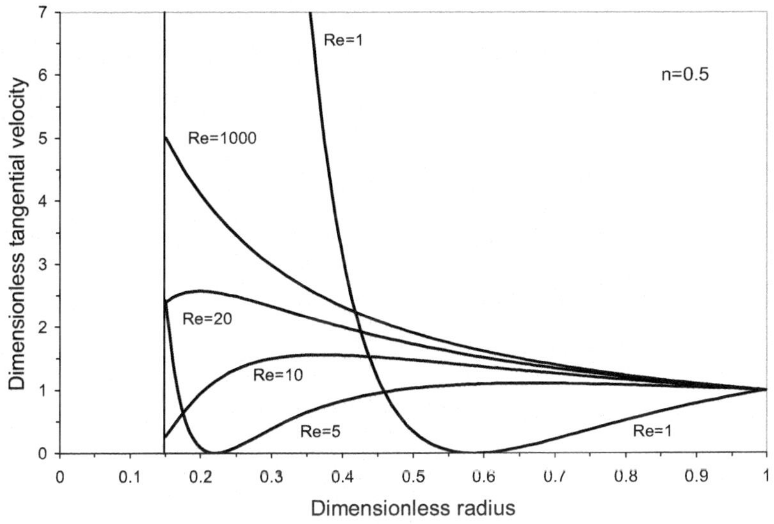

Fig. 5.14 Tangential velocity profile according to Eq. (5.20) for a power law fluid with $n' = 0.5$ and $R_a^* = 0.15$

and

$$C_2(\mathrm{Re}) = 1 + C_1(\mathrm{Re}R_a^*)\exp(-\mathrm{Re})\left(\frac{1}{\mathrm{Re}} + \frac{1}{\mathrm{Re}^2}\right) \tag{5.17}$$

In 1962, Rietema solved the same set of equations and boundary conditions $v_r = (v_R/R)r$, in accordance with Kelsall's measurements. His solution was (see Figs. 5.14 and 5.15):

$$v_\theta^* = \frac{1}{r^*} + \frac{C_3(\mathrm{Re}, R_a^*)}{r^*}\left(\exp(-(1/2)\mathrm{Re}) - \exp\left(-(1/2)\mathrm{Re}r^{*2}\right)\right) \tag{5.18}$$

with

$$C_3(\mathrm{Re}, R_a^*) = \frac{1}{\left(1 + (1/2)\mathrm{Re}R_a^{*2}\right)\exp\left(\mathrm{Re}R_a^{*2}\right) - \exp(-(1/2)\mathrm{Re})} \tag{5.19}$$

In a final work, Van Duijin and Rietema [114] went back to their 1958 solution, but now using the equations for solid/fluid mixture.

Two additional works were done on the same line as Rietema's [110] and Upadrashta et al. (1982) which analyzed the flow of a pseudo-plastic fluid in a hydrocyclone. Their analysis was justified since concentrated suspensions, such as those used in mineral industries, are usually non-Newtonian and of the pseudo-plastic (power law) type. Based on the measurements of Kelsall [71], (which were made with water) they assumed that the radial velocity had the form $v_r = (v_R/R)r$. By

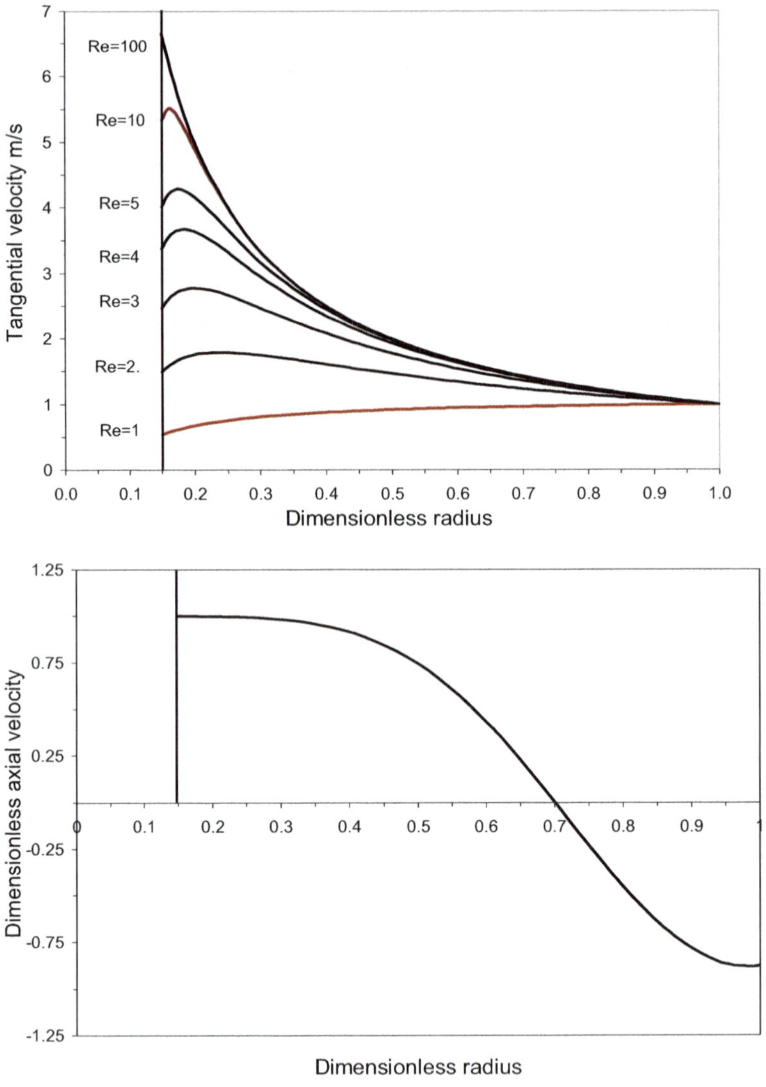

Fig. 5.15 Tangential and axial velocity profile with v_z^* given by (5.26)

introducing this term and a constitutive equation for the non-Newtonian fluid, they obtained the following expression for the dimensionless tangential velocity and the dimensionless air core radius, respectively:

$$v_\theta^* = \frac{1}{r^*}\left(1 - \frac{1}{\mathrm{Re_m}} \frac{2^{n'}(1-n')}{n'}\left(\left(\frac{1}{r^*}\right)^{2n'} - 1\right)\right)^2 \tag{5.20}$$

$$R_a^* = \left(\frac{1}{\mathrm{Re_m}} \frac{2n'(2n'+1)(1-n')}{2n'(1-n') + n'\mathrm{Re_m}} \right)^{1/2n'} \tag{5.21}$$

For the case of a Newtonian fluid, $n' = 1$, (5.20) reduces to a free vortex.

(b) **Models with $v_r = v_r(r), v_\theta(r)$ and $v_z(r,z)$**

The problem with the previous models is that they do not explicitly consider the axial velocity. Furthermore, their solutions serve only to calculate the centrifugal force, but give no clue as to how the particles are separated and how the particles and fluid are evacuated from the device. Furthermore, the authors were not concerned about whether their models satisfied the continuity equation. The conclusion is that it is essential to incorporate the axial component of the Reynolds equation into the dynamic process.

Integrating the continuity Eq. (5.6) yields:

$$r v_r(r) = -\int \frac{\partial v_z}{\partial z} r dr + K \tag{5.22}$$

Since, by assumption, the radial component of the velocity is a function or r only, it is clear from (5.22) that the axial velocity gradient $\partial v_z / \partial z$ must be independent of z; that is, the most general form of the axial velocity should be [55]:

$$v_z = f(r)z + g(r) \tag{5.23}$$

Case 1: $f = 0$; $v_z = g(r)$.

First consider the case in which the axial velocity is an arbitrary function of r independent of z, that is:

$$v_z = g(r), \ \frac{\partial v_z}{\partial z} = 0 \text{ and } v_r = \frac{R v_R}{r} \tag{5.24}$$

Equation (5.24) implies that radial flow becomes infinite at the cylinder axis; therefore, a singularity is introduced into the equations if the axis of symmetry is included in the domain. In that case, no solution exists for which the tangential velocity diminishes as the radius diminishes, reaching a finite value at the axis. On the other hand, for a hydrocyclone with an air core of radius R_a, the tangential velocity distribution has the following form:

$$v_\theta^* = \frac{1}{r^*} \frac{(2/\mathrm{Re})(r/R_a)^{-(R_a-2)} - 1}{(2/\mathrm{Re})(R_a/R)^{-(R_a-2)} - 1} \tag{5.25}$$

Since the function g(r) is arbitrary, assume the following form for the dimensionless axial velocity, which compares favorably with experimental data:

$$v_z^* = 1 - 1.8 \sin^6((r - 0.5)/R) \tag{5.26}$$

Figure 5.17 shows the tangential and axial velocity distributions given by Eqs. (5.25) and (5.26) (Fig. 5.16).

Case 2: $v_z = f(r)z + g(r)$.

Consider the case of an axial velocity as a linear function of z, with $g(r)$ an arbitrary function of r. The problems analyzed by Rietema [102] and Upadrashta et al. [111] correspond to this case. Rietema assumed a constant radial velocity ($v_r = v_R$), which

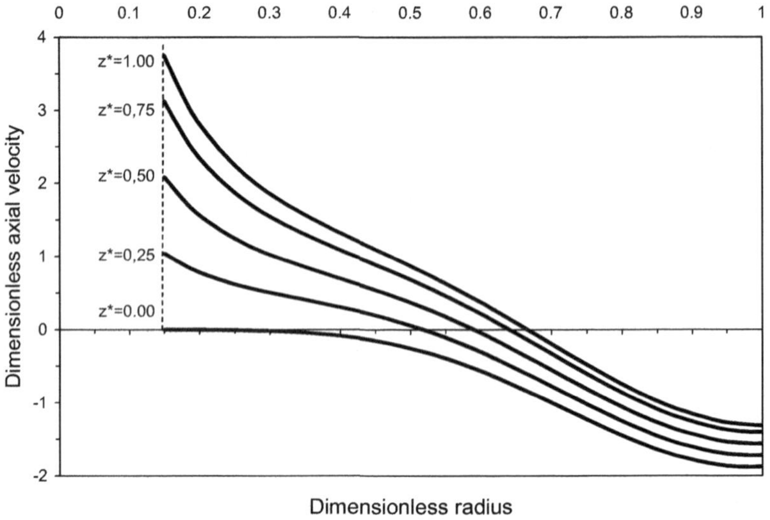

Fig. 5.16 Axial velocity profile according to Eq. (5.27)

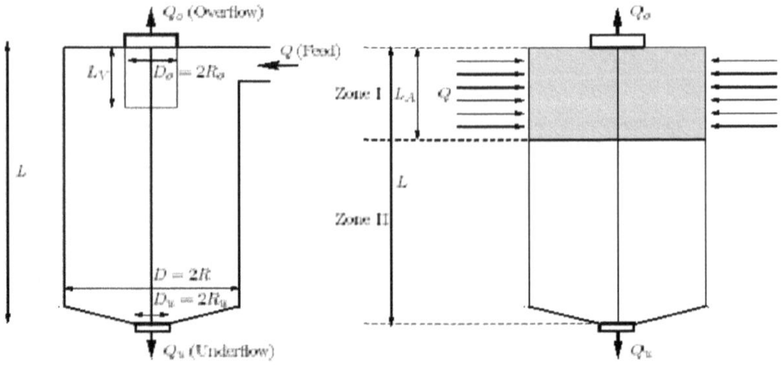

a) Physical model b) Axial symmetrical model

Fig. 5.17 Schematic representation of the flat-bottom hydrocyclone implies that $f(r) = v_R/r$. For this case, $g(r)$ can be an arbitrary function of r. Thus:

$$v_r = v_R,\ v_\theta = v_\theta(r) \text{ and } v_z = \frac{v_R}{r}z - 1.8\sin^6\frac{r-0.5}{R} \tag{5.27}$$

Their axial velocity distribution is given in Fig. 5.18.

Collantes et al. [31] applied this case to a flat-bottom hydrocyclone, with f and g as arbitrary functions of r to be determined. Introducing the general form of the axial velocity into the field equations, they wrote the following equations for continuity and the radial, tangential and axial components of the velocity:

$$\text{Continuity equation:} \quad \frac{1}{r}\frac{d}{dr}(rv_r(r)) + f(r) = 0 \tag{5.28}$$

$$\text{Radial component:} \quad v_r(r)\frac{dv_r(r)}{dr} - \frac{v_\theta^2}{r} - \nu_\varepsilon\frac{d}{dr}\left(\frac{1}{r}\frac{d}{dr}(rv_r(r))\right) = -\frac{1}{\rho}\frac{\partial p(r,z)}{\partial r} \tag{5.29}$$

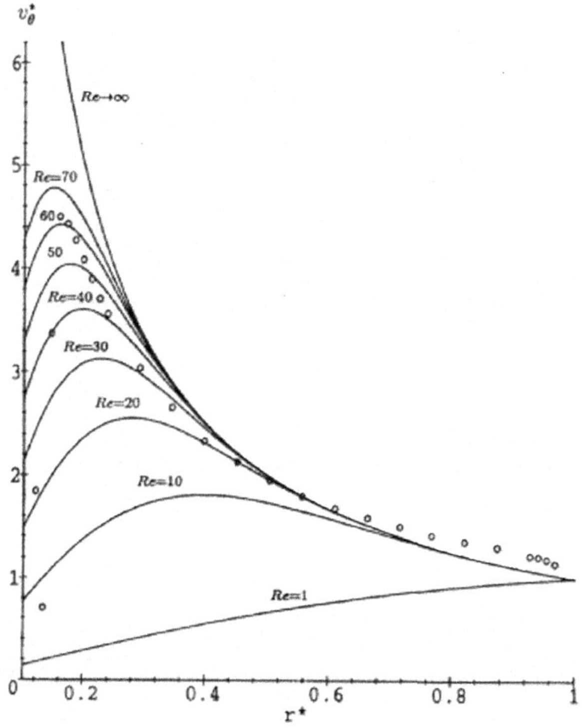

Fig. 5.18 Simulated tangential velocity distribution according to Collantes et al. [31]. Comparison with experimental data from Chiné and Concha [28]

$$\text{Tangential component:} \quad \frac{v_r(r)}{r}\frac{d}{dr}(r v_\theta(r)) = \nu_\varepsilon \frac{d}{dr}\left(\frac{1}{r}\frac{d}{dr}(r v_\theta(r))\right) \tag{5.30}$$

$$\text{Axial component:} \left(v_r(r)\frac{df(r)}{dr} + f^2(r) - \frac{\nu_\varepsilon}{r}\frac{d}{dr}\left(r\frac{df(r)}{dr}\right)\right)z + \left(v_r(r)\frac{dg(r)}{dr} + f(r)g(r) - \frac{\nu_\varepsilon}{r}\frac{d}{dr}\left(r\frac{dg(r)}{dr}\right)\right) = -\frac{1}{r}\frac{\partial p(r,z)}{\partial r} \tag{5.31}$$

Eliminating the pressure from Eqs. (5.29) and (5.31), the following set of equations is obtained for $f(r)$ and $g(r)$:

$$\begin{aligned} &\frac{d}{dr}\left(v_r(r)\frac{df(r)}{dr} + f^2(r) - \nu_\varepsilon \frac{1}{r}\frac{d}{dr}\left(r\frac{df}{dr}\right)\right) = 0 \\ &\frac{d}{dr}\left(v_r(r)\frac{dg(r)}{dr} + f(r)g(r) - \nu_\varepsilon \frac{1}{r}\frac{d}{dr}\left(r\frac{dg}{dr}\right)\right) = 0 \end{aligned} \tag{5.32}$$

To satisfy the continuity equation a stream function ψ is defined, such that:

$$v_r(r) = \frac{1}{r}\frac{\partial \psi}{\partial z} \equiv \frac{q(r)}{r} \text{ and } v_z = -\frac{1}{r}\frac{\partial \psi}{\partial r} \equiv -\frac{1}{r}\frac{dq}{dr}z + g(r) \tag{5.33}$$

Defining the dimensionless variable $\eta = (r/R)^2$ and introducing these expressions into Eq. (5.32), we obtain for $q(\eta)$ and $g(\eta)$:

$$\begin{aligned} &\frac{dq(\eta)}{d\eta}\frac{d^2q(\eta)}{d\eta^2} - q(\eta)\frac{d^3q(\eta)}{d\eta^3} = 0 \\ &\frac{d}{d\eta}\left(q(\eta)\frac{dg(\eta)}{d\eta}\right) - \frac{d}{d\eta}\left(\frac{dq(\eta)}{d\eta}g(\eta)\right) = 0 \end{aligned} \tag{5.34}$$

To obtain this equation, the viscous terms of Eq. (5.32) were neglected. This is the same assumption made by Bloor and Ingham [14], which gave good results for the axial velocity (see previous section). A solution to this ordinary differential equation system, compatible with the boundary conditions in a flat-bottom hydrocyclone, where the feed enters the equipment through a ring of height L_A in the upper part of the cylinder (see Fig. 5.19), is (Fig. 5.20):

Zone I

$$v_r(r) = -Q/2\pi L_A r \equiv v_R R/r\left(\frac{Q_0}{Q}\left(1-\exp\left(-5.6\left(\frac{r}{R_o}\right)^2\right)\right) - \sin\left(\beta\left(\frac{r}{R}\right)^2\right)\right) \tag{5.35}$$

$$v_z(r,z) = \frac{Q}{\pi R^2}\left(\beta\cos\left(\beta\left(\frac{r}{R}\right)^2\right) + \left(5.6\frac{Q_o}{Q}\left(\frac{R}{R_o}\right)^2\exp\left(-5.6\left(\frac{r}{R}\right)^2\right) - \beta\left(\frac{r}{R}\right)^2\right)\frac{z}{L_A}\right) \tag{5.36}$$

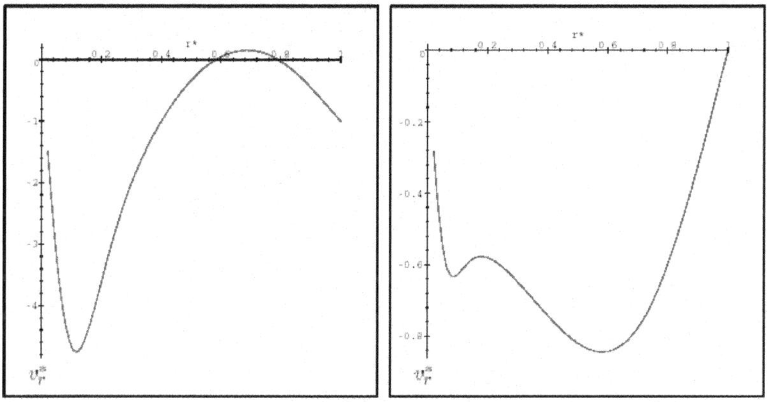

a) Zone I b) Zone II

Fig. 5.19 Radial velocity distribution for zones I and II, according to Collantes et al. [31]

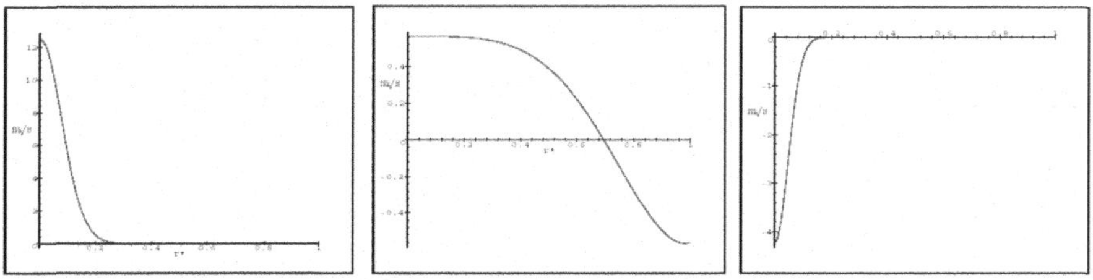

a) At the top wall. b) At the zone I/zone II interface. c) At the bottom wall.

Fig. 5.20 Simulated axial velocity distribution, according to Collantes et al. [31]

$$v_\theta(r) = \frac{\Gamma_R}{2\pi r}\left(\frac{1-\exp\left(-\frac{1}{2}\mathrm{Re}\beta\left(\frac{r}{R}\right)^2\right)}{1-\exp\left(-\frac{1}{2}\mathrm{Re}\beta\right)}\right) \tag{5.37}$$

where $\beta = \pi + \arcsin(Q_u/Q)$, Q, Q_u and Q_o are the volume feed and the overflow rate, respectively, and R is the hydrocyclone radius. All the volume flow rates are taken positive. Note that in (5.35) the factor $-Q/2\pi L_A r \equiv v_R R/r$ and in (5.36) and (5.37) $Q/\pi R^2 \equiv 2v_R L_A/R$ and $\Gamma_R/2\pi r = v_{\theta R} R/r$.

Zone II

$$v_r(r) = -\frac{Q}{2\pi(L-L_A)r}\left(\sin\left(\beta\left(\frac{r}{R}\right)^2\right) + \frac{Q_u}{Q}\left(1-\exp\left(-5.6\left(\frac{r}{R}\right)^2\right)\right)\right) \tag{5.38}$$

$$v_z(r,z) = -\frac{Q}{\pi R^2}\begin{pmatrix} 5.6\frac{Q_u}{Q}\left(\frac{R}{R_u}\right)^2 \exp\left(-5.6\left(\frac{r}{R}\right)^2\right) - \left(\beta\cos\left(\frac{r}{R}\right)^2\right) \\ +5.6\frac{Q_u}{Q}\left(\frac{R}{R_u}\right)^2 \exp\left(-5.6\left(\frac{r}{R}\right)^2\right)\frac{z}{L-L_A} \end{pmatrix} \quad (5.39)$$

$$v_\theta(r) = \frac{\Gamma_R}{2\pi r}\left(\frac{1-\exp\left(\frac{\mathrm{Re}}{2}\beta\frac{L_A}{L-L_A}\left(\frac{r}{R}\right)^2\right)}{1-\exp\left(\frac{\mathrm{Re}}{2}\beta\frac{L_A}{L-L_A}\right)}\right) \quad (5.40)$$

In (5.38) $Q/2\pi(L-L_A)r = v_R R L_A / r(L-L_A)$ and in (5.39) $Q/\pi R^2 \equiv 2v_R L_A/R$. In both zones the pressure distribution can be calculated from:

$$\begin{aligned} p(r,z) = p_0 &- \frac{\rho}{2}\left(v_r(r)f'(r) + f^2(r)\right)z^2 \\ &- \rho\left(v_r(r)g'(r) + f(r)g(r)\right)z + \rho\int\frac{v_\theta^2}{r}\mathrm{d}r - \frac{1}{2}v_r^2(r) \end{aligned} \quad (5.41)$$

From the exact solutions of the Reynolds equations, as given in this section, and from the conclusions of the various researchers who have modeled the hydrocyclone in this manner, we can conclude that the radial velocity depends highly on the shape of the axial velocity. One can obtain completely different radial velocity distributions by choosing appropriate axial flows. Nevertheless, the shape of the tangential velocity, as a combination of a free vortex and a rigid motion, is insensitive to the form of these axial and radial velocities. Only the maximum value of the tangential velocity and its position vary as the radial velocity changes. It is important to note that none of the radial velocities chosen by Rietema or Upadrashta have associated axial velocities that resemble the experimental ones.

For the axial velocity to be a consequence of the radial velocity, it must be a function of r and at least a linear function of z. Finally, we can indicate that, in modeling the hydrocyclone, selecting among the several solutions of the field equations by comparing the value of the maximum tangential velocity with experimental values is not the proper way to go. The three, or at least two, velocity components must be compared, and the axial velocity in particular must be reasonable.

(c) Models with $v_\theta = v_\theta(r), v_r = v_r(r,z)$ and $v_z = v_z(r,z)$

The more recent experimental data of Hsieh and Rajamani [99] and Quian et al. [98] confirm that the radial and axial velocities are functions of the radial and axial coordinates.

Bloor and Ingham ([16], 1983) analyzed the inviscid isochoric flow within a hydrocyclone, as we already saw. Assuming a free vortex for the tangential velocity, they obtained the radial and axial components as functions of the radial and axial coordinates. Since these velocities compared favorably with Kelsall's experimental information, Bloor and Ingham used the resulting axial and radial velocities to recalculate the tangential velocity from Reynolds's equations with variable eddy viscosity. Consider the tangential component of Reynolds's equation:

$$\frac{v_r}{r}\frac{\partial}{\partial r}(r v_\theta) + v_z\frac{\partial v_z}{\partial z} = \frac{1}{r^2}\frac{\partial}{\partial r}\left(v_\varepsilon r^3 \frac{\partial}{\partial r}\left(\frac{v_\theta}{r}\right) + \frac{\partial}{\partial z}\left(v_\varepsilon \frac{\partial v_z}{\partial z}\right)\right) \tag{5.42}$$

Using dimensionless coordinates: $r^* = r/R$, $v_\theta^* = v_\theta / v_{\theta R}$, radial and axial velocities given by $v_r = -A\frac{r^{3/2}}{Z^2}$ and $v_z = (1/2)A\left(1/r^{1/2}\right)(3\alpha - 5(r/z))$, and assuming that the eddy viscosity has the form $\nu_\varepsilon = \nu_t/Z^2$ (necessary for the tangential velocity to be independent of z), Bloor and Ingham obtained from (5.68) with boundary conditions $v_\theta^*(1) = 1$ and $v_\theta^*(0) = 0$:

$$v_\theta^* = \frac{1}{r^*}\gamma\left(\frac{4}{5}, \frac{2}{5}, r^{*5/2}\lambda_{\mathrm{Re}}\right) \Big/ \gamma\left(\frac{4}{5}, \frac{2}{5}, \lambda_{\mathrm{Re}}\right) \tag{5.43}$$

where λ_{Re} is a type of Reynolds number defined by $\lambda_{\mathrm{Re}} = AR^{5/2}/\nu_t$, γ is the incomplete gamma function and ν_t is fluid kinematic viscosity. The assumption that $\nu_\varepsilon = \nu_t/Z^2$ implies that the length scale ℓ of the turbulence in Prandtl's mixing length theory, $\nu_\varepsilon = \ell^2\left|K + \frac{\partial v_\theta}{\partial r} - \frac{v_\theta}{r}\right|$, varies as $\ell = h(r)/z$ (Bloor and Ingham 1983). Figure 5.23 gives the new tangential velocity distribution (Figs. 5.21, 5.22and 5.23).

In 1987 Bloor and Ingham further analyzed the swirling flow in hydrocyclones without underflow in terms of the vorticity function. They started from the continuity equation in spherical coordinates and the Bernouilli equation:

$$\nabla \cdot \boldsymbol{v} = \mathbf{0} \quad \nabla B = \boldsymbol{v} \times \boldsymbol{w} \tag{5.44}$$

where B is the Bernouilli and $\boldsymbol{w}$ is the vorticity defined by $\boldsymbol{w} = \nabla \times \boldsymbol{v}\, \mathbf{w} = \nabla \times \mathbf{v}$. In component form:

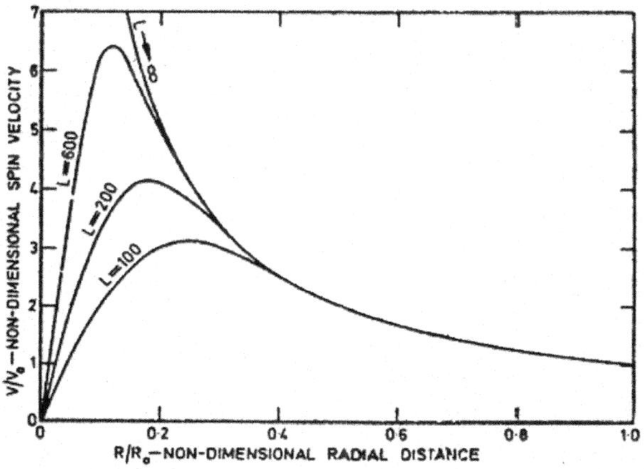

Fig. 5.21 Tangential velocity distribution according to Bloor and Ingahm [16]

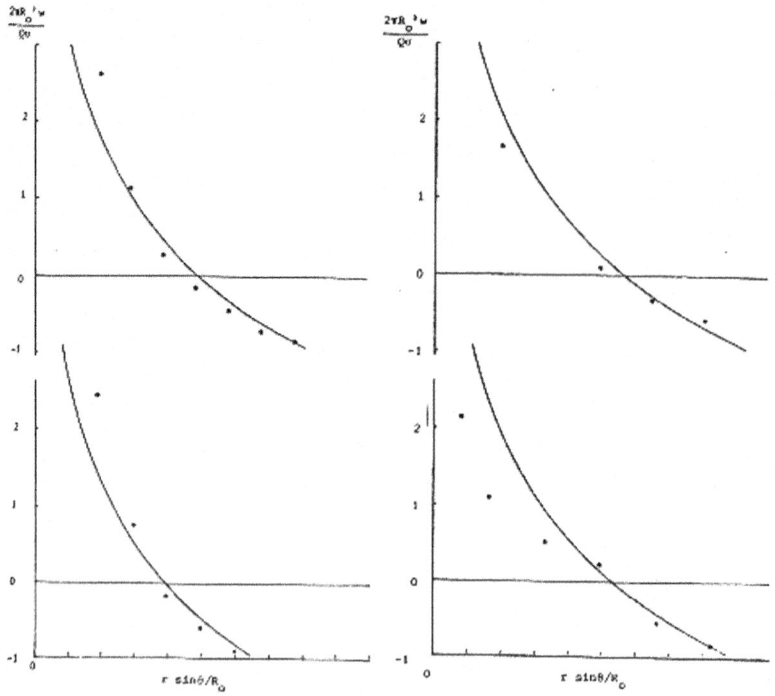

Fig. 5.22 Velocity distribution according to Bloor and Ingham (1987b) compared with experimental values of Kelsall [71] and Knowles et al. [72]

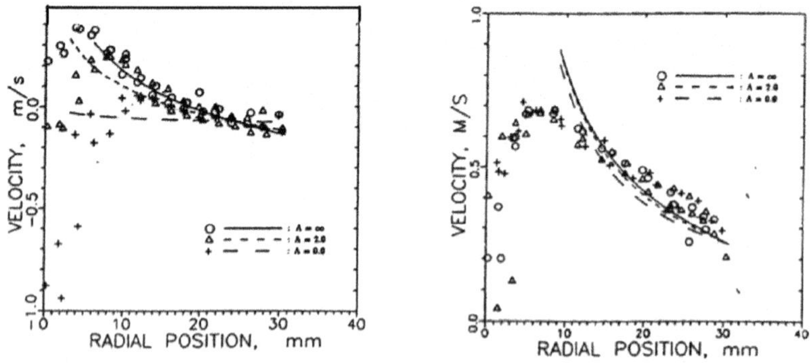

a) Tangential velocity distribution

b) Axial velocity distribution

Fig. 5.23 Simulated velocity distribution compared with measurements according to Hwang et al. [64]

$$\frac{\partial}{\partial \xi}\left(\xi^2 v_\xi \sin\phi\right) + \frac{\partial}{\partial \phi}\left(\xi v_\phi \sin\phi\right) = 0 \tag{5.45}$$

$$w_\xi = \frac{1}{\xi \sin\phi}\frac{\partial}{\partial\phi}(v_\theta \sin\phi);\ w_\phi = -\frac{1}{\xi}\frac{\partial}{\partial\xi}(\xi v_\theta);\ w_\theta = \frac{1}{\xi}\left(\frac{\partial}{\partial\xi}(\xi v_\phi) - \frac{\partial v_\xi}{\partial\phi}\right) \tag{5.46}$$

To satisfy the continuity equation a stream function is used with components:

$$v_\xi = \frac{1}{\xi^2 \sin\phi}\frac{\partial\psi}{\partial\phi} \text{ and } v_\phi = -\frac{1}{\xi\sin\phi}\frac{\partial\psi}{\partial\xi} \tag{5.47}$$

From the ϕ and θ components of Eq. (5.46), the θ component of the velocity and the ξ and θ components of vorticity were obtained, and the solution to the stream function led to the following velocity distributions:

$$v_\theta = \frac{R}{r}v_{\theta R}\left(1 - \frac{2\psi}{R^2 v_z}\right)^{1/2} \tag{5.48}$$

$$\begin{aligned} v_r = {} & 2\sigma \sin\phi\Big(\big(\mathrm{cosec}^2\alpha + \ln(\tan\alpha/2) - \mathrm{cosec}\,\alpha\cot\alpha\big)\cos\phi - (\cos\phi\ln(\tan\alpha/2))\Big) \\ & + \frac{r^2\cos\phi(\cos^2\phi - \cos^2\alpha - \sin^2\phi)}{\sin^2\phi(1 - 5\cos^2\phi)} \end{aligned} \tag{5.49}$$

$$\begin{aligned} v_z = {} & 2\sigma \sin\phi\Big(\big(\mathrm{cosec}^2\alpha + \ln(\tan\alpha/2) - \mathrm{cosec}\alpha\cot\alpha\big)\cos\phi - (\cos\phi\ln(\tan\alpha/2))\Big) \\ & + \frac{r^2\cos\phi(\cos^2\phi - \cos^2\alpha - \sin^2\phi)}{\sin^2\phi(1 - 5\cos^2\phi)} \\ & + 2\sigma\left(\begin{array}{l}\left(\begin{array}{l}\big(\mathrm{cosec}^2\alpha + \ln(\tan\alpha/2) - \mathrm{cosec}\alpha\cot\alpha\big)\sin^2\phi \\ -\sin^2\phi\ln(\tan\phi/2) + \cos\phi - 1\end{array}\right) \\ +2r^2\frac{\cos^2\phi - \cos^2\alpha}{1 - 5\cos^2\alpha}\end{array}\right) \end{aligned} \tag{5.50}$$

where

$$\sigma = -\pi R^2 v_{\theta R}^2 / Q v_z. \tag{5.51}$$

Bloor and Ingham ([11], 1987b) showed that their solution compared well with experimental values of Kelsall [71] and Knowles et al. [72]. They indicated that the way the entry conditions are modeled has a profound effect on the nature of the flow (see Fig. 5.24).

Hwang et al. [67] extended the work of Bloor and Ingham [15] to include underflow through the apex. To do so, they used a stream function for the sink in the form:

$$\psi_1 = \frac{Q\cos\phi\cos\alpha}{2\pi(1 + Q_o/Q_u)(1 - \cos\alpha)} \tag{5.52}$$

With this modification the velocity distributions become (see Fig. 5.25):

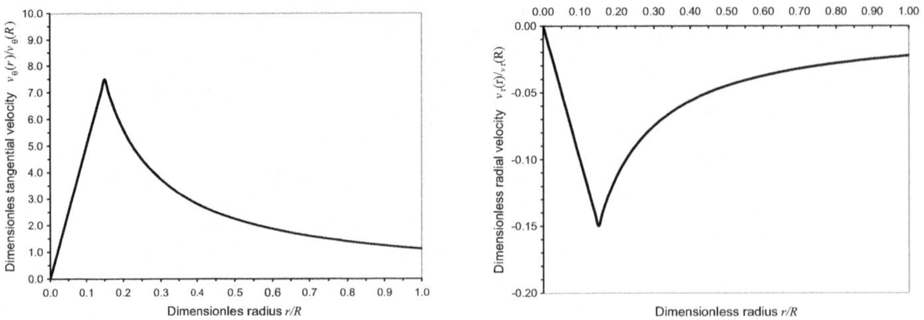

Fig. 5.24 Tangential and radial velocity profiles for a potential flow model of a hydrocyclone for $\Gamma = 7.065$, $Q = 0.1413$ and $\omega = 50$

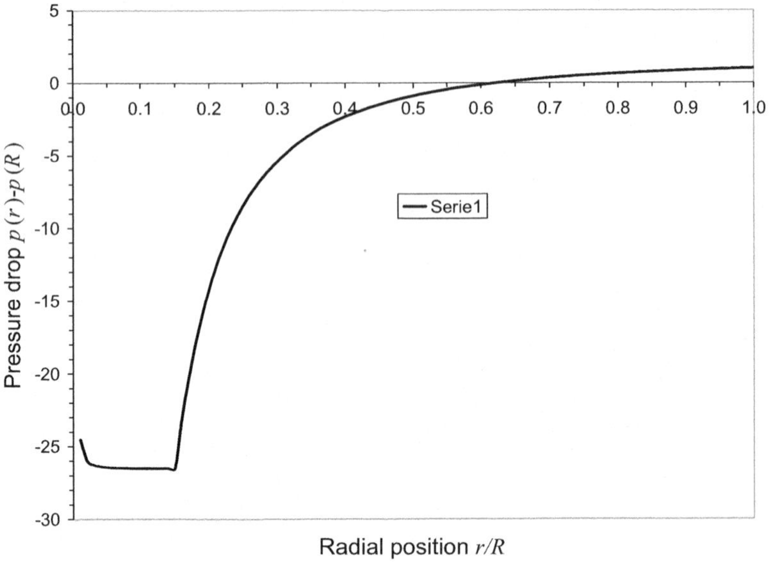

Fig. 5.25 Pressure distribution for a potential flow model of a hydrocyclone for $\Gamma = 7.065$, $Q = 0.1413$ and $\omega = 50$

$$v_\theta = \frac{R}{r} v_{\theta R} \left(1 - \frac{2\psi}{R^2 v_z}\right)^{1/2} \tag{5.53}$$

$$v_r = -\frac{Q\sin\phi}{2\pi R^2}\left(\frac{1+\Lambda}{\eta^2}\frac{1}{1-\cos\alpha} + 2\sigma\left(\begin{array}{l}\frac{D(\alpha)}{\sin^2\alpha}\cos\phi - 1 - \cos\phi\ln(\tan(\phi/2))\\ +\frac{Q\cos\phi}{2\pi R^2}\left(\frac{D(\alpha)}{\sin^2\alpha}\sin\phi - \frac{D(\alpha)}{\sin\phi}\right)\end{array}\right)\right)$$

$$\begin{aligned} v_z = &-\frac{Q\cos\phi}{2\pi R^2}\left(\frac{1+\Lambda}{\eta^2}\frac{1}{1-\cos\alpha} + 2\sigma\left(\frac{D(\alpha)}{\sin^2\alpha}\cos\phi - 1 - \cos\phi\ln(\tan(\phi/2))\right)\right)\\ &-\frac{Q\sin\phi}{2\pi R^2}\left(2\sigma\left(\frac{D(\alpha)}{\sin^2\alpha}\sin\phi - \frac{D(\alpha)}{\sin\phi}\right)\right)\end{aligned} \tag{5.54}$$

where Q is the feed volume flow rate, R is the radius of the cylindrical part of the hydrocyclone, σ is given by (5.51), $\Lambda = Q_o/Q_u$ is the flow partition between overflow and under flow streams and where $D(o)$ is the operator $D(\circ) = 1 - \cos(\circ) + \sin^2(\circ)\ln(\tan(\circ)/2)$.

5.4 Extreme Solutions

5.4.1 Euler's Inviscid Isochoric Models

The difficulties encountered in obtaining general solutions to the flow equations have shown that it is advantageous to divide the flow into categories depending on the relative value of the several terms of the equations and to neglect the terms of lower orders of magnitude. To perform this task the Navier Stokes equations are written in dimensionless form:

$$\frac{1}{St}\frac{\partial \boldsymbol{v}^*}{\partial t^*} + \nabla^* \boldsymbol{v}^* \cdot \boldsymbol{v}^* = -\frac{1}{Ru}\nabla^* p^* + \frac{1}{\mathrm{Re}}\nabla^{*2}\boldsymbol{v}^* + \frac{1}{Fr}\boldsymbol{b}^*$$
$$\nabla^* \boldsymbol{v}^* = 0 \tag{5.55}$$

with

$$St = \frac{t_0 u_0}{L}, \quad Ru = \frac{\rho u_0^2}{p_0}, \quad \mathrm{Re} = \frac{\rho u_0 L}{\mu}, \quad Fr = \frac{u_0^2}{Lg}$$

where the stress tensor $\boldsymbol{T}$ is expressed as the sum of a pressure p and an extra stress modeled as a Newtonian fluid, L, t_0, u_0 and p_0 are characteristic parameters of the flow, *St, Ru,* Re and *Fr* are the Strouhal, Ruark, Reynolds and Froude dimensionless numbers and the asterisk indicates dimensionless field variables. The Reynolds number is the ratio of the convective to the diffusive forces in the flow; therefore, it is a good parameter to divide the flow into three categories.

Euler's isochoric flows, or inviscid flows, are obtained if the convective forces are much greater than the diffusive, or viscous, forces. In this case the Reynolds Number is infinite, making the second term on the right side of (5.55) vanish:

$$\rho\frac{\partial \boldsymbol{v}}{\partial t} + \nabla \boldsymbol{v} \cdot \boldsymbol{v} = -\nabla p + \boldsymbol{b}$$
$$\nabla \cdot \boldsymbol{v} = 0 \tag{5.56}$$

Many researchers have suggested an inviscid isochoric flow for the hydrocyclone, that is, the flow of an ideal fluid. The is because, for hydrocyclones, the tangential velocity shape is very close to a free vortex, as can be seen from the experimental observations. Within ideal flows, we can distinguish rotational and irrotational flows.

The field equations for inviscid isochoric flows are:

$$\nabla \cdot \boldsymbol{v} = 0 \quad \nabla B = \boldsymbol{v} x \boldsymbol{w} \quad B = \frac{p}{\rho} + \frac{1}{2} v^2 \tag{5.57}$$

where $\boldsymbol{v}$, $\boldsymbol{w}$ and p are the velocity, vorticity and pressure fields and B is the Bernouilli. Body forces have been neglected.

(a) **Irrotational inviscid isochoric flow (Potential flow model)**

For potential flow the field equations become:

$$\nabla \cdot \boldsymbol{v} = 0 \quad \nabla \times \boldsymbol{v} = 0 \quad B = 0 \tag{5.58}$$

Combining these equations results in:

$$\boldsymbol{v} = -\nabla \phi, \quad \nabla^2 \phi = 0, \quad \frac{1}{2} v^2 + \frac{p}{\rho} = \text{constant} \tag{5.59}$$

where ϕ is the velocity potential. For cylindrical plane flows the Laplace equation is valid:

$$\frac{\partial^2 \phi}{\partial r^2} + \frac{1}{r}\frac{\partial \phi}{\partial r} + \frac{1}{r^2}\frac{\partial^2 \phi}{\partial \theta^2} = 0, \quad p(r) = p(R) - \frac{1}{2}\rho\left(v_R^2 + v_{\theta R}^2\right)\left(1 - \frac{v_r^2 + v_\theta^2}{v_R^2 + v_{\theta R}^2}\right) \tag{5.60}$$

$$v_r = -\frac{\partial \phi}{\partial r}, \quad v_\theta = -\frac{1}{r}\frac{\partial \phi}{\partial \theta} \tag{5.61}$$

Measurements of the flow patterns in hydrocyclones [71], 63, 98 have shown that the tangential velocity has the form of a free vortex from the cyclone wall in the direction of the axis. Near the axis, the tangential velocity resembles that of a solid body rotation. There is no agreement on radial velocity measurements, as can be seen in Figs. 5.1, 5.2 and 5.3. The best assumption that can be made, from Hsieh and Rajamani's measurement, is that the radial flow is similar to a sink close to the cyclone wall but is proportional to the radius closer to the cyclone axis. Taking these observations into consideration, the simplest model for the hydrocyclone is a combination of a free vortex and a rigid motion with a sink. This combination is also called a *forced vortex with a sink.*

Consider the flow in a hydrocyclone, working without an air core, divided into two regions, the vortex core $0 \leq r \leq R_c$, where the flow is a rigid body motion with constant angular velocity $\omega = v_\theta(R_c)/R_c$, and the exterior region $R_c \leq r \leq R$, where the flow is a free vortex with a constant circulation $\Gamma = 2\pi R v_{\theta R}$. The total flow rate into the cyclone per unit length is $\widehat{Q} = 2\pi R v_R$.

Region $R_c \le r \le R$:

The solution of Eqs. (5.60) and (5.61) in this region yields a free vortex and a sink for the tangential and radial velocity, respectively:

$$v_\theta = \frac{\Gamma}{2\pi r}, \qquad v_r = -\frac{\widehat{Q}}{2\pi r} \tag{5.62}$$

From Bernouilli's equation:

$$p(r) - p(R) = \frac{1}{2}\rho\left(v^2(R) - v^2(r)\right) \tag{5.63}$$

Since $v^2 = v_\theta^2 + v_r^2$, substituting the values of (10) at r and R, yields:

$$p(r) = p(R) - \frac{\rho}{2}\left(\frac{\Gamma}{2\pi R}\right)^2\left(1 + \left(\frac{\widehat{Q}}{\Gamma}\right)^2\right)\left(1 - \left(\frac{R}{r}\right)^2\right) \tag{5.64}$$

Region $0 \le r \le R_c$.

In this region a forced vortex and a sink represent the tangential and radial velocity, respectively:

$$v_\theta(r) = \omega r \quad v_r = -\frac{\widehat{Q}}{2\pi r} = -\left(\frac{\widehat{Q}}{\Gamma}\right)\left(\frac{R}{r}\right) \tag{5.65}$$

where ω is the angular velocity of the rigid motion. At $r = R_c$, the velocity fields of both regions intersect, that is, $\omega R_c = \Gamma / 2\pi R_c$, from which R_c can be calculated as:

$$R_c = \left(\frac{\Gamma}{2\pi\omega}\right)^{1/2} \tag{5.66}$$

Fig. 5.4 shows the tangential and radial velocity profiles for the *forced vortex with a sink* model.

For a rigid body motion the pressure can be calculated from $\nabla B = 2\rho\omega^2 r\mathbf{e}_r$. Integrating from R_c to r yields:

$$p(\xi) + \frac{1}{2}\rho v^2(\xi)\Big|_{\xi=R_c}^{\xi=r} = \rho\omega^2\eta^2\Big|_{\eta=R_c}^{\eta=r}$$

$$p(r) = p(R_c) + \frac{1}{2}\rho v^2(R_c) - \frac{1}{2}\rho v^2(r) + \rho\omega^2 r^2 - \rho\omega^2 R_c^2$$

Replacing v from (12) gives:

$$p(r) = p(R_c) + \frac{1}{2}\rho\omega^2\left(r^2 - R_c^2\right) + \frac{1}{2}\rho\left(\frac{\widehat{Q}}{\Gamma}\right)^2\left(\left(\frac{R}{R_c}\right)^2 - \left(\frac{R}{r}\right)^2\right) \quad \text{for } 0 \le r \le R_c \tag{5.67}$$

Fig. 5.5 shows the pressure distribution.

Several authors, for example Fontain and Dijsman [60], suggested models of this type.

(b) **Rotational inviscid isochoric model**

For steady flow $\dot{B} = \nabla B \cdot \boldsymbol{v}$, therefore, from Eq. (5.57) $\dot{B} = \boldsymbol{v} \times \boldsymbol{w} \cdot \boldsymbol{v} \equiv 0$, which indicates that Bernouilli is constant along streamlines. Thus, in *spherical coordinates* (ξ, ϕ, θ) we can write:

$$v_\xi \frac{\partial}{\partial \xi}(\xi \sin\phi v_\theta) + \frac{v_\phi}{\xi}\frac{\partial}{\partial \phi}(\xi \sin\phi v_\theta) = 0 \tag{5.68}$$

therefore, $\xi \sin\phi v_\theta$ is constant along a streamline:

$$\xi \sin\phi v_\theta = f(\psi) \tag{5.69}$$

where $f(\psi)$ is the vorticity function. At the wall the streamline must be a constant, $f(\psi)|_{\phi=\alpha} = k$. References [16, 19] assumed that the tangential velocity was a free vortex everywhere, that is $v_\theta = f(\psi)/\xi \sin\phi$. Using the stream function to satisfy the continuity equation, the velocity components may be written in the form:

$$v_\xi = \frac{1}{\xi^2 \sin\phi}\frac{\partial \psi}{\partial \phi} \text{ and } v_\phi = -\frac{1}{\xi \sin\phi}\frac{\partial \psi}{\partial \xi} \tag{5.70}$$

Substituting Eqs. (5.69) and (5.70) into Eq. (5.68), neglecting the first term and using the fact that for small ϕ, sen$\phi \approx \phi$, yields:

$$\frac{\partial}{\partial \phi}\left(\frac{1}{\phi}\frac{\partial \psi}{\partial \phi}\right) = -\xi^4 f(\psi) \tag{5.71}$$

Assuming the following form for the vorticity function,

$$f(\psi) = F\psi^{-5/3} \tag{5.72}$$

the solution of (5.71) by Bloor and Ingham [11] yields $\psi(\xi, \phi) = K\xi^{3/2}\phi^{3/2}(\alpha - \phi)$, so that:

$$v_\xi = \frac{3}{2}\frac{A\phi^{1/2}}{\xi\phi}(3\alpha - 5\phi) \quad v_\phi = -\frac{3}{2}\frac{A\phi^{1/2}}{\xi}(\alpha - \phi) \tag{5.73}$$

and the radial, tangential and axial velocities in cylindrical coordinates become [12, 13]:

$$v_r = -A\frac{r^{3/2}}{z^2} \quad v_\phi = \frac{k}{r} \quad v_z = \frac{1}{2}Ar^{-1/2}(3\alpha - r(r/z)) \tag{5.74}$$

where

$$k = \frac{0.8QR}{A_i} \text{ and } A = \frac{Q}{2\pi\left(d_v/2\right)^{3/2}\left(\alpha - d_v/2\ell\right)} \tag{5.75}$$

And Q is the volume flow rate to the hydrocyclone, A_i is the feed inlet area, d_v is the external diameter of the vortex finder tube and ℓ is the distance from the apex to the lower part of the vortex finder tube. Figure 5.6 shows the tangential and radial velocity, while Fig. 5.7 shows the axial velocity distribution, according to [16].

Brayshaw [23] explored other vorticity functions for Eq. (5.71). Using *cylindrical coordinates*, he derived the following boundary conditions for the stream function $f(\psi)$ at the entry section, underflow and overflow, respectively:

$$\frac{1}{R^2}\frac{d^2\psi}{dz^2}, \quad \frac{1}{r^2}\frac{d^2\psi}{dr^2}\bigg|_{r=R_u} - \frac{1}{r^3}\frac{d\psi}{dr}\bigg|_{r=R_u}, \quad -\frac{1}{r^2}\frac{d^2\psi}{dr^2}\bigg|_{r=R_o} + \frac{1}{r^3}\frac{d\psi}{dr}\bigg|_{r=R_u} \tag{5.76}$$

He found that a gamma function and an exponential function were suitable for $f(\psi)$ and gave rise to different particle selectivity curves, with the gamma function showing a sharper separation. Figure 5.8 shows a plot of different vorticity function as given by Brayshaw [23].

Davidson [47] indicated that Boor and Ingahm's solution of Eq. (5.71) could be generalized in the form:

$$\psi = (\xi\phi)^m\left(\alpha^\beta - \phi^\beta\right); \text{ with } \beta = 2(2 - \text{m}) \tag{5.77}$$

Unfortunately, no analytical solution of this form can be obtained. Instead, Davidson [47] proposed a similar solution, in cylindrical coordinates with the tangential velocity independent of the axial coordinate, of the form:

$$v_z = z^n F(r/z), \quad v_r = z^n G(r/z), \quad p = P(r) + \rho z^{2n} p_0(r/z) \tag{5.78}$$

$$\psi = z^{n+2}H(r/z), \quad \frac{dH}{d(r/z)} = F(r/z), \quad \text{G}(r/z) = (r/z)^2 F - (n+1)H \tag{5.79}$$

$$H = \gamma^{1/2}(r/z)_c^2 \times \left(d^2 - \phi^2\right), \quad F = a\gamma^{1/2}\phi\left(d^2 - \phi^2\right)^{-1/2} \tag{5.80}$$

where ϕ is a function obtained by solving the equation

$$\sum_{i=0}^{k} a_i\phi^{2k-2i+1} = \frac{1}{4}(-1)^{k+1}\left[\left(\frac{r/z}{r/z_c}\right) - 0.5\right] \tag{5.81}$$

Davidson solved these equations numerically, obtaining excellent agreement for the velocity distributions with $n = 0.5$ (corresponding to $m = 3/2$) compared to the experimental values of Knowles et al. [72] and [71]. In this case the axial velocity is:

$$v_z = Ar^{-1/2}[1.5\alpha - 2.5(r/z)] \tag{5.82}$$

where A is a constant. Figure 5.9 shows this axial velocity.

5.5 Boundary Layer Models

Prandtl's flow, or boundary layer flow, is obtained when the Reynolds number is high and convective forces are on the same order of magnitude as the diffusive forces.

In previous sections we learned that the tangential and axial velocities have an order of magnitude of several m/s and that the tangential velocity is insensitive to the axial coordinate, while the axial velocity depends on z. The radial velocity is two orders of magnitude smaller. Since the axial velocity is on the order of 1 m/s, the length of the hydrocyclone is on the order of 1 m, the kinematic viscosity is on the order of 10^{-4} m^2/s and the axial Reynolds number is on the order of 10^4. With this high Reynolds number, it is only natural to think of the flow inside the hydrocyclone as divided into a viscous boundary layer close to the walls and an inviscid flow in the central region. The inviscid flow was thoroughly analyzed by Bloor and Ingham in several papers ([11, 16, 12, 13, 18, 19]), yielding the velocity distribution. In later papers, Bloor and Ingham [13, 14] studied the boundary layer along the lateral walls and at the top wall of a conical hydrocyclone, giving the velocity distribution and volume flow within the boundary layer and the boundary layer thickness.

Consider the Reynolds equations for isotropic turbulent swirling flow in a cone, characterized with a turbulent eddy viscosity ν_ε:

$$\frac{\partial}{\partial r}(rv_r) + \frac{\partial}{\partial z}(rv_z) = 0 \tag{5.83}$$

$$v_r\frac{\partial v_r}{\partial r} + v_z\frac{\partial v_r}{\partial z} - \frac{v_\theta^2}{r} = -\frac{1}{\rho}\frac{\partial p}{\partial r} + v_z\left(\frac{\partial}{\partial r}\left(\frac{1}{r}\right)\frac{\partial}{\partial r}(rv_r) + \frac{\partial^2 v_r}{\partial z^2}\right) \tag{5.84}$$

$$\frac{v_r}{r}\frac{\partial}{\partial r}(rv_\theta) + v_z\frac{\partial v_\theta}{\partial z} = \nu_\varepsilon\left(\frac{\partial}{\partial r}\left(\frac{1}{r}\frac{\partial}{\partial r}(rv_\theta)\right) + \frac{\partial^2 v_\theta}{\partial z^2}\right) \tag{5.85}$$

$$v_r\frac{\partial v_z}{\partial r} + v_z\frac{\partial v_z}{\partial z} = -\frac{1}{\rho}\frac{\partial p}{\partial z} + \nu_\varepsilon\left(\frac{1}{r}\frac{\partial}{\partial r}\left(r\frac{\partial v_z}{\partial r}\right) + \frac{\partial^2 v_z}{\partial z^2}\right) \tag{5.86}$$

(a) **Boundary layer at the lateral walls of a hydrocyclone**

Since the length scale in the direction "y" perpendicular to the lateral wall is much smaller than in the direction "x" along the wall, it is convenient to make a coordinate transformation from (z, r) to (x, y) by means of:

$$\begin{aligned} z &= x\cos\alpha + y\sin\alpha \\ r &= x\sin\alpha - y\cos\alpha \\ \theta &= \theta \end{aligned} \tag{5.87}$$

where α is the semi-angle of the hydrocyclone. In addition, it is convenient to define dimensionless variables:

$$\begin{aligned} &x^* = \frac{x}{L_x},\ y^* = \frac{y}{L_y},\ v_x^* = \frac{v_x}{u_x},\ v_y^* = \frac{v_y}{u_y},\ v_\theta^* = \frac{v_\theta}{u_x}, \\ &p^* = \frac{p}{p_0},\ \frac{\partial}{\partial x^*}(\circ) = L_x\frac{\partial}{\partial x}(\circ),\ \frac{\partial}{\partial y^*}(\circ) = L_y\frac{\partial}{\partial y}(\circ) \end{aligned} \tag{5.88}$$

With these considerations the field equations become:
Continuity equation:

$$\begin{aligned} &\frac{\partial}{\partial x^*}\left(x^* v_x^*\right) + \left\{\frac{L_x}{L_y}\frac{u_y}{u_x}\right\}\frac{\partial}{\partial y^*}\left(x^* v_y^*\right) \\ &\quad - \cot\alpha\left\{\frac{u_y}{u_x}\right\}\frac{\partial}{\partial y^*}\left(y^* v_y^*\right) + \left\{\frac{L_y}{L_x}\right\}\frac{\partial}{\partial x^*}\left(y^* v_x^*\right) = 0 \end{aligned} \tag{5.89}$$

x^* component:

$$\begin{aligned} v_x^*\frac{\partial v_x^*}{\partial x^*} + \left\{\frac{L_x}{L_y}\frac{u_y}{u_x}\right\}v_y^*\frac{\partial v_x^*}{\partial y^*} - \frac{\sin\alpha v_\theta^{*2}}{x^*\sin\alpha - \left\{\frac{L_y}{L_x}\right\}y^*\cos\alpha} &= -\left\{\frac{p_0}{\rho u_x^{*2}}\right\}\frac{\partial p}{\partial x^*} + \left\{\frac{L_x}{L_y}\frac{u_y}{u_x}\right\}\frac{\partial^2 v_x^*}{\partial x^{*2}} \\ &\quad + \frac{1}{\text{Re}}\left\{\frac{L_y}{L_x}\right\}\frac{\partial^2 v_x^*}{\partial y^{*2}} \\ &\quad + \frac{1}{\text{Re}}\frac{\sin\alpha\frac{\partial v_x^*}{\partial x^*} - \cos\alpha\left\{\frac{L_y}{L_x}\right\}\frac{\partial v_x^*}{\partial y^*}}{x^*\sin\alpha - \left\{\frac{L_y}{L_x}\right\}y^*\cos\alpha} \\ &\quad + \frac{1}{\text{Re}}\frac{\sin\alpha\cos\alpha\frac{u_x}{u_y}v_y^*}{\left(x^*\sin\alpha - \left\{\frac{L_y}{L_x}\right\}y^*\cos\alpha\right)^2} \\ &\quad - \frac{1}{\text{Re}}\frac{\sin^2\alpha v_x^*}{\left(x^*\sin\alpha - \left\{\frac{L_y}{L_x}\right\}y^*\cos\alpha\right)^2} \end{aligned} \tag{5.90}$$

θ component:

$$v_x^*\frac{\partial v_\theta^*}{\partial x^*} + \left\{\frac{L_x}{L_y}\frac{u_y}{u_x}\right\}v_y^*\frac{\partial v_\theta^*}{\partial y^*} - \frac{\sin\alpha v_\theta^* - \left\{\frac{u_y}{u_x}\right\}\cos\alpha v_y^*}{x^*\sin\alpha - \left\{\frac{L_y}{L_x}\right\}y^*\cos\alpha} = \frac{1}{\text{Re}}\frac{\partial^2 v_\theta^*}{\partial x^{*2}} + \frac{1}{\text{Re}}\left\{\frac{L_x}{L_y}\right\}\frac{\partial^2 v_\theta^*}{\partial y^{*2}}$$

$$+\frac{1}{\mathrm{Re}}\frac{\cos\alpha\left\{\frac{L_x}{L_y}\right\}v_\theta^*}{\left(x^*\sin\alpha-\left\{\frac{L_y}{L_x}\right\}y^*\cos\alpha\right)^2}$$
$$+\frac{1}{\mathrm{Re}}\frac{\sin\alpha\frac{\partial v_\theta^*}{\partial x^*}-\cos\alpha\left\{\frac{L_x}{L_y}\right\}\frac{\partial v_\theta^*}{\partial y^*}}{\left(x^*\sin\alpha-\left\{\frac{L_y}{L_x}\right\}y^*\cos\alpha\right)^2} \tag{5.91}$$

y^* component:

$$v_x^*\frac{\partial v_y^*}{\partial x^*}+\left\{\frac{L_x}{L_y}\frac{u_y}{u_x}\right\}v_y^*\frac{\partial v_y^*}{\partial y^*}-\frac{\left\{\frac{u_y}{u_x}\right\}\cos\alpha\, v_\theta^*}{x^*\sin\alpha-\left\{\frac{L_y}{L_x}\right\}y^*\cos\alpha}=-\left\{\frac{p_0}{\rho u_x^2}\right\}\left\{\frac{L_x}{L_y}\frac{u_y}{u_x}\right\}\frac{\partial p^*}{\partial y^*}$$
$$\frac{1}{\mathrm{Re}}\frac{\partial^2 v_y^*}{\partial x^{*2}}+\frac{1}{\mathrm{Re}}\left\{\frac{L_x}{L_y}\right\}\frac{\partial^2 v_y^*}{\partial y^{*2}}$$
$$+\frac{1}{\mathrm{Re}}\frac{\sin\alpha\frac{\partial v_y^*}{\partial x^*}-\cos\alpha\left\{\frac{L_x}{L_y}\right\}\frac{\partial v_y^*}{\partial y^*}}{x^*\sin\alpha-\left\{\frac{L_y}{L_x}\right\}y^*\cos\alpha}$$
$$+\frac{1}{\mathrm{Re}}\frac{\sin\alpha\cos\alpha\left\{\frac{u_x}{u_y}\right\}v_x^*}{\left(x^*\sin\alpha-\left\{\frac{L_y}{L_x}\right\}y^*\cos\alpha\right)^2} \tag{5.92}$$

A study of the order of magnitude of the terms shows that:

$$\left\{\frac{L_x}{L_y}\right\}^2=\mathrm{Re}\,,\quad \left\{\frac{u_x}{u_y}\right\}^2=\mathrm{Re}\,,\quad \left\{\frac{L_x}{L_y}\frac{u_y}{u_x}\right\}=1 \tag{5.93}$$

Substituting into the previous equations and neglecting all terms with powers of Re^{-1}, the boundary layer equations in dimensional form for the swirling flow in a cone are:

$$\frac{\partial}{\partial x}(xv_x)+\frac{\partial}{\partial y}(xv_y)=0 \tag{5.94}$$

$$v_x\frac{\partial v_\theta}{\partial x}+v_y\frac{\partial v_\theta}{\partial y}=\nu_\varepsilon\frac{\partial^2 v_\theta}{\partial y^2} \tag{5.95}$$

$$v_x\frac{\partial v_x}{\partial x}+v_y\frac{\partial v_x}{\partial y}=-\frac{1}{\rho}\frac{\partial p}{\partial x}+\nu_\varepsilon\frac{\partial^2 v_x}{\partial y^2} \tag{5.96}$$

$$\frac{\partial p}{\partial y}=0 \tag{5.97}$$

Since the pressure is independent of y, the pressure gradient in Eq. (5.96) can be written in terms of the velocity outside the boundary layer, according to Bernoulli's equation.

Bloor and Ingham used the velocities obtained for the inviscid flow (see previous analysis) as boundary conditions at the boundary layer limit; thus:

$$\frac{\partial}{\partial r}(r v_r) + \frac{\partial}{\partial z}(r v_z) = 0 \; v_x = v_x(R)\left(\frac{x}{x_0}\right)^{-1/2}, \; v_\theta = \frac{R v_\theta(R)}{r} \frac{\gamma\left(\frac{4}{5}, \frac{2}{5}\frac{Lr}{R^{5/2}}\right)}{\gamma\left(\frac{4}{5}, \frac{2}{5}L\right)} \tag{5.98}$$

where $x_0 = \ell/\tan\alpha$ and R is the radius of the cylindrical portion of the hydrocyclone. Bloor and Ingham [14] obtained a solution to these equations using the approximate integral method of von Polhausen. Defining the new similar variable $\eta = y/\delta(x)$, where $\delta(x)$ is the thickness of the boundary layer, they described the velocity distributions in the form:

$$\frac{v_x}{v_x(R)} = 2\eta - \eta^2 + a(x)\eta(1-\eta)^2, \quad \frac{v_\theta}{v_\theta(R)} = 2\eta - \eta^2 \tag{5.99}$$

Substituting these expressions into the von Karman integrals of the field equations results in two ordinary differential equations, the result of which indicates that the dimensionless thickness of the boundary layer $\delta^*(x^*)$ increases from $x^* = 0$ to $x^* = 0.5$, then stays constant and finally diminishes to 0 at $x^* = 1$. At the points of maximum thickness, the boundary layer has the form:

$$\delta_m = \frac{1.88 x_0}{\mathrm{Re}^{1/2}} \tag{5.100}$$

where $x_0 = \ell/\tan\alpha$.

(b) **Boundary layer at the top wall of a hydrocyclone**

A dimensional analysis of Eqs. (5.83) to (5.86) yields the following set of field equations for the boundary layer at the top wall of the hydrocyclone:

$$\frac{\partial}{\partial r}(r v_r) + \frac{\partial}{\partial z}(r v_z) = 0 \tag{5.101}$$

$$v_r \frac{\partial v_r}{\partial r} + v_z \frac{\partial v_r}{\partial z} - \frac{v_\theta^2}{r} = -\frac{1}{\rho}\frac{\partial p}{\partial r} + \frac{1}{\rho}\frac{\partial T_{rz}^E}{\partial z} \tag{5.102}$$

$$\frac{v_r}{r}\frac{\partial}{\partial r}(r v_\theta) + v_z \frac{\partial v_\theta}{\partial z} = \frac{1}{\rho}\frac{\partial T_{\theta z}^E}{\partial z} \tag{5.103}$$

$$0 = \frac{1}{\rho}\frac{\partial p}{\partial z} \tag{5.104}$$

with boundary conditions:

$$\begin{array}{ll} v_r(r,0) = 0 & v_\theta(r,0) = 0 \\ v_r(r,\delta(r)) = V_r & v_\theta(r,\delta(r)) = V_\theta \end{array} \tag{5.105}$$

Since the pressure is independent of z, the pressure gradient in Eq. (5.102) can be written in terms of the velocity outside the boundary layer, according to Bernoulli's equation: $V_r \partial V_r / \partial r - V_\theta^2 / r = -(1/\rho)\partial p / \partial r$. Using the von Karman integral method, Bloor and Ingham [13] approximated the velocity distribution with the following functions of a similar variable ξ:

$$v_r = V_r \xi^{1/7}\left(\gamma\left(\frac{15}{7} - 2\gamma\right)\xi + \left(\gamma - \frac{8}{7}\right)\xi^2\right); \quad v_\theta = \frac{1}{7} V_\theta \xi^{1/7}(8 - \xi) \tag{5.106}$$

where $\xi = z/\delta(r)$.

Smith [105] analyzed the stability of the boundary layer in the vortex flow of a pneumatic cyclone. He explains that the flow is naturally unstable close to the wall, since in this region the angular momentum diminishes with increasing radius. The instability manifests itself in the intermittent bursts of fluid from the boundary layer moving radially inwards, where it diffuses and loses identity. These bursts make the flow in the cyclone very turbulent even if there is no turbulence convected into the device by the incoming fluid or produced by the interaction of the incoming fluid with the flow in the cyclone. This turbulence resulting from the unstable boundary layer is inherent in any device employing a rotating flow within a stationary container. On a conical wall, the secondary flows and the generation of turbulence occur in combination. During a burst, the heavy particles are left behind, forming the streaks that are commonly observed in dust separation cyclones. The lighter particles are dragged by the fluid and remix into the main flow. This is a mechanism of redispersion of small particles trapped in the boundary layer. Although these mechanisms were observed in pneumatic cyclones, they can also explain the high turbulent motion in hydrocyclones and the redispersion of small particles.

Kumari and Nath [74] studied the unsteady laminar motion in the boundary layer of nozzles, diffusers and hydrocyclones. They concluded that the tangential velocity at the edge of the boundary layer is of a non-free vortex type, in contrast to the steady flow. In the flow through hydrocyclones the swirl flow dominates the axial flow, which increases approaching the apex.

5.6 Similarity Solution of Reynolds Equations

Mikaylov and Romensky [76] solved Reynolds equations for the flow within a hydrocyclone assuming axial symmetry and self-similarity, with the similar variable $\eta = z/r \equiv z^*/r^*$, independence of v_θ from z and an air core modeled given by the following equation:

$$\frac{d_a}{R} = 0.64 \frac{d_o}{R} \tan\left(\frac{\alpha}{2}\right)^{0.3} \tag{5.107}$$

Defining the following dimensionless variables, $r^* = r/R_{\max}$ and $v_\theta^* = v_\theta / v_{\theta\,\max}$, where the characteristic variable $R_{\max}$ is the value of r where the tangential velocity has its maximum $v_{\theta\,\max}$, they obtained the following set of field equations:

$$\frac{v_\theta^*}{r^*} = \frac{1}{2} \frac{\mathrm{d}p^*}{\mathrm{d}r^*} \tag{5.108}$$

$$\frac{v_r^*}{r^*} \frac{\mathrm{d}}{\mathrm{d}r^*}\left(r^* v_r^*\right) - \frac{r^*}{z^*} v_z^* \frac{\mathrm{d}v_\theta^*}{\mathrm{d}r^*} = \frac{1}{2r^{*2}} \frac{\mathrm{d}}{\mathrm{d}r^*}\left(r^* T_{r\theta}^{*E}\right) \tag{5.109}$$

$$\frac{1}{r^*} \frac{\mathrm{d}}{\mathrm{d}r^*}\left(r^* v_r^*\right) - \frac{r^*}{z^*} \frac{\mathrm{d}v_z^*}{\mathrm{d}r^*} = 0 \tag{5.110}$$

The solutions to these equations have the form:

$$v_\theta^* = \frac{1}{\left(1 + \frac{1}{b^2}\left(r^* - \frac{1}{r^*}\right)^2\right)^{0.5}} \tag{5.111}$$

$$v_z^* = C v_\theta^* \frac{\mathrm{d}}{\mathrm{d}r^*}\left(r^* v_\theta^*\right) + \frac{0.5}{r^* v_\theta^*} \frac{z^*}{r^*} \frac{\mathrm{d}}{\mathrm{d}r^*}\left(r^* T_{r\theta}^{*E}\right) + f\left(r^*, z^*\right) \tag{5.112}$$

$$v_r^* = \left(v_\theta^* \frac{r^*}{z^*} - \frac{\left(r^* v_\theta^*\right)^2}{r^*}\right)\left(\frac{C_1}{z^*} + g\left(r^*\right)\right) + C_2 \tag{5.113}$$

$$\Delta p = 2 \int \frac{v_\theta^*}{r^*} \mathrm{d}r^* + C_3 \tag{5.114}$$

The constants depend on the geometry of the hydrocyclone and the boundary conditions. To obtain the solution they modeled the extra stress $T_{r\theta}^{*E}$ empirically from measurements of the velocity distributions:

$$\frac{\mathrm{d}}{\mathrm{d}r^*}\left(r^{*2} T_{r\theta}^{*E}\right) = C_4 r^* v_\theta^* h\left(r^*\right) \tag{5.115}$$

The authors claimed good agreement with their own experimental information. However, in setting up the equations, they arbitrarily eliminated terms in the dimensionless field equations. The reasons for some eliminations are not clear since in the r and z components of the field equations they eliminated terms on the order of $O(10^{-2})$, while in the θ component they eliminated terms on the order of $O(10^{-3})$. Furthermore, in the z component, they eliminated the pressure gradient $\partial p^*/\partial z^*$, which is on the order of $O(1)$. Their results are given in Fig. 5.26.

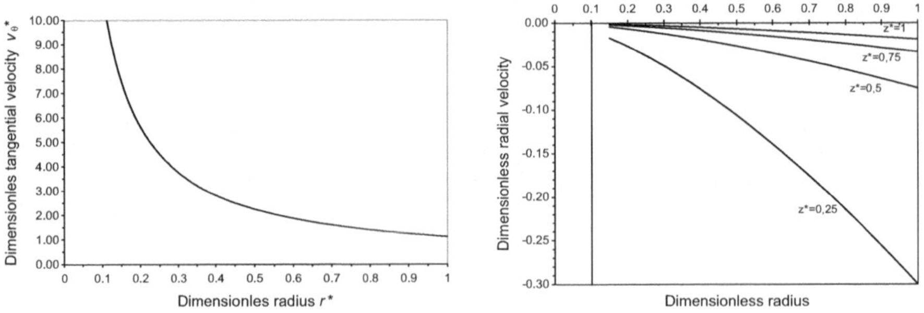

Fig. 5.26 Velocity distribution for the isochoric rotational flow model of a hydrocyclone with A = 0.01875, according to Bloor and Ingham [16]

Nepomnyashchii and Pavlovskii [82, 83], made a dimensional analysis of the integral angular momentum equations and, using the similarity theory, concluded that the velocity components and the pressure within a hydrocyclone could be expressed as an expansion in series of functions of the similarity variables:

$$v_\theta^* = z^{*-3/2}\varphi_1(\eta) + \mathrm{Re}^{-1} z^{*-1}\varphi_2(\eta) + \ldots \tag{5.116}$$

$$v_z^* = z^{*-3/2} f_1(\eta) + \mathrm{Re}^{-1} z^{*-1} f_2(\eta) + \ldots \tag{5.117}$$

$$v_r^* = \mathrm{Re}^{-1} z^{*-1} F_1(\eta) + \ldots \tag{5.118}$$

$$p^* = z^{*-3} g_1(\eta) + \mathrm{Re}^{-1} z^{*-5/2} + \ldots \tag{5.119}$$

where $v_\theta^* = v_\theta / v_{\theta R}$, $v_z^* = v_z / v_{zR}$, $v_r^* = v_r / v_{rR}$ and $p^* = (p - p_0)/\rho v_\theta^2$.

In this case, the Reynolds number is defined in terms of the tangential characteristic velocity $v_{\theta R}$. Inserting the expressions for the velocities into the field equations yields as the first term in the expansion:

$$\varphi_1(\eta) = C_1\eta^{-3/2},\; f_1(\eta) = C_2\eta^{-3/3},\; g_1(\eta) = -\frac{1}{3}C_1^2\eta^{-3},\; F_1(\eta) = A_1\eta^{-1},$$

$$\varphi_2(\eta) = D_1\eta^{-2} + D_2\eta^{-1},\; f_2(\eta) = B_2\eta^{-1},\; g_2(\eta) = E_1\eta^{-7/2} + E_2\eta^{-5/2} \tag{5.120}$$

$$A_1 = -\frac{20C_1^2 + 63C_2^2}{8C_1^2 + 42C_2^2},\; D_1 = -\frac{21C_1C_2}{8C_1^2 + 42C_2^2},$$
$$E_1 = -\frac{6C_1^2C_2}{4C_1^2 + 21C_2^2},\; E_2 = -\frac{4}{5}C_1D_2 \tag{5.121}$$

$$I = \frac{(R - R_a)H_1}{RR_a} + \frac{H_2}{R_a} - \cot\alpha \ln\left(\frac{R}{R - H_2 \tan\alpha}\right) \tag{5.122}$$

$$C_1 = \left(\frac{2}{2I}\right)^{1/2}, C_2 = \frac{Q_o}{2\pi R^{3/2} R_v v_{\theta R}\left(\left(R/R_v\right)^{1/2} - 1\right)}$$

$$\text{and } D_2 = \frac{\frac{15C_1^2C_2}{4C_1^2+21C_2^2}\int_{S_I}\frac{dr}{r^{1/2}}dz}{\frac{6}{5}C_1\int_{S_I}\frac{dr}{r^{1/2}}dz} \tag{5.123}$$

where H_1 and H_2 are the height of the cylindrical and conical sections, respectively, R_v is the outer radius of the vortex finder tube and R_a is the radius of the air core. With these values the velocity distributions are:

$$v_z = \frac{Q}{2\pi}\frac{\frac{R}{r} - \left(\frac{R_o}{R}\right)^{1/2}\left(\frac{R}{r}\right)^{3/2}}{\left(1 - \left(\frac{R_o}{R}\right)^{1/2}\right)^2} \quad v_\theta^* = v_F\left(\frac{R_F}{3}\right)^{3/2} \quad v_r = v_F\left(\frac{R_o}{R}\right)\frac{A_1}{C_1^{2/3}\mathrm{Re}}\frac{R}{r} \tag{5.124}$$

$$p = p_0 - \frac{1}{3}\rho v_F^2\left(\frac{R_F}{r}\right)^3 \quad R_a = \frac{R_F}{R}\left(\frac{\rho v_F^2}{3(p_{\max} - p_o)}\right)^{1/3} \tag{5.125}$$

In these equations $p_{\max}$ is the maximum pressure in the apparatus and R_F is the eccentricity of the hydrocyclone feed, that is, the distance from the axis to the center of the inlet pipe.

This model does not satisfy the boundary conditions at the cyclone wall; therefore, it can be deemed valid outside the boundary layer only.

5.7 Asymptotic Solutions of the Reynolds Equations

According to [32, 33], the most general axial-symmetric model of a hydrocyclone with inflow through a cylindrical mantle may be expressed in terms of a dimensionless axial-symmetric stream function ψ and a dimensionless circulation Γ, with the radial Reynolds number $\mathrm{Re} = Rv_R/\nu_\varepsilon$, the dimensionless number $R_0 = Lv_R/Rv_{\theta R}$, representing the ratio of the total flow Q to the hydrocyclone to the circulation $\Gamma(R)$ at the cyclone's outer edge, and the characteristic length $\lambda = \left(R/L\right)^2$ as parameters. The following two equations represent the flow field [55]:

$$\frac{\partial\psi}{\partial\xi}\frac{\partial\Gamma}{\partial\eta} - \frac{\partial\psi}{\partial\eta}\frac{\partial\Gamma}{\partial\xi} = \frac{2\eta}{\mathrm{Re}}\frac{\partial^2\Gamma}{\partial\eta^2} + \frac{\lambda}{2\mathrm{Re}}\frac{\partial^2\Gamma}{\partial\xi^2} \tag{5.126}$$

$$\Gamma\frac{\partial\Gamma}{\partial\xi} = R_o\left(4\eta^4\left(\frac{\partial\psi}{\partial\xi}\frac{\partial^3\psi}{\partial\eta^3} - \frac{\partial\psi}{\partial\eta}\frac{\partial^3\psi}{\partial\xi\partial\eta^2} - \frac{2}{\mathrm{Re}}\left(2\frac{\partial^3\psi}{\partial\eta^3} + \eta\frac{\partial^4\psi}{\partial\eta^4}\right)\right)\right)$$

$$+\lambda\left(-\frac{\partial\psi}{\partial\xi}\frac{\partial^2\psi}{\partial\xi^2}+\eta\frac{\partial\psi}{\partial\xi}\frac{\partial^3\psi}{\partial\eta\partial\xi^2}-\eta\frac{\partial\psi}{\partial\eta}\frac{\partial^3\psi}{\partial\xi^3}-\frac{1}{\mathrm{Re}}\left(4\eta^2\frac{\partial^4\psi}{\partial\eta^2\partial\xi^2}+\frac{\alpha}{2}\eta\frac{\partial^4\psi}{\partial\xi^4}\right)\right) \tag{5.127}$$

where $\eta = (r/R)^2, \xi = z/L, \Gamma = rv_\theta/Rv_{\theta R}, \psi = \widehat{\psi}/(Q/2\pi)$, L is the length of the hydrocyclone, $\widehat{\psi}$ is the axial-symmetric stream function and Q is the total feed flow.

In hydrocyclone flow, the tangential velocity is two orders of magnitude greater than the radial velocity; therefore, the dimensionless number R_0 is always small (of the order of 10^{-2}). For that reason it is convenient to solve Eqs. (5.126) and (5.127) as a series expansion in R_0:

$$\Gamma = \sum_{n=0}^{\infty}\Gamma_n(\eta,\xi)R_0^n \quad \text{and} \quad \psi = \sum_{n=0}^{\infty}\psi_n(\eta,\xi)R_0^n \tag{5.128}$$

Substituting these terms into Eq. (5.55) yields:

$$\left(\Gamma_o\frac{\partial\Gamma_o}{\partial\xi}\right)R_o^0+\left(\Gamma_1\frac{\partial\Gamma_1}{\partial\xi}\right)R_o^1+\ldots=\left(\Gamma_1\frac{\partial\Gamma_1}{\partial\xi}\right)R_o^1+\left(\Gamma_2\frac{\partial\Gamma_2}{\partial\xi}\right)R_o^2+\ldots \tag{5.129}$$

Equation (5.129) implies that Γ_0 is independent of ξ, that is,

$$\Gamma_0 = \Gamma_0(\eta) \tag{5.130}$$

Then, for the zero-order approximation the following relationship is obtained from Eq. (5.127):

$$\frac{\partial\psi_0}{\partial\xi} = \frac{2\ell}{NL_A}\eta\frac{\Gamma_0''(\eta)}{\Gamma_0'(\eta)} \equiv f_{01}(\eta) \tag{5.131}$$

Assuming that:

$$\psi_0 = f_{00}(\eta)+\xi(\eta) \tag{5.132}$$

and substituting into (5.131), the following ordinary second-order differential equation is obtained:

$$2\eta\Gamma_0''(\eta)-\frac{NL_A}{\ell}f_{01}(\eta)\Gamma_0'(\eta)=0 \tag{5.133}$$

It is convenient to divide the hydrocyclone into two zones: the inlet zone, which will be called zone I, and zone II below, (see Fig. 5.17). Boundary conditions for Eq. (5.133) are obtained from the values of the stream functions $\xi = 1$ in zone I and

$\xi = 0$ in zone II:

$$\begin{bmatrix} \psi(\eta, 0) = g_0(\eta) \\ \psi(\eta, 1) = g_1(\eta) - g_0(\eta) \end{bmatrix} \Rightarrow \begin{bmatrix} f_{00}(\eta, 0) = g_0(\eta) \\ f_{00}(\eta) + f_{01}(\eta) = g_1(\eta) \end{bmatrix}$$

therefore:

$$\begin{aligned} f_{00}(\eta) &= g_0(\eta) \\ f_{01}(\eta) &= g_1(\eta) - g_o(\eta) \end{aligned} \tag{5.134}$$

Physically, this means that for small values of R_0, the stream function is specified entirely by their boundary conditions. For a flat-bottom hydrocyclone such as that shown in Fig. 5.17, the boundary conditions equivalent to setting $v_r(R) = v_R$, $v_r(0) = 0$ $v_\theta(R) = v_{\theta R}$, $v_z(r, 0) = Q/2\pi R_u^2$ and $\mathrm{v_z(r, L) = (Q - Q_D)/2\pi R_u^2}$ were found to be:

$$\text{In zone I:} \quad g_0(\eta) = j(\eta), \qquad g_1(\eta) = \frac{Q_o}{Q}(1 - \exp(-100\eta)) \tag{5.135}$$

$$\text{In zone II:} \quad g_0(\eta) = -\frac{Q_o}{Q}(1 - \exp(-150\eta)), \quad g_1(\eta) = j(\eta), \tag{5.136}$$

$$\text{Where:} \quad \begin{aligned} j(\eta) = &+1445.32\eta^5 - 6166.26\eta^{4.5} + 11025.86\eta^4 - 1067.60\eta^{3.5} \\ &+5964.80\eta^3 - 1829.91\eta^{2.5} + 220.59\eta^2 + 15.53\eta^{1.5} \end{aligned} \tag{5.137}$$

In the zero-order approximation, which is with $n = 0$, the circulation and stream functions are independent of R_0. Γ is a function of η only and ψ is a function of η and a linear function of ξ. This result coincides with the previously analyzed solution, where $v_z = f(r) + g(r)$. The results for the zero-order approximation of velocity distributions are shown in Figures 5.27 and 5.28 [33].

5.8 Modeling the Air Core

The role of the air core in the performance of a hydrocyclone was not recognized until 1987, when Plitt and colleagues commented that it is related to the flow transition from a spray discharge to roping. That same year Davidson et al. [46] indicated that as the slurry viscosity increases, the angular velocity in the hydrocyclone drops and the air core contracts. Castro [26] found that the water split is mainly controlled by the air core size for a given apex/vortex ratio and, therefore, the water split is governed by the operating conditions and variables that affect the air core. Bustamante [25] found that operating a hydrocyclone with water only might be used to predict roping conditions when operating with a pulp. He performed tests in which, for a fixed

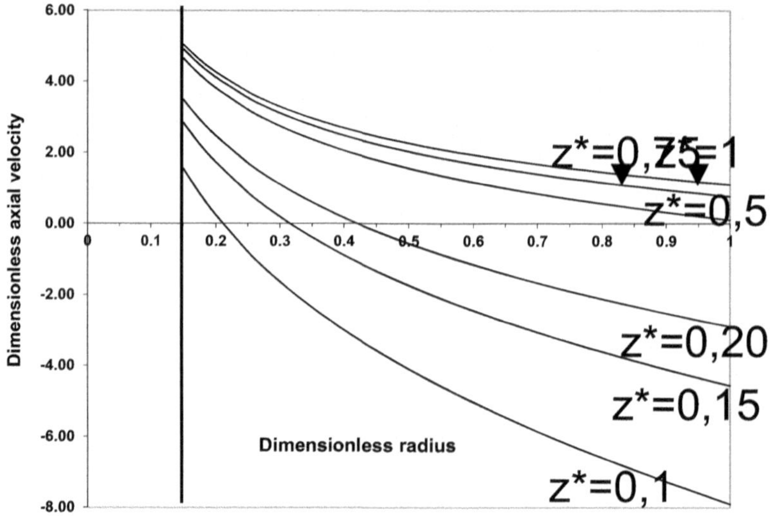

Fig. 5.27 Dimensionless axial velocity distributions for the isochoric rotational flow model of a hydrocyclone according to Bloor and Ingham [16]

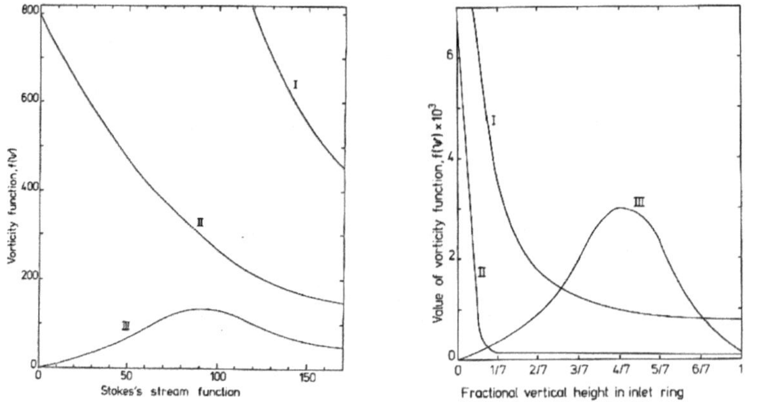

a) Vorticity versus stream function b) Vorticity versus height in inlet ring

Fig. 5.28 Three different vorticity functions according to Brayshaw [22]

apex/vortex ratio, he measured the underflow rate for increasing feed rate values, observing that the air core increased as the feed rate increased. At a given condition, the air core completely obstructed the apex and there was no underflow. The same condition when no underflow occurred when operating with only water coincided with the onset of roping when the hydrocyclone was operated with pulp [3, 34]. This late recognition of the importance of the air core in hydrocyclone operation is strange, since researchers had developed models for the air core size in vortex motions long before.

Binnie and Hookings [10] analyzed water drainage from an open cylindrical tank with controlled axial and tangential inflows and a free discharge at the bottom and center of the tank. They observed a relationship between the total head and the discharge rate that depended on the ratio of tangential to radial inflow. They also observed a critical pressure head that could maintain an air core. Figure 5.29 shows such curves.

When a combination of tangential and axial inflow of a constant ratio was used, they observed an interesting surging phenomenon. This effect was related to the formation and breakdown of the air core. Consider the tank full up to a total head of 0.16 m, with a tangential inflow rate of 0.24×10^{-3} m^3/s and an axial inflow rate of 0.66×10^{-3} m^3/s. The total inflow was maintained at $0.9 \times 10^{-3} m^3/s$. The curve for the total head versus the discharge flow rate is also given in Fig. 5.29. At $H = 0.16$m, the ratio of tangential to axial inflow rates of 0.37 is not sufficient to maintain an air core and the entire discharge leaves through the exit pipe with a flow rate of 2.3×10^{-3} m^3/s. Since this outflow is larger than the inflow of 0.90×10^{-3} m^3/s, the container begins to loose water and the total head and discharge rate diminishes until H becomes $H = 0.06\,m$. This is the lowest total head at which a coreless discharge is possible with the given inflow ratio of 0.37 and, once again, an air core forms suddenly, abruptly diminishing the discharge rate to 0.57×10^{-3} m^3/s. This outflow is now smaller than the inflow of 0.90×10^{-3} m^3/s, and the water level will rise until the discharge rate is equal to the inflow rate. At this point, with a total head of $H = 0.16$m, the air core again becomes unstable, collapsing to a coreless outflow of 2.3×10^{-3} m^3/s. Then a new cycle begins.

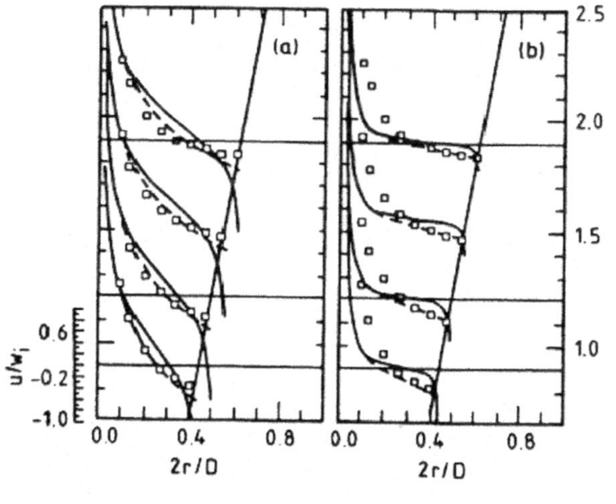

Fig. 5.29 Axial velocity for the isochoric rotational flow model of a hydrocyclone according to Davidson [47], compared with experimental data of [71]. **a** n = − 0.5; **b** n = − 1.8

Binnie and Hookings [10] obtained an approximate expression for the discharge rate in the above problem by considering a potential flow. With the Bernouilli given by $B = p/\rho + (1/2)(v_\theta^2 + v_z^2)$, the total head is $H = B/g$, and the axial velocity at the interface, where $p = 0$, is given by $v_z = (2gH - v_\theta^2)^{1/2}$. If the underflow and the air core radii are denoted by R_u and R_a respectively, the outflow volume is given by:

$$Q_u = \pi\left(R_u^2 - R_a^2\right)v_z \tag{5.138}$$

and therefore:

$$Q_u = \pi\left(R_u^2 - R_a^2\right)\left(2gH - V_\theta^2\right)^{1/2} \tag{5.139}$$

For an inviscid flow, the tangential velocity is a free vortex of the form $v_\theta = V_{\theta u}/r$, where $V_{\theta u}$ is the tangential velocity at $r = R_u$, and, therefore, for $r = R_a$:

$$Q_u = Q_u = \pi\left(R_u^2 - R_a^2\right)\left(2gH - \left(V_{\theta u}/R_a\right)^2\right)^{1/2} \tag{5.140}$$

Binnie and Hookings argued that, for the stability of the flow, R_a should adjust itself so that the discharge is maximum, which means that $\partial Q_u/\partial R = 0$, leading to the following condition:

$$4gHR_a^4 - V_{\theta u}^2 R_a^2 - V_{\theta u}^2 R_u^2 = 0 \tag{5.141}$$

from which the radius of the air core is:

$$R_a = \left(\frac{V_{\theta u}^4 + \left(V_{\theta u}^4 + 16gHV_{\theta u}^2 R_u^2\right)^{1/2}}{8gH}\right)^{1/2} \tag{5.142}$$

Equation (5.141) can also be written in the form:

$$\begin{aligned} 4gHR_a^4 - 2V_{\theta u}^2 R_a^2 &= V_{\theta u}^2 R_a^2 + V_{\theta u}^2 R_u^2 - 2V_{\theta u}^2 R_a^2 = 0 \\ 2gH - V_{\theta u}^2 &= \frac{V_{\theta u}^2}{2R_a^2}\left(\frac{R_u^2}{R_a^2} - 1\right) \equiv v_z^2(R_a) \end{aligned} \tag{5.143}$$

This is the value of the axial velocity in the discharge tube. Introducing this value into Eq. (5.138), the flow rate through the discharge can be found:

$$Q_u^2 = \frac{1}{2}\pi R_a^2 V_{\theta u}^2\left(\left(\frac{R_u}{R_a}\right)^2 - 1\right)^3 \tag{5.144}$$

or

$$\left(\frac{Q_u}{\Gamma_u}\right)^2 = \frac{1}{8}\left(\frac{R_a}{R_u}\right)^2\left(\left(\frac{R_u}{R_a}\right)^2 - 1\right)^3 \tag{5.145}$$

where $\Gamma_u = 2\pi R_u V_{\theta u}$ is the circulation at $r = R$. This equation was derived by [105] for the outflow of a hydrocyclone.

Davidson [48] followed Binnie and Hookings' approach to calculate the air core size for one outlet at the underflow and two outlets at the underflow and overflow. For one outlet they described the tangential velocity with the equation $v_\theta = A/r^n$. The Bernouilli can now be written in the form $B = p/\rho + v_z^2 + A^2/r^{2n}$. At the air core interface the pressure is zero, $r = R_a$ and $A = V_{\theta u}$,thus:

$$v_z = \left(2B - V_{\theta u}^2/R_a^{2n}\right)^{1/2} \tag{5.146}$$

and the volume flow results:

$$Q_u = \pi\left(R_u^2 - R_a^2\right)\left(2B - V_{\theta u}^2/R_a^{2n}\right)^{1/2} \tag{5.147}$$

Differentiating this expression with respect to R_a and equating the result to zero yields:

$$\frac{R_a}{R_u} = \left(\frac{nV_{\theta u}^2}{nV_{\theta u}^2 + 2V_z^2}\right)^{1/2} \tag{5.148}$$

For $n = 1$, Eq. (5.145) derived by Smith [105] is recovered:

For two outlets, Davidson [48] described the axial velocities with the following equations $V_{zu} = 2B - V_{\theta u}^2/R_a^{2n}$ and $V_{zo} = 2B - V_{\theta o}^2/R_a^{2m}$. The total flow rate is given by:

$$Q = \pi\left(R_u^2 - R_a^2\right)\left(2B - V_{\theta u}^2/R_a^{2n}\right)^{1/2} + \pi\left(R_o^2 - R_a^2\right)\left(2B - V_{\theta o}^2/R_a^{2m}\right)^{1/2} \tag{5.149}$$

After differentiating with respect to R_a, the radius of the air core becomes:

$$R_a = \left(\frac{nR_u^2 V_{\theta u}^2 V_{zu} + mR_o^2 V_{\theta o}^2 V_{zo}}{\left(2V_{zu}^2 + nV_{\theta u}^2\right)V_{zo} + \left(2V_{zo}^2 + mV_{\theta o}^2\right)V_{zu}}\right)^{1/2} \tag{5.150}$$

For large $V_{\theta u}$, $R_a \to R_u$ and for large $V_{\theta o}$, $R_a \to R_o$. Since R_u is smaller than R_o the apex will be blocked first.

Davidson [48] indicated that, using inviscid flow to calculate the air core size, Eq. (5.150) gives correct results when the fluid is water. For a suspension he assumed that a correction factor proportional to the ratio of the kinematic viscosity of the suspension ν_s to the kinematic viscosity of the water ν_w can be used for the air core radius, as follows:

$$R_a|_{suspension} \approx R_a|_{water}\left(\frac{\nu_s}{\nu_w}\right)^{0.3} \tag{5.151}$$

Davidson [48] further indicated that his equation for the air core size compares well with the experiments of [71, 96], Hsieh and Rajamani [64] and Svarovsky and Marasinghe [108]. The agreement with the experiments of [106] is only fair.

Mikhaylov and Romenskiy [76] proposed the following equation for the air core in a conical hydrocyclone:

$$R_a = k_a d_o (\tan\alpha)^{0.30} \tag{5.152}$$

where 2α is the angle of the conical section of the hydrocyclone and k_a is a constant.

Nepomnyashchii and Pavlobskii [82] derived the following equation for the air core radius [see Eq. (5.125)]:

$$R_a = \frac{R_F}{R}\left(\frac{\rho v_F^2}{3\Delta p}\right)^{1/3} \tag{5.153}$$

where R_F and R are the eccentricity of the hydrocyclone feed and the hydrocyclone radius, respectively, and v_F is the inlet velocity of the feed.

Svarovsky [107] presented this equation in dimensionless form:

$$\frac{d_a}{d_c} = \frac{d_F}{d_c}\left(\frac{1}{3Eu}\left(\frac{d_c}{d_F}\right)^4\right)^{1/3} \tag{5.154}$$

where d_a , d_F and d_c are the air core diameter, twice the inlet eccentricity and the hydrocyclone diameter, respectively, and Eu $= \Delta p/(1/2)\rho v^2$ is Euler's number.

Barrientos et al. [3] calculated the air core radius using the momentum jump balance for the normal stresses. Across the interface, a jump in the normal stresses exists and is given by the Young–Laplace equation:

$$[\boldsymbol{Tn}\cdot\boldsymbol{n}] = -2H\gamma \tag{5.155}$$

where the square brackets represent the jump of a property across the interface, $\boldsymbol{T}$ is the stress tensor, $\boldsymbol{n}$ is the normal unit vector to the interface, H is the main curvature of the air core and γ is the liquid surface tension. Assuming the air behaves as an ideal fluid rotating as a rigid body with density ρ_a and angular velocity ω_a and that the liquid is a Newtonian fluid with viscosity μ, Eq. (5.72) can be written in the form:

$$\frac{1}{2}\rho_a\omega_a^2 R_a^3 + \left(2\mu\left.\frac{\partial v_r}{\partial r}\right|_{r=R_a} - \Delta p_a\right)R_a - \gamma = 0 \tag{5.156}$$

Introducing the dimensionless variables and making a study of order of magnitude of the resulting equation yields the following equation for the air core:

$$R_a \approx \frac{\gamma}{2\mu v - \Delta p_a} \tag{5.157}$$

Based on this equation, Concha et al. [34] and Castro et al. [27] developed empirical equations for the air core diameter in terms of the operating variables in a laboratory and industrial hydrocyclone, respectively. Concha et al. [34] obtained the following equation for the air core:

$$d_a = \frac{3.03\gamma d_o}{270.3\mu_w\left(1 + \frac{1.18\times10^5}{\Delta P^{2.61}}\right)\left(\frac{d_u}{d_o}\right)^{-0.20} - 1.38 \times 10^{-3}} \tag{5.158}$$

In this equation γ is measured in N/m, μ_w is in kg/ms, ΔP is in Pa and d_a , d_o and d_u are in m. In this case ΔP is the total pressure drop in the hydrocyclone.

Castro et al. [27] obtained the following relationship:

$$d_a = \frac{103.6(d_u/d_o)^{0.89}}{6.3(\mu/\mu_w)(1 + \frac{0.98}{\Delta P^{1.15}} - \frac{1.41}{\Delta P^{1.57}})} \tag{5.159}$$

where μ and μ_w are the viscosities of the pulp and the water, respectively.

Ovalle and Concha [91] studied the generation and propagation of waves at the air core-water interface of a conical hydrocyclone. They showed that inside the hydrocyclone the flow is sub-critical (Froude number less than 1) and therefore in a favorable condition for the existence of undulatory motion in the fluid. They used a hydraulic analogy consisting of a launder with a ramp, where the water enters through the bottom at the left side and overflows both at the left and the right of the upper part of the physical model (see Fig. 5.30). The centrifugal force of the hydrocyclone is substituted by the gravitational force in the hydraulic model.

The flow in this hydraulic analogy was modeled as inviscid and irrotational, with the form of the free surface obtained as:

$$\eta(x, y, t) = 2A\frac{k}{\omega}\sinh(k(y+h))\cos(kx - \omega t)$$

where A is a constant and $\omega(t)$ and $k(x)$ represent the temporal and spatial behavior, respectively. Figure 5.31shows some isobaric lines resulting from the calculations of the temporal evolution of the waves imposed over the free surface (Figs. 5.32, 5.33 and 5.34).

The authors concluded that the form of the free surface depends mainly on the global geometrical characteristics of the flow and possibly to a minor extent on physicochemical properties of the surface. They showed that the waves propagating through the water–air interface of the air core will increase their amplitude as the

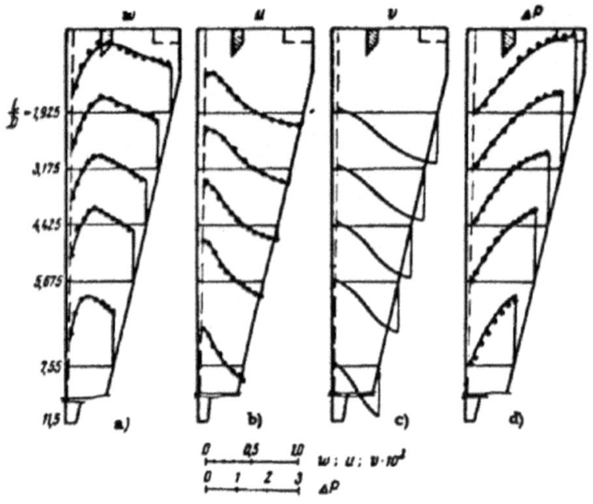

Fig. 5.30 Velocities and pressure distributions according to Mikhailov and Romenskiy [73]

hydrocyclone diameter becomes smaller, so that the air core may become of the size of the apex, in which case instability is induced in the underflow. The apex will intermittently become blocked by the air core, inducing roping. Unfortunately, the presence of waves in a conical hydrocyclone is inherent to the physics of the flow.

Although most of the experimental evidence on the flow pattern in a hydrocyclone is based on studies with water, one can say that the flow pattern in a hydrocyclone treating mineral suspensions is well understood. What is not yet sufficiently clear is a model of the air core and how it affects the flow pattern, the fluctuating velocity component and hydrocyclone operation. Nevertheless, our knowledge of the flow pattern in a hydrocyclone is sufficient to develop classification models of the device.

5.9 Type of Hydrocyclone Discharge and Roping

Although roping in hydrocyclones is a frequent problem in industries and has been studied by many researchers, we do not yet have a theory that relates all the variables involved [97]. Several experimental works with different approaches such as mechanical energy balance, work on hydrocyclones with water only and hydrocyclone air core measurements with different instruments in the laboratory and plant have attempted to explain the roping phenomenon. They have addressed design variables such as apex-to-vortex diameters and different cone angles, as well as operating variables such as inlet solid concentration, particle size, pressure and overflow and underflow flow rates and concentrations in order to understand their effect on the roping condition.

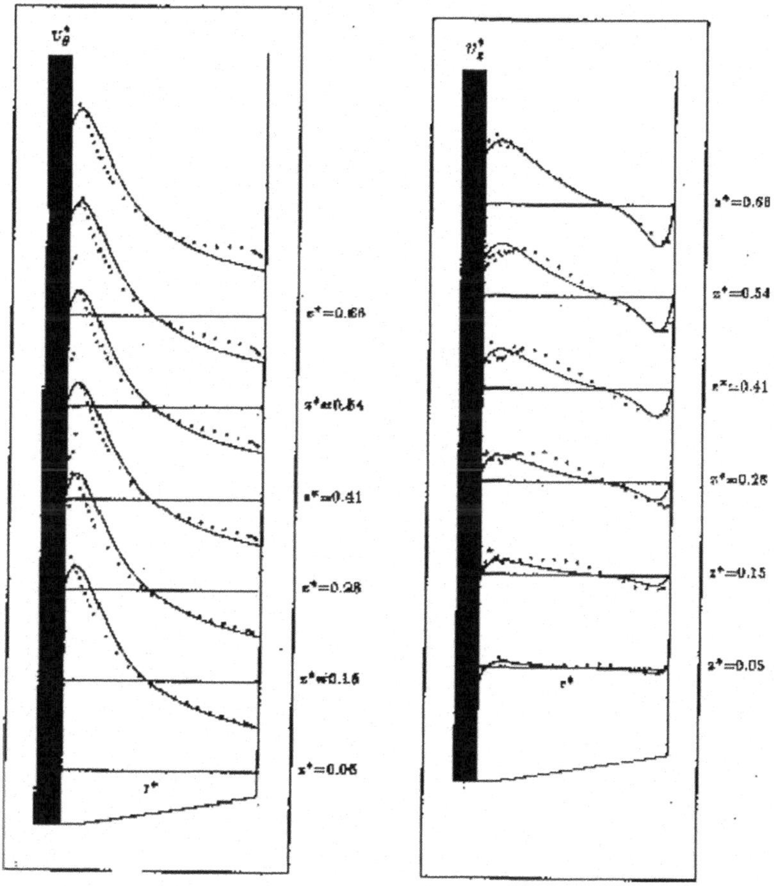

Fig. 5.31 Tangential velocity distribution for several values of the radial Reynolds number for the zero-order approximation [33]

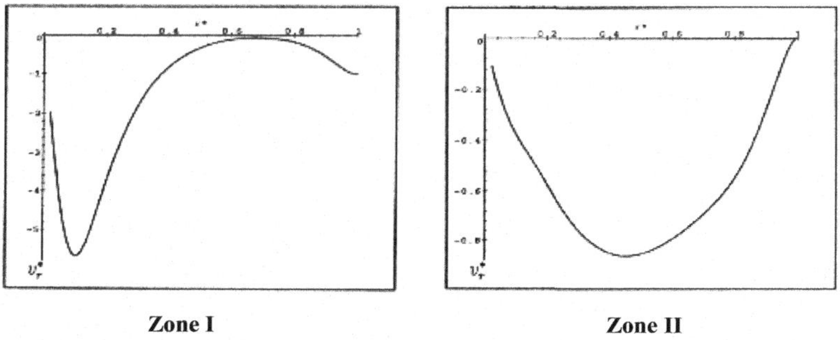

Fig. 5.32 Predicted radial velocity profile in a flat-bottom hydrocyclone for the zero-order approximations in zones I and II [33]

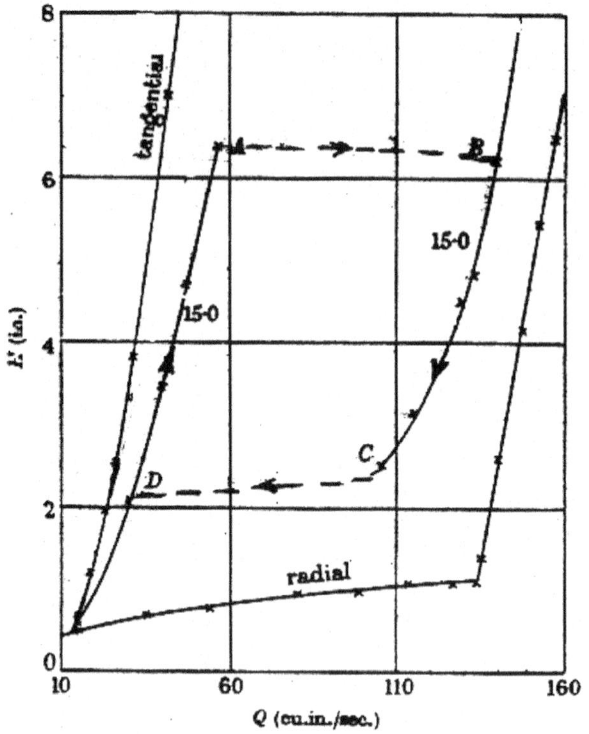

Fig. 5.33 Discharge volume flow rate Q_D versus total head H for axial or tangential water inlet of 0.9×10^{-3} m³/s. Also shown is the curve for a tangential inflow of 0.24×10^{-3} m³/s and axial inflow of 0.66×10^{-3} m³/s

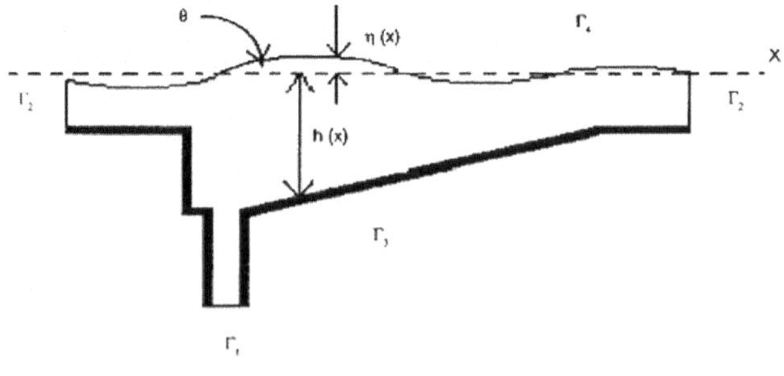

Fig. 5.34 Hydraulic analogy of a conical hydrocyclone

Bustamante [25], operating a 6-inch Krebs hydrocyclone with water only measured the air core diameter increase as the inlet pressure was increased. He showed that a certain point the air core equaled the apex diameter and no water leaves through the underflow. He established that when operating the hydrocyclone with pulp at the same operating conditions the transition from spray to roping

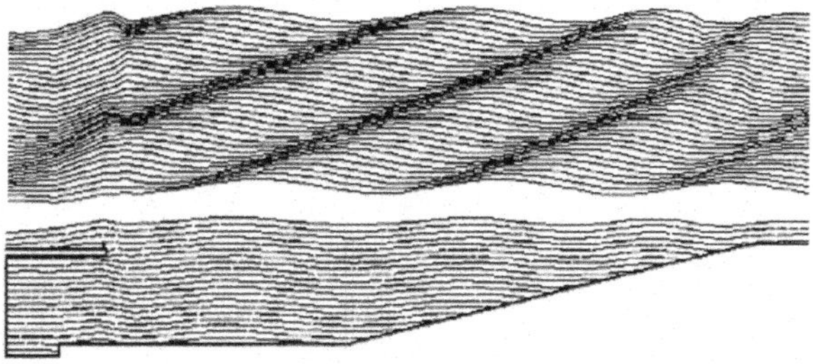

Fig. 5.35 Isobaric lines in the hydraulic model of a conical hydrocyclone

occurred when the apex was blocked and no the air came in to forming the air core. Using the Young Laplace equation, Concha et al. [34] showed for that this hydrocyclone the apex to vortex diameter $d_a/d_v < 0.34$ roping always exists and for $d_a/d_v < 0.50$ there is no roping under any operating conditions. In the intermediate range $0.34 < d_a/d_v < 0.50$, spray and roping can coexist depending on other variables such as pressure, feed concentration, particle feed size and size distribution. The conclusion of this work was that the precondition for roping is the absence of the air core in the apex and the formation of a bed of particles in the bottom of the hydrocyclone.

This same year [81] pointed out that a hydrocyclone can operate with a spray, semi-rope and rope discharge depending on several variables but mainly on the particle concentration of the feed. Figure 5.35 shows the representation of these phenomena. In dense flow separation, more solids are stored in the conical part of the hydrocyclone and slowly moved upward reducing solids recovery to the underflow. The air core does not extend over the entire length of the hydrocyclone and oscillates intensively. The discharge assumes the shape of a rope and is characterized by high solids content and fewer fines, which is the advantage of this variant. The transition state between spray and rope discharge presents the advantages of both high solid recovery and high solids content in the underflow.

Using CFD Perez [97], confirmed that the incremental increase in inlet pressure produces an expansion of the air core diameter leading to roping which is characterized by a change in the frequency spectrum of the pressure oscillations at the hydrocyclone wall with high amplitudes and low frequencies for spray discharge and noisy signals in all spectrums and damped low frequencies for roping. They also indicated that transition from spray to roping is generated by an increase in particle diameter at a constant solid concentration in the feed flow. With 20 μm diameter particles in the feed an increase in pressure results in a decrease in the discharge angle, although it always maintaining a spray shape, while with 34 μm diameter particles generates a semi-rope discharge and 70 μm diameter particles always roping condition. Figure 5.36 shows slurry-air contours and discharge angles as the inlet pressure increases for the following three cases:

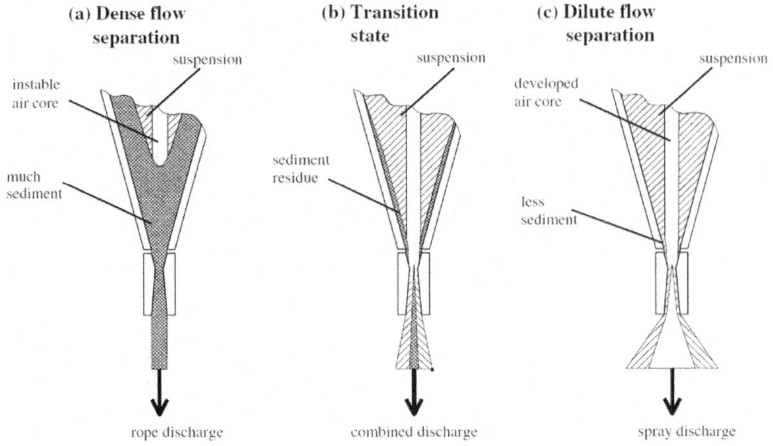

Fig. 5.36 Operating state of a hydrocyclone

	Case I	Case II	Case III
Particle size (μm)	20	34	70
Pressure (psi)	8.6, 19.2, 33.2, 50.2, 70.5	7.4,19.1, 25.8, 50.2, 58.8	4.7, 4.8, 8.6, 14

In a recent work, Daza et al. [49] examined the air core and material distribution inside the hydrocyclone, the underflow shape and spray angle, the flow and solid concentrations, the efficiency curve and separation size and the particle size distribution in the underflow sizes. The study confirmed that in a feed with particle size distribution the feed has a direct impact on the underflow behavior. When the device operates with coarser material it tends to rope with a decrease in the spray angle. The same happens with an increase in the feed pressure inlet. The roping condition generated by the coarse material directly affects the efficiency of the hydrocyclone, triggering an increase in the separation size. The underflow particle size distribution tends to be coarser with an increase in the inlet pressure. Figure 5.38 shows the experimental setup, Fig. 5.39 the material size distribution while Fig. 5.40 shows the CFD simulation (Fig. 5.37).

CUI et al. [39] studied the flow field in a 50 mm hydrocyclone with CFD simulation and experimental studies based on particle image velocimetry (PIV). Figure 5.41a shows a photo of the experimental model setup and Fig. 5.41b shows the PIV measurement system.

The procedure consisted of fluid circulation system and a PIV system. Polystyrene powder (5–10 μm) was used as tracer particle in the tests. A total of 200 transient velocity frames were recorded to obtain a time to obtain an average value.

Vakamalla et al. [113] setup consisting of an electrical resistance tomography and a high speed camera to measure the air core size at different operating conditions of a 75 mm diameter hydrocyclone. See Fig. 5.41. A comparison with CFD measurement are shown in Figs. 5.42 and 5.43.

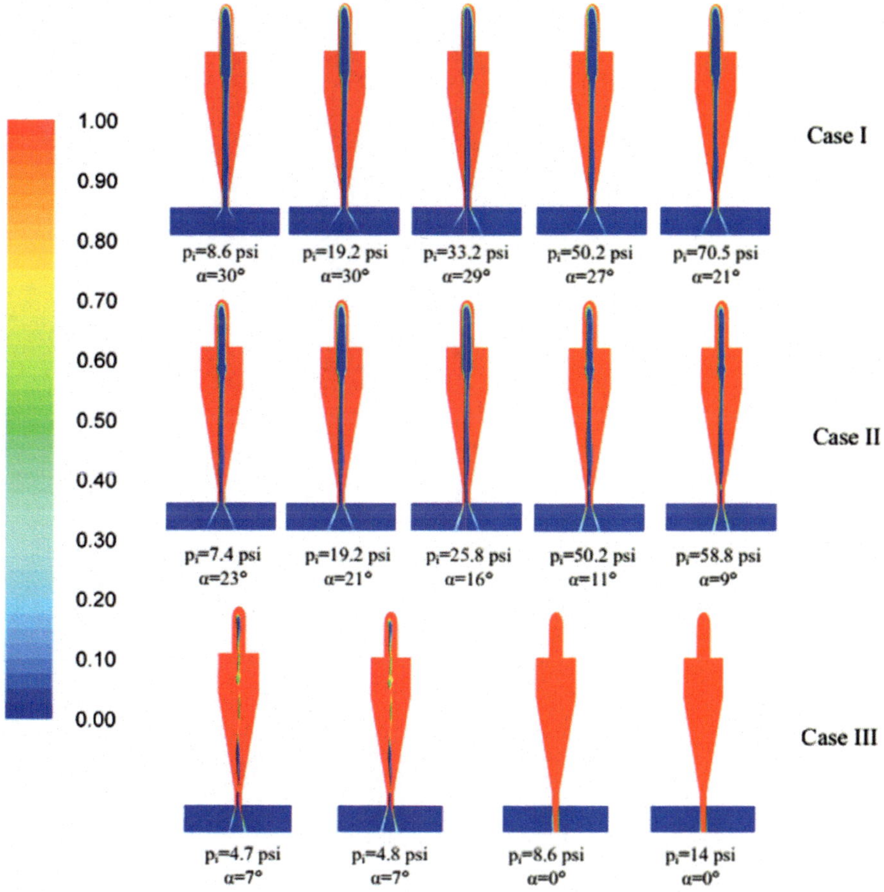

Fig. 5.37 Slurry-air contours and discharge angles as the inlet pressure increases

Fig. 5.38 Experimental setup

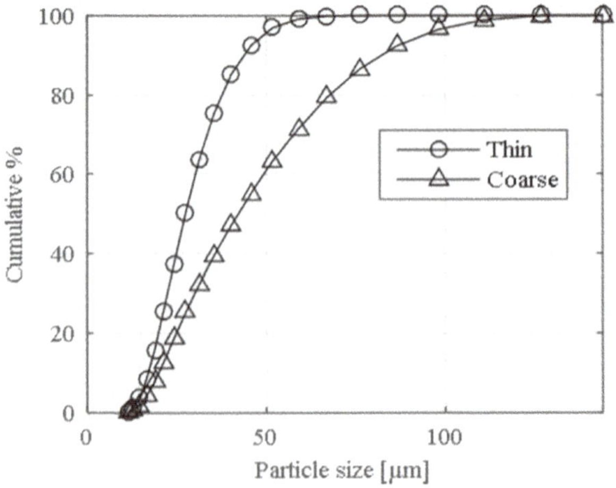

Fig. 5.39 Material size distribution

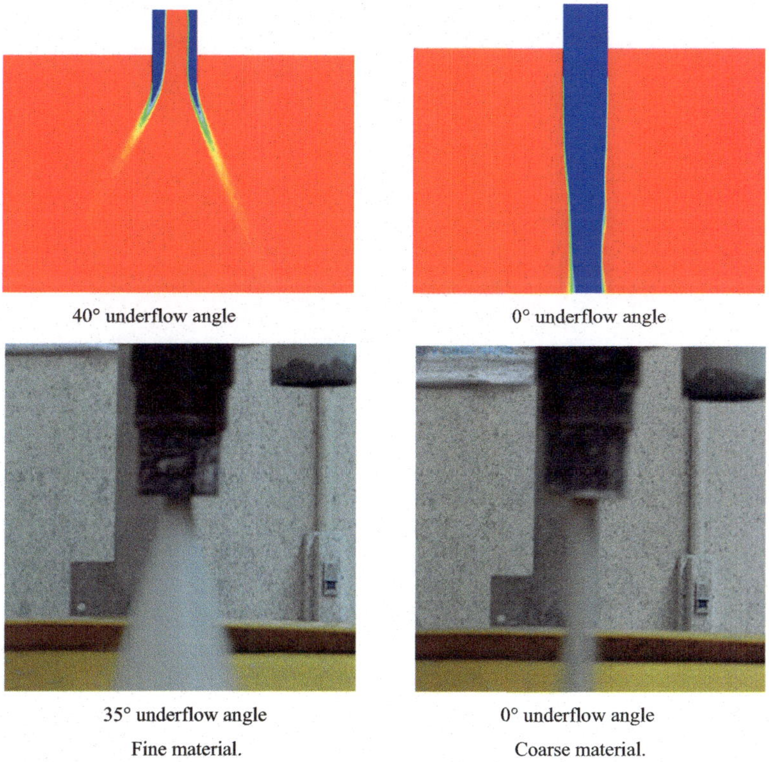

Fig. 5.40 Simulated and experimental underflow at 1 bar pressure

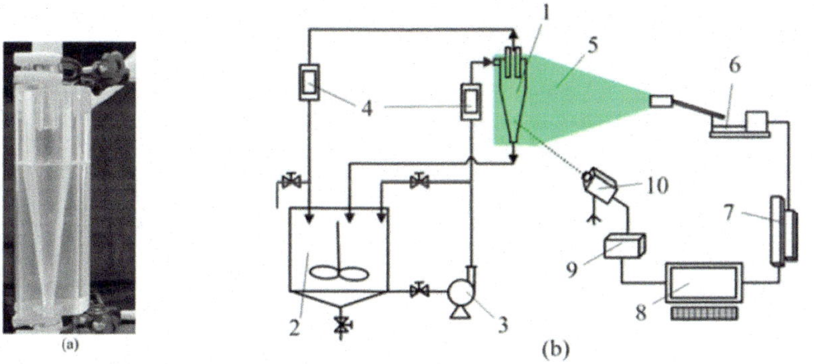

Fig. 5.41 Experimental model setup: **a** Photo of hydrocyclone; **b** PIV measurement system [39]

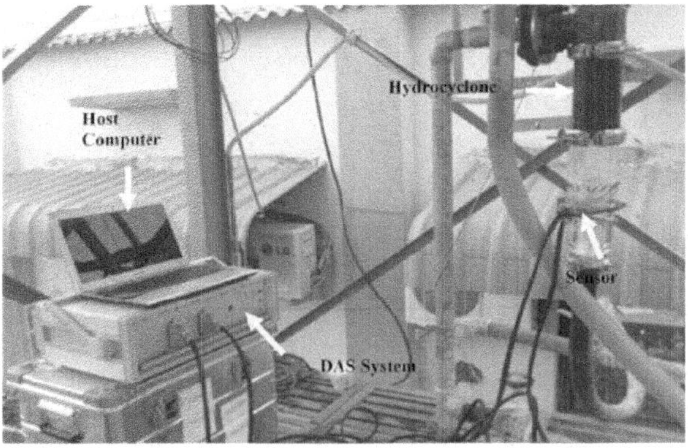

Fig. 5.42 Setup of an electrical conductivity tomograhpy system to measure the air-core size [112]

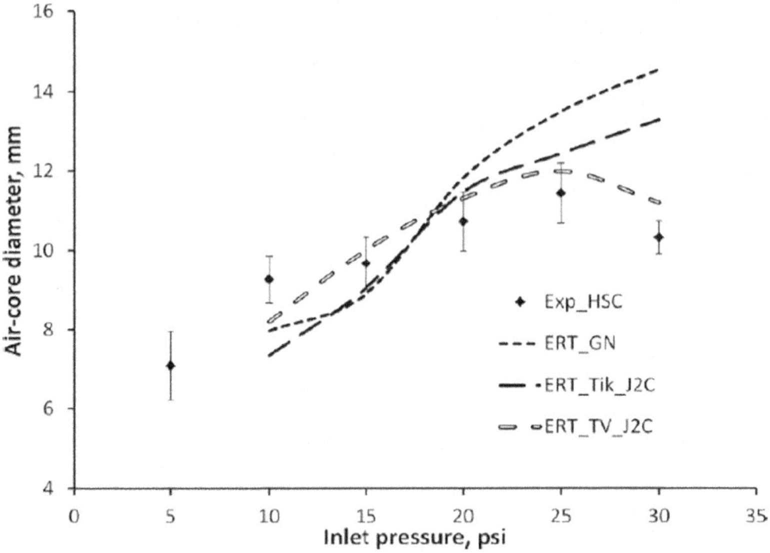

Fig. 5.43 Comparison of air-core diameter by CFD against a high-speed camera and electrical Resistance tomography of a 76 mm hydrocyclone operating at different pressure

References

1. Aguilera GF, Arismendy JE (2014) Development of a hydrodynamic model of a hydrocyclone including the simulation of air-core effect, using CFD, Holos 2014, 30(3):184–192
2. Bank N, Gauvin WH (1977) Measurement of flow characteristics in a confined vortex flow. Can J Chem Eng 55:397–402
3. Barrientos A, Sampaio R, Concha F (1993) Effect of the air core on the performance of a hydrocyclone. In: XVIII international mineral processing congress, Sydney, pp 23–28, 267–270

4. Bhaskar KU, Murthy YR, Raju MR, Tivari S, Srivastava JK, Ramakrishnan N (2007) CFD simulation and experimental validation studies on hydrocyclone. Miner Eng 20:60–71
5. Bhattacharyya P (1980) Theoretical study of the flow field inside a hydrocyclone with vortex finder diameter greater than the apex opening; I. Laminar Case, Appl Sci Res 36:213–225
6. Bhattacharyya P (1980) Theoretical study of the flow field inside a hydrocyclone with vortex finder diameter greater than the apex opening; II. Turbulent Case, Appl Sci Res 36:197–212
7. Bhattacharyya P (1984) The flow field inside a conventional hydrocyclone. In: Proceeding of 2nd International Conference Hydrocyclones, BATH, Cranfield, Paper H2, pps 323–330
8. Binnie AM, Hookings GA (1957) Laboratory experiments on whirlpools. Proc Roy Soc A 236:261–274
9. Binnie AM, Hookings GA, Kamel MYM (1948) The flow of swirling water through a convergent-divergent nozzle. Proc Roy Soc A 194:398–415
10. Binnie AM, Hookings GA (1948) Laboratory experiments on whirlpools. Proc Roy Soc A 194:5
11. Bloor MIG, Ingham DB (1973) Theoretical investigation of the flow in a conical hydrocyclone. Trans Inst Chem Eng 51(1):36–41
12. Bloor MIG, Ingham DB (1975) Turbulent spin in a cyclone. Trans Instn Chem Eng 53:1–6
13. Bloor MIG, Ingham DB (1975) The leakage effect in the industrial cyclone. Trans Instn Chem Eng 53:7–11
14. Bloor MIG, Ingham DB (1976) Boundary layer flows on the side walls of conical cyclones. Trans Instn Chem Eng 54:276
15. Bloor MIG, Ingham DB (1987) The flow in industrial cyclones. J Fluid Mech 178:507–519
16. Bloor MIG, Ingham DB (1974) Theoretical analysis of the conical cyclone. In: First European conference on mixing and centrifugal separation, Cambridge, Sept 1974, paper E-6, 95–106, BHRA, Cranfield, Bedford, England
17. Bloor MIG, Ingham DB (2015) Theoretical aspects of hydrocyclone flow. In: Wakeman RJ (ed) Progress in filtration and separation, Part. 3. Elsevier, Amsterdam, pp 57–148
18. Bloor MIG, Ingham DB (1984a). The influence of vorticity on the efficiency of the hydrocyclone. In: Proceeding of 2nd International Conference Hydrocyclones, BATH, Cranfield, Paper B2, 51–66
19. Bloor MIG, Ingham DB (1984b) A theoretical investigation of the fluid mechanics of the hydrocyclone. Filtr Sep July/August, 266–269
20. Bloor MIG (1987) On axially symmetric flow models for hydrocyclones. In: 3rd International conference on hydrocyclones, Oxford, England, Sept 1987, Paper D-2, 83–89
21. Bond CJ, Climpson J, Dyakowski T , Faulks I, Jia X, Kostuch JA, Payton D, Wang M, Wang, SJ, West RM, Williams RA (1999) World congress on industrial process tomography, Buxton, Greater Manchester1999. Industrial monitoring of hydrocyclone operation using electrical resistance tomography, 1st Apr 14–17
22. Bond J, Cullivan JC, Climpsos N, Dyakowski T, Faulks I, Jia X, Kostuch J, Payton D, Wang M, Wang SJ, West RM, Williams RA (1999) Industrial monitoring of hydrocyclone operation using electrical, resistance tomography. In: 1st World congress on industrial process tomography, Buxton, Greater Manchester, 14–17 Apr
23. Brayshaw MD (1990) A numerical model for inviscid flow of fluid in a hydrocyclone. Int J Mineral Process 29:51–75
24. Brennan M (2006) CFD simulations of hydrocyclones with an air core. Chem Eng Res Des 84:495–505. https://doi.org/10.1205/cherd.05111
25. Bustamante MO (1991) Effect of the hydrocyclone geometry on normal operation condition. M.Sc. Thesis, University of Concepción (in Spanish)
26. Castro O (1990) Pulp Rheology effects for hydrocyclone models. M.Sc. Thesis, University of Queensland
27. Castro O, Concha F, Montero J, Miranda J, Castro J, Urizar D (1996) Air core modelling for an industrial hydrocyclone, Hydrocyclones-96. In: Claxton D, Svarovsky L, Thew MT (eds). MEP Ltd., London

28. Chiné B, Concha F (2000) Flow patterns in conical and flat bottom hydrocyclones. Chem Eng J 80:267–273
29. Chiné B, Concha F (2000) Flow patterns in conical and cylindrical hydrocyclones. Chem Eng J 80(1–3):267–273. https://doi.org/10.1016/S1383-5866(00)00101-5
30. Chu L-Y, Chen W (1999) Numerical simulation of turbulence and its structure. 9:128–136
31. Collantes J, Concha F, Chiné B (2000) Axial symmetric model for a flat bottom hydrocyclone. Chem Eng J 80:257–265
32. Collantes J, Concha F (1998) Asymptotic solutions of the Navier-Stokes equation for the swirling flow in a flat bottom hydrocyclone, XXVI Congresso Brasileiro de Sistemas Particulados, Teresópolis, RJ, Brasil, pp 19–22
33. Collantes J, Concha F (2000) Vortex flow in flat-bottomed hydrocyclones: leading order steady solution, XXI IMPC, Development in Mineral Processing 13. In: Massacci P (ed). Elsevier, Amsterdam, pp A4-77, A4-86
34. Concha F, Barrientos A, Montero J, Sampaio R (1996) Air core and roping in hydrocyclones. Int J Miner Process 44–45:743–749
35. Concha F (2007) Flow pattern in Hydrocyclones. KONA Powder and Particle 25:97–132
36. Concha F (2009) Settling velocities of particulate Systems. KONA Powder and Particle 27:18–37
37. Concha F, Collantes J (1999) Solution to the Navier-stokes equation for the flow in a flat-bottom hydrocyclone with the axial velocity dependent on the radial and axial coordinate. In: Conference on solid-liquid separation, Hawaii
38. Concha F, Castro B, Ovalle E, Romero J (1998) Numerical simulation of the flow pattern in a hydrocyclone. In: Innovation in physical separation technology, pp 35–60
39. Cui BB, Wei W, Gao S, Liu W, Feng T (2014) Numerical and experimental studies of flow field in hydrocyclone with air core. Trans Nonferrous Met Soc China 24:2642–2649
40. Cullivan JC, Williams RA, Cross CR (1998) Understanding the hydrocyclone separator through computational fluid dynamics. Trans I Chem E, Part A, Chem Eng Res Des 81(A4):455–466
41. Cullivan JC, Williams RA, Dyakowski T, Cross CR (2004) New understanding of a hydrocyclone flow field and separation mechanism from computational fluid dynamics. Mineral Eng 17(5):651–660
42. Dabir B, Petty CA (1984) Laser Doppler anemometry measurements of tangential and axial velocities in a hydrocyclone operating without an air core. In: Proceeding 2nd International Conference Hydrocyclones, BATH, Cranfield, Paper A2, 15–26
43. Dai GQ, Li JM, Chen WM. Numerical prediction of liquid flow within a hydrocyclone. Chem Eng J 74:211–216
44. Datta A, Som SK (2000) Numerical prediction of air core diameter, coefficient of discharge and spray cone angle of a swirl spray pressure nozzle. Int J Heat Fluid Flow 21:412–419
45. Davailles A, Climent E, Bourgeois F (2012) Fundamental understanding of swirling flow pattern in hydrocyclones. Sep Purif Technol 92:152–160
46. Davidson MR (1988) Similarity solutions for the flow in hydrocyclones. Chem Eng Sci 43(7):1499–1505
47. Davidson MR (1988) Numerical calculations of flow in a hydrocyclone operating without an air core. Appl Math Model 12:119–128
48. Davidson MR (1994) An adaptive method of predicting the air core diameter for numerical models of hydrocyclone flow. CSIRO Div Min Process Eng 1–21
49. Daza J, Cornejo P, Rodriguez C, Betancourt F, Concha F (2018) Transition from spray to roping in hydrocyclones: influence of the feed particle size distribution. Miner Eng 123:71–78
50. Delgadillo JA, Rajamani RK (2005) A comparative study of three turbulence-closure models for the hydrocyclone problem. Int J Mineral Process 77:217–230
51. Delgadillo JA, Rajamani RK (2007) Exploration of hydrocyclone designs using computational fluid dynamics. Int J Miner Process 84:252–261
52. Delgadillo JA, Rajamani RK (2009) Computational fluid dynamics prediction of the air-core in hydrocyclones. Int J Comut Fluid Dyn 23:189–197

53. Doby MJ, Kraipech W, Nowakowski AF (2005) Numerical prediction of outlet velocity patterns in solid–liquid separators. Chem Eng J 111(2–3):173–180
54. Doby MJ, Nowakowski AF, Dyakowski T (2008) Understanding air core formation in hydrocyclones by studying pressure distribution as a function of viscosity. Int J Mineral Process Processing 86:18–25
55. Donaldson C, Sullivan RD (1960) Behavior of solutions of the Navier-Stokes equations for a complete class of three-dimensional viscous vortices. In: Proceedings of the Heat transfer and Fluid mechanics Institute, pp 16–30
56. Drew DA, Passman SL (1998) Theory of multicomponent fluids. In: Mardsen JE, Sirovich L (eds) Applied mathematical sciences 135. Springer, New York
57. Dyakowski T, Williams RA (1993) Modeling turbulent flow within a small-diameter hydrocyclone. Chem Eng Sci 6:1143–1152
58. Dyakowski T, Williams RA (1995) Prediction of air-core size and shape in a hydrocyclone. Int J Miner Process 43(1–2):1–14
59. Fisher MJ, Flack RD (2002) Velocity distributions in a hydrocyclone separator. Exp Fluids 32:302–312
60. Fontain FJ, Dijksman C (1953) Recent developments in mineral dressing. Inst Min Met, London 229
61. Gu F, Li W (1987) Measurement and study of velocity field in various cyclones by use of laser Doppler anemometry. In: Proceeding of 3rd International Conference Hydrocyclones, Oxford, BHRA, Paper C2, pp 65–74
62. He P, Salcudean M, Gartshore IS (1999) A numerical simulation of hydrocyclones. Chem Eng Res Des 77(A5):429–441
63. Hsieh KT, Rajamani K (1988a) Phenomenological model of hydrocyclone: model development and verification for single phase flow. Int J Mineral Process 22:223–310
64. Hsieh KT, Rajamani RK (1991) Mathematical model of the hydrocyclone based on physics of fluid flow. AIChE J 37
65. Hsieh KT, Rajamani K (1988b) Phenomenological model of hydrocyclone: numerical solution and verification of particle movements. In: Forssberg E (ed) Proceedings of the XVI IMPC. Amsterdam, pp 377–388
66. Hsieh KT, Rajamani K (1991) AIChE J 37:735
67. Hwang CC, Shen HQ, Zhu G, Khonsari MM (1993) On the main flow pattern in hydrocyclones. J Fluids Eng 155:21–25
68. Ingham M, Wen X (2000) Numerical modelling of the fluid and particle penetration through small sampling cyclones. J Aerosol Sci 31(9):1097–1119
69. Karimi M, Akdogan G, Dellimore KH, Bradshaw SM (2012) Quantification of numerical uncertainty in computational fluid dynamics modelling of hydrocyclones. Comput Chem Eng 43:45–54
70. Kelsall DF (1963) Some applications of hydraulic cyclones in hydrometallurgical processes. Trans AIME 226:225–231
71. Kelsall DF (1952) A study of the motion of solid particles in a hydraulic cyclone. In: Recent development in mineral processing, 1st IMPC, London, pp 209–228
72. Knowles SR, Woods DR, Feurstein IA (1973) The velocity distribution within a hydrocyclone operating without an air core, Can J Chem Engng 51:263–271
73. Knowles SR, Woods DR, Fuerstein IA (1973) The velocity distribution within a hydrocyclone operating without an air core. Can J Chem Eng 51:263
74. Kumari K, Nath G (1985) Unsteady swirling boundary layer flow with heat and mass transfer in nozzles, diffusers and hydrocyclones. Int J Heat Mass Trans 28(7):1377–1383
75. Lim EWC, Chen Y-R, Wang C-HR-M (2010) Experimental and computational studies of multiphase hydrodynamics in a hydrocyclone separator system. Chem Eng Sci 65:6415–6424
76. Mikaylov PM, Romensky AA (1974) On the calculation of the flow dynamics in liquid cyclones. Fluid Mech Soviet Res 3(1):154–159
77. Murthy YR, Bhaskar KU (2012) Parametric CFD studies on hydrocyclone. Powder Technol 230:36–47

78. Narashimha M, Spripriya R, Banerjee CFD modelling of hydrocyclone—prediction of cut size. Int J Miner Process 75:53–68
79. Narasimha M, Brennan M, Holtham MPN (2006) Large eddy simulation of hydrocyclone prediction of air-core diameter and shape. Int J Mineral Process 80:1–14
80. Neesse T, Schneider M, Dueck J, Golyk V, Buntenbach S, Tiefel H (2004) Hydrocyclone operation at the transition point rope/spray discharge. Miner Eng 17(733–737):2004
81. Neesse Th, Schubert H, Graichen K (1991) Practical and theoretical aspects of dense-flow separation. Aufbereitungstechnik 32(9):459–472
82. Nepomnyashchii EA, Pavlobskii VV (1977) Hydrodynamic design of a hydrocyclone. Theor Found Chem Eng (Engl Transl) 11(1):83–88
83. Nepomnyashchii EA, Pavlobskii VV (1980) Calculating the velocity field in a hydrocyclone on the basis of a laminar analog of the averaged turbulent flow. Theor Found Chem Eng (Engl Transl) 13(5):671–674
84. Nepomnyashchii EA, Pavlobskii VV (1986) Hydrodynamic calculation of a pressurized hydrocyclone. Theor Found Chem Eng (Engl Transl) 20(2):151–156
85. Nowakowski AF, Kraipech W, Williams RA, Dyakowski T (2000) The hydrodynamics of a hydrocyclone based on a three-dimensional multi-continuum model. Chem Eng J 80(1–3):275–282
86. Nowakowski AF, Doby MJ (2008) The numerical modelling of the flow in hydrocyclones. Kona Powder Particle J 26:66–80
87. Nowakowski AF, Dyakowski T (2003) Investigation of swirling flow structure in hydrocyclones. Chemical Eng Res Des 81(A):862–873
88. Nowakowski AF, Cullivan JC, Williams RA, Dyakowski T (2005) Application of CFD to modeling the flow in hydrocyclones. Is this a realizable option or still a research challenge? Minerals Eng 17(5):661–669
89. Ohasi H, Maeda S (1958) Motiuon of water in a hydraulic cyclone. Trans Chem Eng (kagaku Kogaku) 22:200–2007
90. Ovalle E, Araya R, Concha F (2005) The role of wave propagation in hydrocyclone operations I: an axisymmetric streamfunction formulation for a conical hydrocyclone. Chem Eng J 111(2–3):205–211
91. Ovalle E, Concha F (2005) The role of wave propagation in hydrocyclone operation II: Wave propagation in the air-water interface of a conical hydrocyclone. Chem Eng J 111:213–223
92. Ovalle E, Concha F (2005) Role of wave propagation in hydrocyclone operations II. Wave propagation in the air-water interface of a conical hydrocyclone. Chem Eng J 111:213–223
93. Pan T-W, Joseph DD, Bai R, Glowinski R, Sarin V (2002) Fluidization of 1204 spheres: simulation and experiment. J Fluid Mech 451:169–191
94. Patankar SV (1980) Numerical heat transfer and fluid flow. Hemisphere Publishing Corporation, Taylor and Francis Group, New York
95. Patankar NA, Joseph DD (2001) Modeling and numerical simulation of particulate flows by the Eulerian-Lagrangian approach. Int J Multiph Flow 27(10):1659–1684
96. Petersen KRP (1993) Vortex chamber modelling for hydrocyclone simulation. M.Sc. Thesis, JKMRC, University of Queensland, Australia
97. Pérez D, Cornejo P, Rodríguez C, Concha F (2018) Transition from spray to roping in hydrocyclones. Miner Eng 123:71–84
98. Quian L, Changlie D, Jirun X, Lixin Y, Guangai X (1989) Comparison of the performance of water-sealed and commercial hydrocyclone. Int J Mineral Process 25:297–310
99. Rajamani K, Hsieh KT (1988) Hydrocyclone model: a fluid mechanical approach. SME Annual meeting, Phoenix Arizona, Preprint 88–163
100. Rajamani RK, Devulapalli B (1994) Hydrodynamic modeling of swirling flow and particle classification in large-scale hydrocyclones, Kona 12:95–104
101. Rietema K, Krajenbrink HJ (1958) Theoretical derivation of tangential velocity profiles in a flat vortex chamber. Influence of turbulence and wall friction. Appl Scio Res, Section A, 8:177–197

102. Rietema K (1961) Performance and design of hydrocyclones: I. General considerations, Chem Eng Sci 15:298–302
103. Rietema K (1962) Liquid-solids separation in a cyclone: the effect of turbulence on separation. In: Symposium on the interaction between fluids and particles, London, Institution of Chemical Engineers, pp 275–281
104. Slack MD, Prasad RO, Bakker A, Boysan F (2000) Advances in cyclone modelling using unstructured grids. Trans IchemE 78:1098–1104
105. Smith JL (1962) An analysis of the vortex flow in the cyclone separator. In: Trans ASME, J Basic Eng Dec 609–618
106. Steffens PR, Whiten WJ, Appleby S, Hitchins J (1993) Prediction of air core diameters for hydrocyclones. Int J Miner Process 39:61–74
107. Svarovsky L (1984) Hydrocyclones. Holt, Rinehart and Winston, London, 198 pp
108. Svarovsky L, Marasinghe BS (1980) Performance of hydrocyclones at high solid concentrations. In: International conference hydrocyclones. BHRA, Cambridge, Paper 10, pp 127–142
109. Swain S, Mohanty S (2013) A 3-dimensional Eulerian-Eulerian CFD simulation of a hydrocyclone. Appl Math Model 37:2921–2932. https://doi.org/10.1016/j.apm.2012.06.007
110. Upadrashta KR, Venkateswarlu D (1983) Tangential flow pattern of power-law fluids in a hydrocyclone. J Powder Bulk Solids Technol 7(4):19–28
111. Upadrasta KR, Ketcham VJ, Miller JD (1987) Tangential velocity profile for pseudo-plastic power-law fluids in the hydrocyclone. A theoretical derivation. Int J Mineral Process 20:309–318
112. Vakamlla TR, Narashima M (2014) Air-core size measurement of operating hydrocyclone by electrical resistance tomography. Chem Eng Technol 37(5)
113. Vakamlla R, Mangadoddy N (2017) Numerical simulation of industrial hydrocyclones performance: Role of turbulence modelling. Separation and Purification Technology 176(4):23–39
114. Van Duijin G, Rietema K (1983) Performance of a large-cone-angle hydrocyclone, I. Hydrodynamics. Chem Eng Sci 38(10):1651–1661
115. Wang B, Chu KW, Yu AB (2007) Numerical study of particle—fluid flow in a hydrocyclone. Ind Eng Chem Res 46:4695–4705
116. Wang B, Yu AB (2006) Numerical study of particle–fluid flow in hydrocyclones with different body dimensions. Miner Eng 19:1022–1033
117. Xu J, Luo Q, Qiu J (1990) Studying the flow field in a hydrocyclone with no forced vortex, Part 1. Average velocity. Filtr Sep July/Aug 276–278
118. Zhang Y, Liu P, Xiao L, Yang X, Yang J (2014) Numerical simulation of multiphase flow inside hydrocyclone based on CFD. Comput Model New Technol 18(11):1374–1379

Chapter 6
Numerical Models of Hydrocyclones

Abstract The previous chapters showed that the problem associated with the mathematical modeling of flow patterns in hydrocyclones involves the solution of strongly coupled, non-linear partial differential equations of mass and momentum balances, which lies well beyond any analytical approach, except when major simplifications are assumed. In this chapter we will study the challenges related to the complex flow behavior arising from laminar and turbulent three-dimensional flows. Numerical simulation of partial differential equations with finite element and finite volume platforms such as Fluent has reached a reasonable level of maturity and is accepted as an analysis tool. Important aspects of the numerical simulation of hydrocyclones are the formulation of the governing equations, selection of the appropriate turbulence model, imposition of boundary conditions and proper representation of the computational domain.

6.1 Introduction

In Chap. 3 we discussed the fact that the exact solution of the Reynolds equations could be solved only for highly idealized conditions using different mathematical techniques such as asymptotic expansions, similarity solutions and boundary layer flows. To solve the complete turbulent Reynolds equations, numerical techniques are necessary.

Laminar flow

In the case of a 5-mm mini-hydrocyclone operating in the laminar-transition region at Reynolds numbers of 100 to 1000, Zhu et al. [30] used a balance for the laminar incompressible flow with direct numerical simulation (DNS) of the steady flow field combined with Lagrangian multiphase flow. The mass and momentum balance for the fluid are:

$$\begin{aligned} \nabla \cdot \rho_f \mathbf{v} &= 0 \\ \nabla \cdot \rho_f \mathbf{v}\mathbf{v} &= -\nabla p + \nabla \cdot \boldsymbol{\tau} + \rho \mathbf{g} \end{aligned} \tag{6.1}$$

F. Concha A. and J. L. Bouso A., *Fluid Mechanics Fundamentals of Hydrocyclones and its Applications in the Mining Industry*, Fluid Mechanics and Its Applications 126,
https://doi.org/10.1007/978-3-030-67913-2_6

where $\boldsymbol{v}$ is the fluid velocity vector, ρ_f the fluid density, p the fluid pressure, $\boldsymbol{\tau} = \mu(\nabla\mathbf{v} + \nabla\mathbf{v}^T)$ the viscous stress tensor, μ the fluid viscosity and $\mathbf{g}$ the gravitational body force. For the particle velocity vector, the authors used:

$$\rho_p\dot{\mathbf{u}}_p = -\nabla p + F_D\rho_p(\mathbf{v} - \mathbf{u}_p) + (\rho_p - \rho_f) + \frac{1}{2}\rho_f(\dot{\mathbf{v}} - \dot{\mathbf{u}}_p) \tag{6.2}$$

where $\mathbf{u}_p$ is the solid particle vector velocity, F_D the drag coefficient, ρ_p the solid particle density and $\dot{\mathbf{v}}$ and $\dot{\mathbf{u}}$ the fluid and solid particle acceleration.

For the drag coefficient F_D, the particle Reynolds number and the drag coefficient the authors used:

$$F_D = \frac{18\mu}{d_p\rho_p}\frac{\mathrm{Re}_p}{24}C_D, \quad \mathrm{Re}_p = \frac{\rho_f d_p|\mathbf{u}_p - \mathbf{v}|}{\mu}, \quad C_D = \frac{24}{\mathrm{Re}_p}\left(1 + b_1\mathrm{Re}_p^{b_2}\right) + \frac{b_3\mathrm{Re}_p}{\mathrm{Re}_p + b_4} \tag{6.3}$$

with $b_1 = 0.1806,\ b_2 = 0.6459,\ b_3 = 0.4251$ and $b_4 = 6880.95$.

They found that the overflow rate became unstable from $\mathrm{Re}_p = 300$ to 600 (Figs. 6.1 and 6.2).

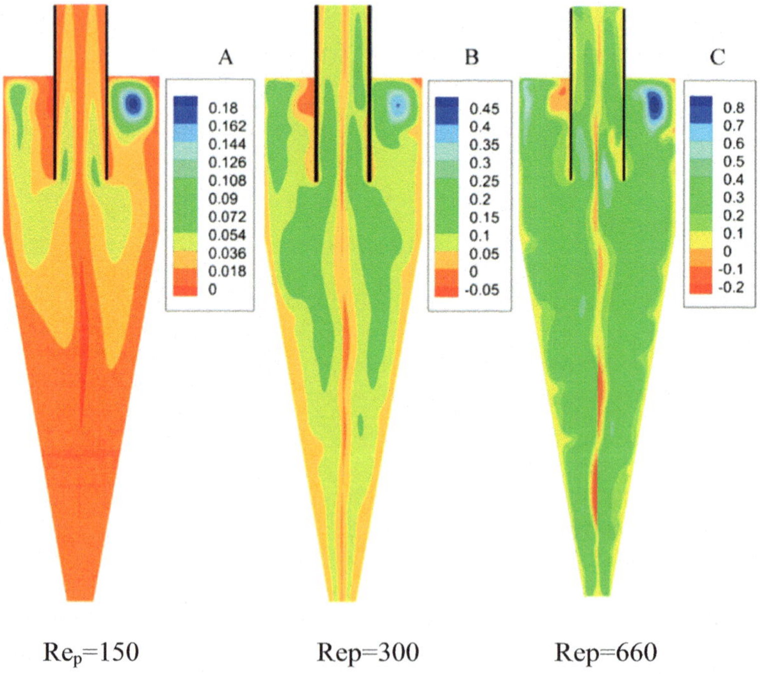

Fig. 6.1 Contours of tangential velocity in the vertical plane for several particle Reynolds numbers

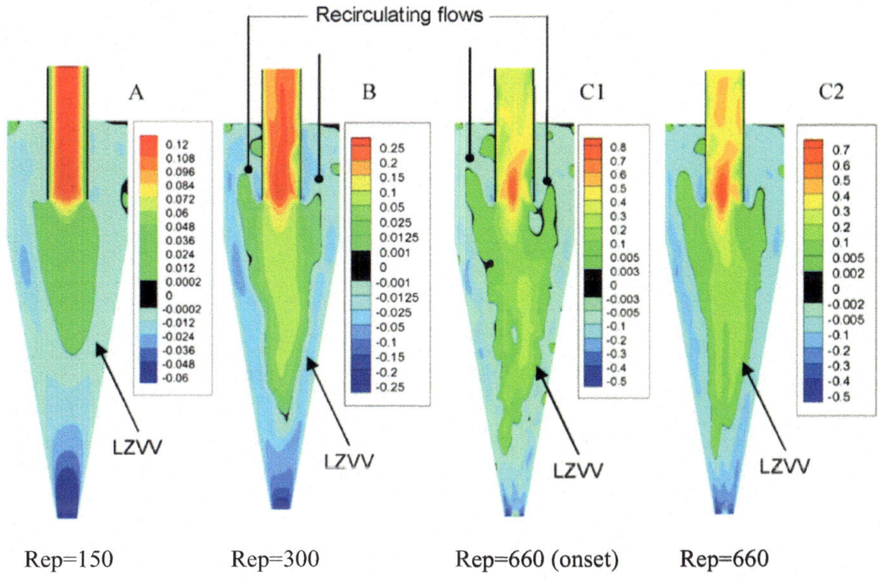

Fig. 6.2 Contours of axial velocity in the vertical plane for several particle Reynolds numbers

Turbulent flow

The turbulent flow of a Newtonian fluid is modeled with the instantaneous and time-averaged velocity, stress and external force fields of the Navier–Stokes equations $\mathbf{v} = \overline{\mathbf{v}} + \mathbf{v}'$, $\mathbf{T} = \overline{\mathbf{T}} + \mathbf{T}'$ and $\mathbf{b} = \overline{\mathbf{b}} + \mathbf{b}'$. Integrating the resulting equations for a short time generates Reynolds equations:

$$\frac{\partial \rho \overline{\mathbf{v}}}{\partial t} + \nabla \bullet \rho \overline{\mathbf{v}\mathbf{v}} = \nabla \bullet \left(\overline{\mathbf{T}} - \rho \overline{\mathbf{v}'\mathbf{v}'} \right) + \overline{\mathbf{b}}$$
$$\nabla \bullet \overline{\mathbf{v}} = 0 \tag{6.4}$$

where $\overline{\mathbf{v}}$**,** $\overline{\mathbf{T}}$ and $\overline{\mathbf{b}}$ are the mean velocity, mean stress tensor and mean body force fields, while $\mathbf{v}'$ is the fluctuating velocity field, ρ is the constant fluid density and $\mathbf{T}_t = -\rho \overline{\mathbf{v}'\mathbf{v}'}$ are the Reynolds stresses. The choice of constitutive equations for the stress tensor, the Reynolds stresses and the numerical mathematical tools used to solve the equations are discussed in this chapter.

Turbulent flow equation

The turbulent flow of a Newtonian fluid is modeled by the time-averaged Navier Stokes equations for the instantaneous velocity, stress and external forces, where the field variables are $\mathbf{v} = \overline{\mathbf{v}} + \mathbf{v}'$, $\mathbf{T} = \overline{\mathbf{T}} + \mathbf{T}'$ and $\mathbf{b} = \overline{\mathbf{b}}$. Integrating these equations for a short time produces the following Reynolds equations:

$$\frac{\partial \rho \overline{\mathbf{v}}}{\partial t} + \rho \nabla \cdot \overline{\mathbf{v}\mathbf{v}} = \nabla \cdot \left(\overline{\mathbf{T}} + \mathbf{T}' \right) + \overline{\mathbf{b}}$$
$$\nabla \cdot \rho \overline{\mathbf{v}} = 0 \tag{6.5}$$

$$\overline{\mathbf{T}} = -\overline{p}\mathbf{I} + \overline{\nu}\left(\nabla \overline{\mathbf{v}} \cdot \nabla \mathbf{v}^T \right)$$
$$\overline{\mathbf{F}} = \overline{\mathbf{T}} \cdot \mathbf{n} \tag{6.6}$$

The boundary conditions are:

$$\overline{\mathbf{v}} = \mathbf{0} \text{ at the solid walls}$$
$$\overline{\mathbf{v}} = \hat{\mathbf{v}} \text{ at the inlet and outlet boundaries} \tag{6.7}$$

6.2 Finite Element Method

The finite element method consists of computing the flow problem in a hydrocyclone modeled by Navier–Stokes equations in the three-dimensional domain Ω, with boundary condition $\partial\Omega$, which represent the solid walls and inlet and outflow boundaries, respectively. The underlying equations, written in primitive variables and stress-divergence form, are:

Nowakowski and Dyakowski [24] used these equations for laminar and turbulent regime with a Eulerian–Lagrangian approach model for fluid–solid simulation. This approach provides a direct description of the particulate flow by tracking the motion of individual particles. Newton's second law, with empirical forms for the hydrodynamic forces, governs particle motion.

$$m_p \frac{d\mathbf{u}_p}{dt} = \mathbf{F}_d + \mathbf{F}_g + \mathbf{F}_C + \mathbf{F}_B + \mathbf{F}_I \tag{6.8}$$

where m_p denotes the mass of the particle with velocity $\mathbf{u}_p$, and $\mathbf{F}_d, \mathbf{F}_g, \mathbf{F}_C, \mathbf{F}_B$ and $\mathbf{F}_I$ are drag, gravitational, centrifugal, buoyant and inertial forces exerted on the particle, respectively.

The particles do not perturb the flow field and the fluid satisfies the continuum equations that are solved in a fixed field in the usual Eulerian way. Figures 6.3 and 6.4 give some of the results for particle trajectories, pressure and velocity fields. Nowakowski et al. [25] presented a 3D finite element code for hydrocyclone modeling. Some important factors in the numerical solution of the model equations, namely proper representation of geometry, imposition of boundary conditions and the choice of turbulence model, are discussed. Numerical results indicate the versatility and effectiveness of the new approaches.

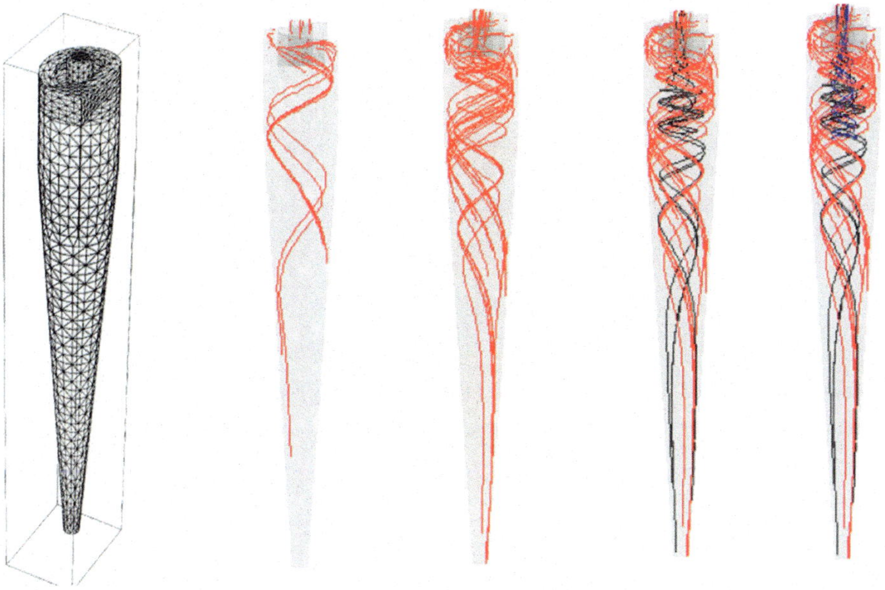

Fig. 6.3 Simulation results obtained from the developed finite element software. Trajectories of fluid particles fed from the hydrocyclone inlet. Nowakowski and Dyakowski [24], Novakowski et al. [25]

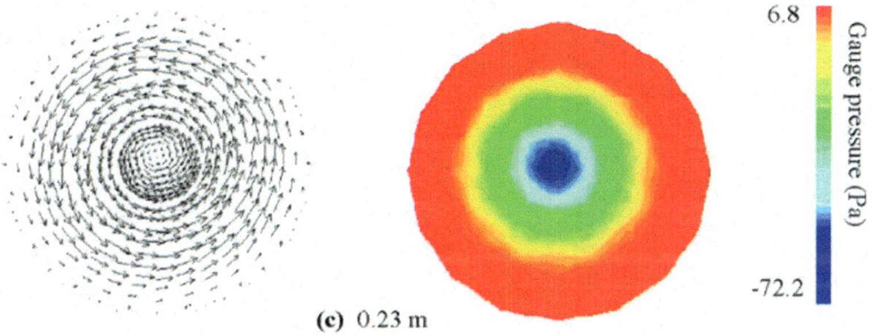

Fig. 6.4 Velocity and pressure field projected on horizontal plane situated 236 cm from the apex. Nowakowski and Dyakowski [24]

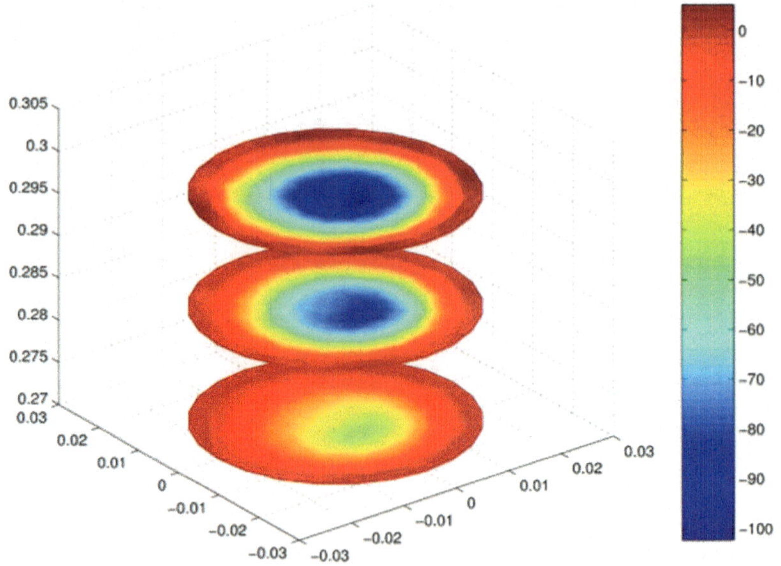

Fig. 6.5 Cross-section of the pressure profile at three different hydrocyclone heights [13]

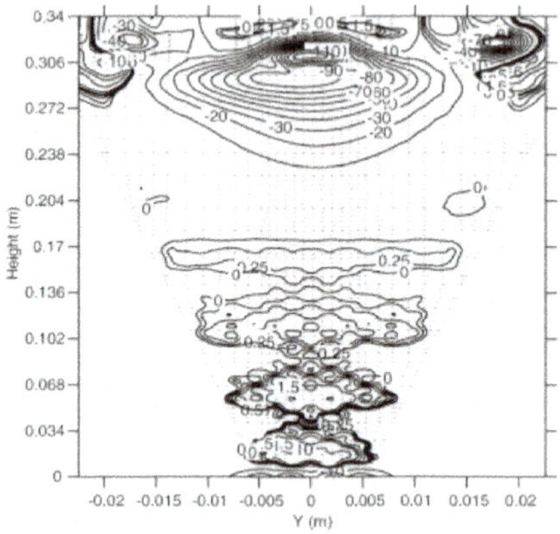

Four-node tetrahedral mesh.

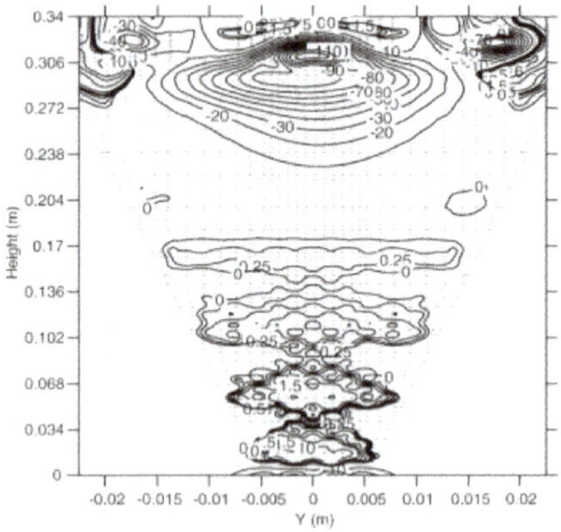

Ten-node tetrahedral mesh.

Fig. 6.6 Pressure contour in a hydrocyclone at Re = 293 [14]

Doby et al. [13, 14] investigated the interaction between the swirling flow and velocity profile at the outlet. The studies used fluids with different properties. The results show that, with proper imposition of boundary conditions, the flow develops naturally. See Figs. 6.5 and 6.6.

6.3 Turbulence Modeling

To solve Reynolds equations, a constitutive equation must be postulated for the Reynolds stresses $\mathbf{T}' \equiv -\rho\overline{\mathbf{v}'\mathbf{v}'}$. The use of an appropriate turbulence model in numerical simulation of a hydrocyclone is particularly important due to the existence of highly rotational fluid flow of an anisotropic nature.

For the hydrocyclone there is the possibility of using different turbulence models such as the Eddy diffusivity model, the $k-\varepsilon$ and $k-\varepsilon$RNG models, the algebraic stress model, Reynolds stress model (RSM), the differential stress model (DSM) and the large-eddy simulation model (Concha et al. 1989).

6.3.1 *Eddy Diffusivity Model (EDM)*

These models are based on the concept of an "eddy," which represents a portion of fluid interchanging momentum and energy with neighboring eddies. Large and small eddies coexist in the same fluid, but the energy of the main flow is extracted by the largest eddies, transferred to smaller neighboring eddies and finally converted into thermal energy. The different numerical models depend on the description of the turbulent viscosity.

The simplest turbulence model is the eddy diffusivity model, which relates the Reynolds stresses to the main shear rate through a turbulence coefficient which, in turn, is a function of the main flow field. Lilley and Chigier [20] indicate that the turbulent viscosity in a hydrocyclone is anisotropic and should be described by a second-order tensor. The model has a characteristic length scale of turbulence as a parameter. For this model, the stress tensor field is given by:

$$
\begin{aligned}
\mathbf{T}-\rho\overline{v'v'} &= -\left(\overline{p}+\frac{2}{3}\rho k\right)\mathbf{I}+2(\mu+\mu_t)\overline{\mathbf{D}} \\
\overline{\mathbf{D}} &= \frac{1}{2}\left(\nabla\mathbf{v}+\nabla\mathbf{v}^T\right) \\
k &= \frac{1}{2}\mathbf{v}'\cdot\mathbf{v}' \\
\mu_t &= \rho\ell_m^2\sqrt{2\mathrm{tr}\mathbf{D}^2}
\end{aligned}
\tag{6.9}
$$

where k, $\mathbf{I}$ and $\overline{\mathbf{D}}$ are the turbulent kinetic energy, the second-order unit tensor and the mean rate-of-deformation tensor, respectively, and μ_t and ℓ_m are the Eddy viscosity and Prandtl mixing length.

Pericleous [27] applied the EDM turbulence model to simulate the flow in a hydrocyclone, including the effect of particles on the turbulence, with the expression (Fig. 6.7):

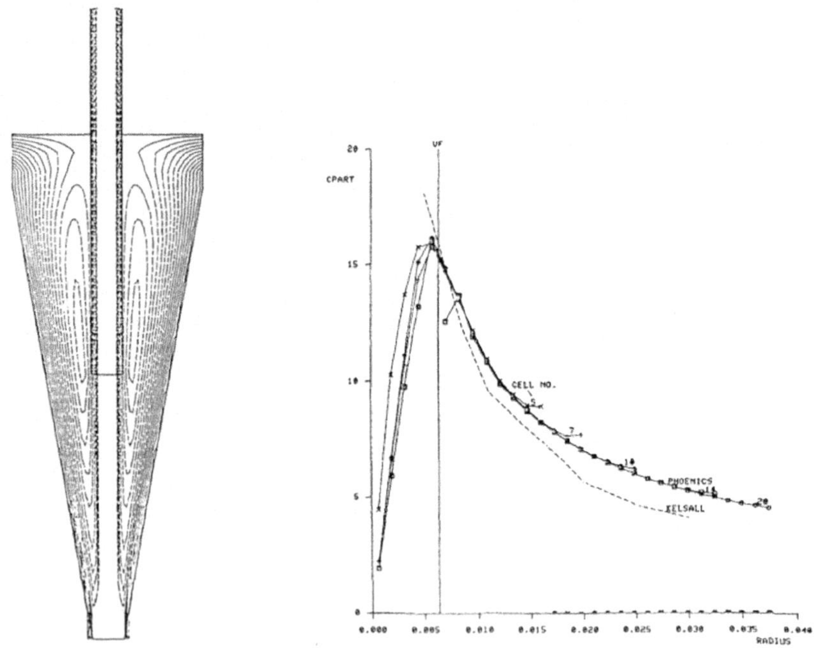

Two-dimensional streamline pattern. Swirl profiles at various levels.

Fig. 6.7 Flow pattern from Pericleous [27]

$$\left(\frac{1 + 0.5 \sum_{i=1}^{n} c_i}{1 - \sum_{i=1}^{n} c_i^4}\right) \tag{6.10}$$

6.3.2 A $k - \varepsilon$ Stress Model

The $k-\varepsilon$ model is a more general model that incorporates the production, convection, diffusion and dissipation of turbulent kinetic energy, where k stands for the kinetic energy, ε is the rate of dissipation and μ_t is the turbulent viscosity of the flow. Therefore, balance of turbulent kinetic energy and dissipation energy is:

$$\underbrace{\begin{bmatrix}\text{Rate of change}\\ \text{of } \varepsilon \text{ or } k \text{ at time } t\end{bmatrix}}_{\Downarrow} = \underbrace{\begin{bmatrix}\text{Flux of } \varepsilon \text{ or } k\\ \text{by convection}\end{bmatrix}}_{\Downarrow} + \underbrace{\begin{bmatrix}\text{Flux of } \varepsilon \text{ or } k\\ \text{by diffusion}\end{bmatrix}}_{\Downarrow} + \underbrace{\begin{bmatrix}\text{Source or sink}\\ \text{of } \varepsilon \text{ or } k\end{bmatrix}}_{\Downarrow}$$

$$\begin{aligned}\frac{\partial}{\partial t}(\rho k) &= -\nabla\cdot(\rho k\overline{\mathbf{v}}) + \nabla\left(\left(\mu+\frac{\mu_t}{\sigma_k}\right)\nabla k\right) + P - \rho\varepsilon\\ \frac{\partial}{\partial t}(\rho \varepsilon) &= -\nabla\cdot(\rho \varepsilon\overline{\mathbf{v}}) + \nabla\left(\left(\mu+\frac{\mu_t}{\sigma_k}\right)\nabla \varepsilon\right) + C_1\frac{\varepsilon}{k}\left(P - C_2\rho\frac{\varepsilon^2}{k}\right)\end{aligned} \tag{6.11}$$

where $\mu_t = C_\mu \rho \frac{k^2}{\varepsilon}$, $P = (\mu+\mu_t)\overline{\nabla\mathbf{v}} : \left(\overline{\nabla\mathbf{v}} + \overline{\nabla\mathbf{v}}^T\right)$, C_μ, C_1, C_2, σ_k and σ_e are constants.

Davidson [10] presented results using this method (Fig. 6.8).

Narashimha et al. [23] used FLUENT with the discrete phase modeling technique, using $\kappa - \varepsilon$ model for turbulence, to study the velocity vectors of water-only hydrocyclone distribution in hydrocyclones with two apex sizes. The geometry of the hydrocyclones is given below (Figs. 6.9 and 6.10).

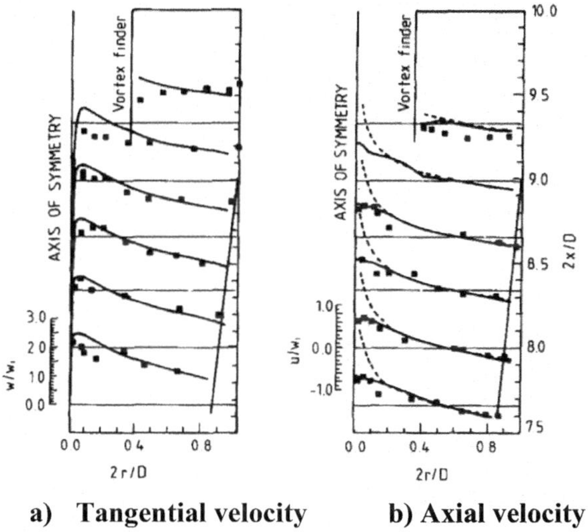

Fig. 6.8 Calculated velocity profiles compared with **a** corresponding experimental data of Knowles et al. (1973) and **b** theoretical work of [3, 10]

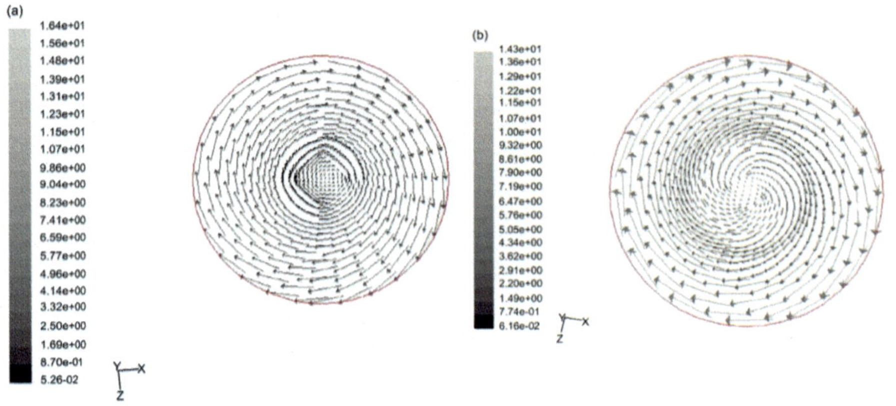

Fig. 6.9 Comparison of velocity vectors in the underflow of the hydrocyclone for an inlet velocity of 9.5 m/s for an apex of **a** 10 and **b** 20 mm

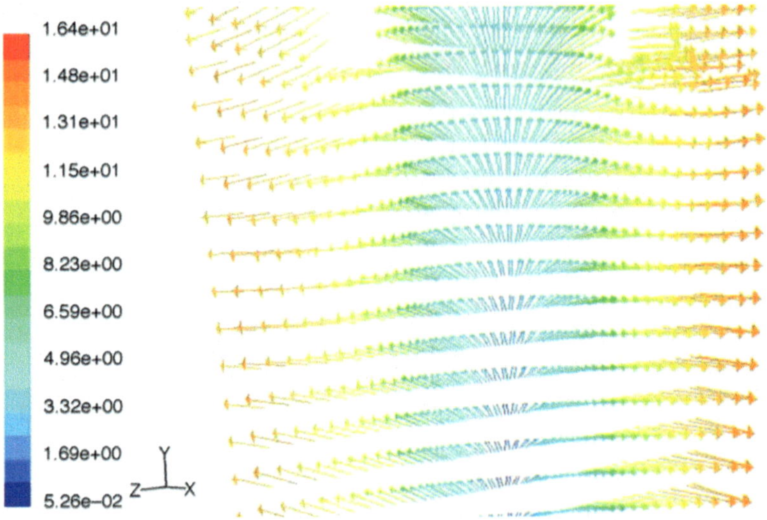

Fig. 6.10 Velocity vectors in the center axial–radial plane

Dimensions	Cyclone 1 (case 1)	Cyclone 2 (case 2)
Cyclone diameter (mm)	101	101
Inlet entry (mm)	25 × 12.5	25 × 12.5
Cylindrical length (mm)	85	85
Vortex finder diameter (mm)	35	35
Vortex finder length (mm)	50	50
Spigot diameter (mm)	10	20
Cone angle (°)	20	20

Cyclone 1 is considered as the case. Cyclone 2 is studied for variation on the Spigot diameter

Fig. 6.11 Contour (left) and vector (right) diagrams of tangential velocity seen from above with a feed pressure of 0.4 MPa

Gao et al. [16] used FLUENT with the $\kappa - \varepsilon$ turbulent model to describe the tangential velocity in a 75-mm hydrocyclone with a feed pressure of 0.4 MPa (Fig. 6.11).

6.3.3 *k*−*ε* RNG Stress Model

A modification to the $k - \varepsilon$ stress model is the introduction of the re-normalization group. This model separates the representation of large and small eddies. The region of small eddies is modeled by the Kolkmogorov law of −5/3, introduced as a perturbation to large-eddy transport. The balance of turbulent kinetic energy is the same as in (0.11), but the balance of dissipation energy becomes:

$$\frac{\partial}{\partial t}(\rho\varepsilon) = -\nabla \cdot (\rho\varepsilon\overline{\mathbf{v}}) + \nabla\left(\left(\mu + \frac{\mu_t}{\sigma_k}\right)\nabla\varepsilon\right) + (C_1 - C_{1RNG})\frac{\varepsilon}{k}P - C_2\rho\frac{\varepsilon^2}{k} \tag{6.12}$$

where

$$C_{1\mathrm{RNG}} = \frac{\eta(1 - \eta/\eta_0)}{(1 - \beta\eta^3)}\frac{k}{\varepsilon} \quad \text{and} \quad \eta = \left(\frac{P}{\mu}\right)^{0.5}\frac{k}{\varepsilon}, \quad \text{where } \eta_0 \text{ and } \beta \text{ are constants} \tag{6.13}$$

Delgadillo and Rajamani [11] proposed the following value for the $C_{1\mathrm{RNG}}$:

$$C_{1\mathrm{RNG}} = \frac{0.0845\rho\eta^2(1 - \eta/\eta_0)}{1 + \beta\eta^3}\frac{k}{\varepsilon} \quad (6.14)$$

Bashaskar [2] presented contours and axial distributions of water in a hydrocyclone using this method (Fig. 6.12).

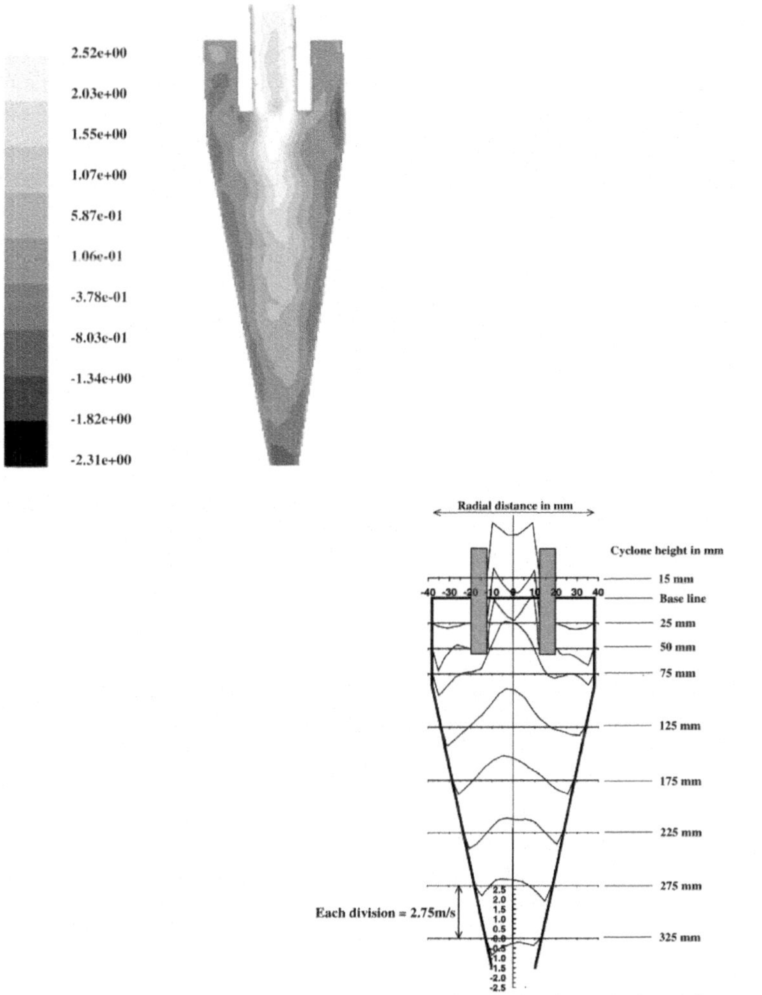

Velocity Contours. Velocity distribution [m/s] in axial plane.

Fig. 6.12 Axial water velocity in a hydrocyclone

6.3.4 Reynolds Stress Model

Next in complexity is the second-order Reynolds stress model (RSM), which consists of transport equations for each of the Reynolds stresses. There are two sub-models: the differential stress model (DSM) and the algebraic stress model (ASM).

(a) **DSM turbulence model**

Through the application of a second-order differential-stress turbulence model (DSM) with fully three-dimensional computational fluid dynamics (CFD) modelling of the hydrocyclone, four key understandings of hydrocyclone operation—symmetry, the air-core development mechanism, air-core precession and the particle separation mechanism—have been challenged. DSM implicitly accounts for local turbulence asymmetry.

The transport equations for the DSM model are:

$$\begin{aligned}
\frac{\partial}{\partial t}\left(\rho\overline{\mathbf{v}'\mathbf{v}'}\right) &= -\nabla\cdot\left(\rho\left(\overline{\mathbf{v}'\mathbf{v}'}\right)\overline{\mathbf{v}}\right) + \nabla\cdot\left(\rho\frac{C_s}{\sigma_{DS}}\frac{k}{\varepsilon}\left(\overline{\mathbf{v}'\mathbf{v}'} + \overline{\mathbf{v}'\mathbf{v}'^{T}}\right)\right) + \mathbf{P} + \mathbf{\Phi} - \frac{2}{3}\rho\varepsilon\mathbf{I} \\
\frac{\partial}{\partial t}(\rho\varepsilon) &= -\nabla\cdot(\rho\varepsilon\overline{\mathbf{v}}) + \nabla\cdot\left(\rho\frac{C_s}{\sigma_\varepsilon}\frac{k}{\varepsilon}\overline{\mathbf{v}'\mathbf{v}'}\cdot\nabla\varepsilon\right) + \mathbf{P} + \mathbf{\Phi} - \frac{2}{3}\rho\varepsilon\mathbf{I}
\end{aligned} \tag{6.15}$$

where

$$\begin{aligned}
\mathbf{P} &= -\rho\left(\nabla\overline{\mathbf{v}}\mathbf{g}\overline{\mathbf{v}'\mathbf{v}''} + \left(\nabla\overline{\mathbf{v}}\mathbf{g}\overline{\mathbf{v}'\mathbf{v}''}\right)^{T} - \boldsymbol{\omega}\times\overline{\mathbf{v}'\mathbf{v}'}\right) \\
\mathbf{\Phi} &= -\rho C_{1S}\frac{\varepsilon}{k}\left(\overline{v'v'} - \frac{2}{3}k\mathbf{I}\right) - C_{2S}\left(\mathbf{P} - \frac{1}{3}(\mathrm{tr}\mathbf{P})\mathbf{I}\right)
\end{aligned} \tag{6.16}$$

Cullivan et al. [8] presented results with this method (Fig. 6.13).

Chih et al. [5], using CFD with RSM and the discrete phase model (DPM), used simulation and experiments to show the presence of a fish hook at several inlet pressures. In the experiments the fish hook starts for 10 μm while the simulation gives a value of 3 μm with a much exaggerated minimum and maximum (Fig. 6.14).

Aguilera and Arismendy [1] used ANSYS FLUENT 12.1 software with turbulence described by RSM to study the size of the air core in a 400-mm hydrocyclone operating at 34–35 psi on a kaolin suspension with concentrations from 3.06 to 12.04 weight %. From the pressure contour, shown in Fig. 6.15, the size of the air core was calculated at 8.6 mm. The vortex and apex diameters were 16.1 and 14.0 mm, respectively, and the surface tension was 73.5 dynes/cm. The ring of pulp flowing from the underflow was thus 2.7-mm wide and increased slightly with the feed concentration.

Cui et al. [7] carried out numerical studies on the flow field with an air core inside a 50-mm hydrocyclone using ANSYS/Fluent with RSM calculation and volume of fluid (VOF) to capture the air–liquid interface. The work was validated with PIV.

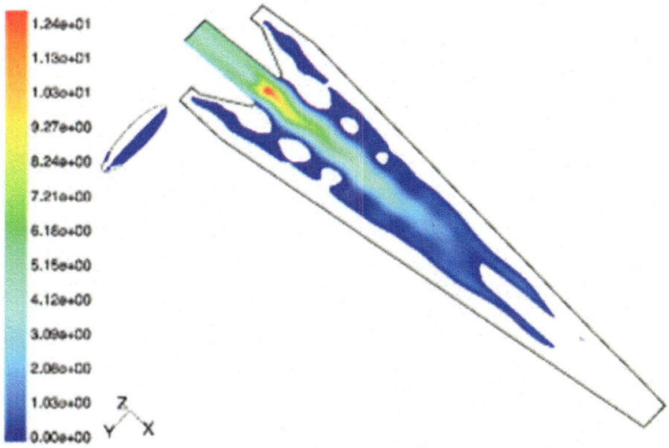

Fig. 6.13 Positive axial-velocity contours during hydrocyclone start-up

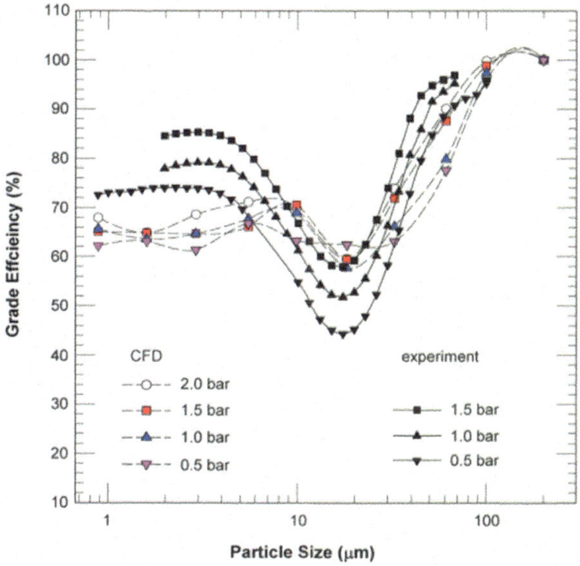

Fig. 6.14 Selectivity functions for inlet pressure values for particles smaller than 3 μm in a 30-mm hydrocyclone and very high short circuit

In Fig. 6.17a, the tangential velocities increase slightly with the increase in feed rate and figure (b) shows that the tangential velocities also increase as the cone angle increases from 11° to 23°. Figures (c) and (d) show that the apex and vortex diameter have little effect on the tangential velocities at the selected location. In all these examples the peak values of the tangential velocity are unchanged (Fig. 6.17).

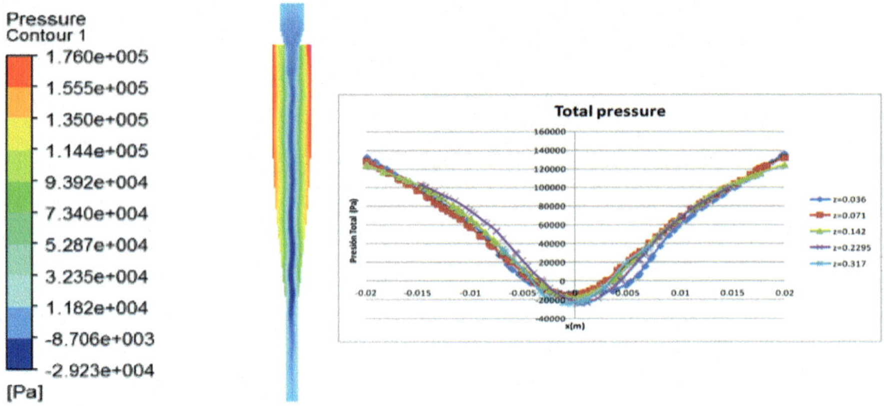

Fig. 6.15 Pressure versus distance from the hydrocyclone axis given by the model, where z is the height in m from the vortex

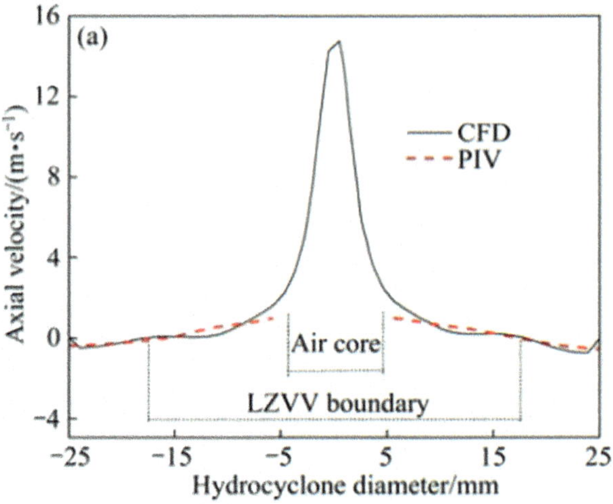

Fig. 6.16 Comparison of the simulated and experimental axial velocity distributions showing the line of zero vertical velocity (LZVV), with an apex diameter of 8 mm and a vortex diameter of 35 mm

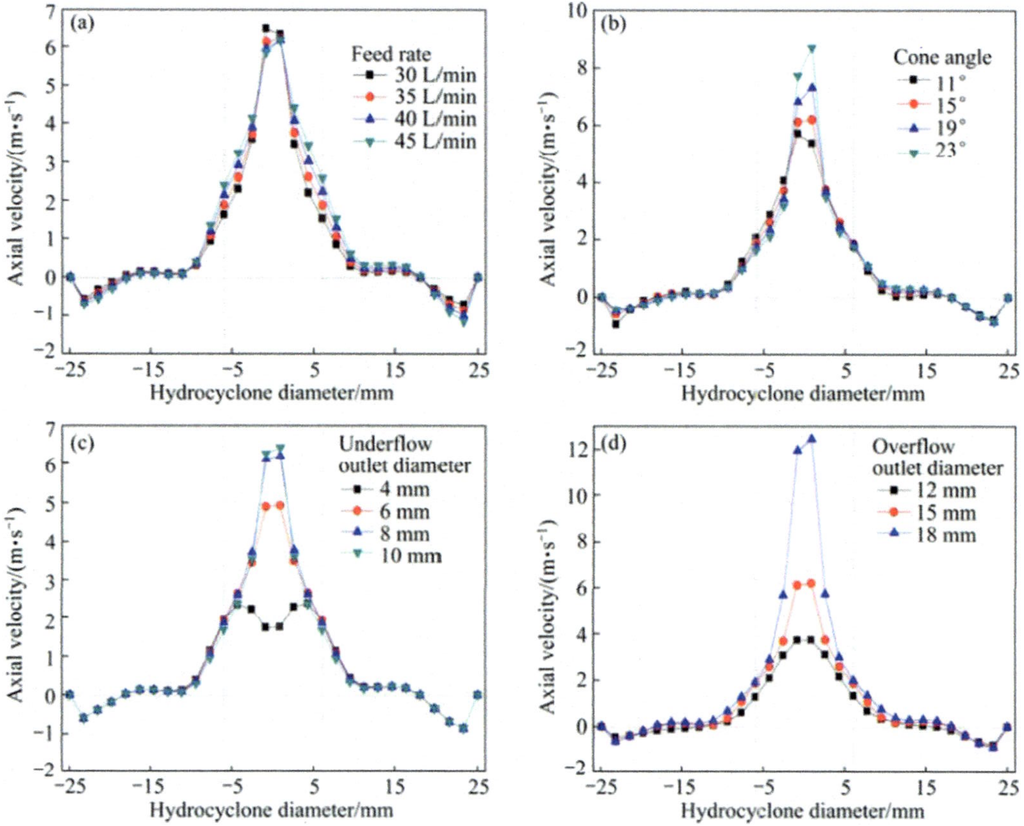

Fig. 6.17 Distributions of axial velocities under different parameters obtained from the numerical method: **a** feed rate, **b** cone angle, **c** apex diameter and **d** vortex diameter

(b) **Algebraic Stress Model**

The following equations represent the ASM:

$$\overline{\mathbf{v}'\mathbf{v}'} = \frac{2}{3}k\mathbf{I} + \frac{(1 - C_{2S})k\left[P - \frac{1}{3}(\mathrm{tr}\mathbf{P})\mathbf{I}\right]}{\left(\rho(C_{1s} - 1)\varepsilon + \frac{1}{2}\mathrm{tr}\mathbf{P}\right)} \tag{6.17}$$

where

$$\frac{\partial}{\partial t}(\rho k) = -\nabla \cdot (\rho k \overline{\mathbf{v}}) + \nabla\left(\rho \frac{C_S}{\sigma_k}\frac{k}{\varepsilon}\overline{\mathbf{v}'\mathbf{v}'} \cdot \nabla k\right) + \frac{1}{2}(\mathrm{tr}\mathbf{P}) - \rho\varepsilon \tag{6.18}$$

An example of this method for hydrocyclones was presented in a paper by [26] on large-eddy simulation (LES) (Figs. 6.18, 6.19, 6.20, 6.21 and 6.22).

(c) **Large-Eddy simulation (LES)**

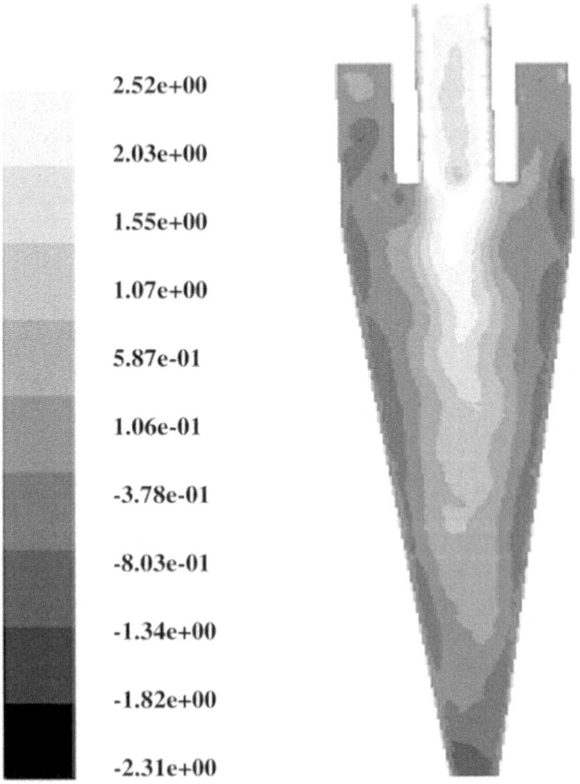

Fig. 6.18 Contours of axial velocity of water with 2.50 [m/s] in a D10 mm hydrocyclone (Udaya et al. 2007)

Fig. 6.19 Contours of tangential velocity. Olson and Van Ommen [26]

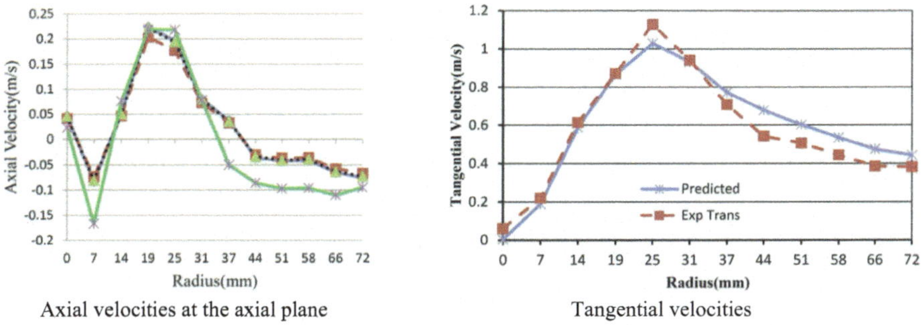

Fig. 6.20 Comparison of experimental and simulated velocities at the axial plane Z = 200 mm [12]

In the LES simulation approach, Reynolds number restrictions are bypassed by directly simulating large scales only and supplying the effect of missing small scales using a sub-grid model. LES retains the full three-dimensionality and time dependence of the fluctuating turbulent field.

As LES is intrinsically a dynamic transient simulation, a three-dimensional body-fitted structured grid was generated in Gambit. It is well known that the accuracy of any numerical simulation depends greatly on the selection of optimum mesh densities; however, sometimes computational time requirements play a dominant role in deciding this optimum mesh density.

Davailles et al. [9] presented an experimental and numerical work using numerical LES to simulate a 100-mm hydrocyclone, studying the effect of apex sizes and feed concentration. They found that dense regions can form inside the hydrocyclone, mainly toward the bottom of the conical section, which dissipate energy through viscous effects. The swirling flow cannot develop throughout the entire hydrocyclone body. At the highest solid concentration tested in this study, the conical section of

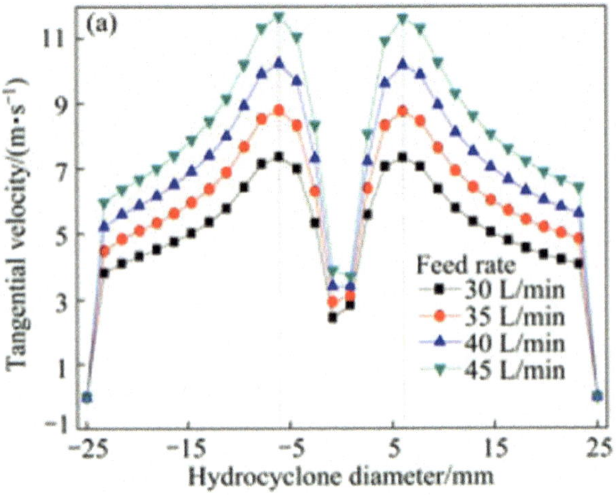

Fig. 6.21 Distributions of tangential velocities under different parameters obtained from the numerical method [7]

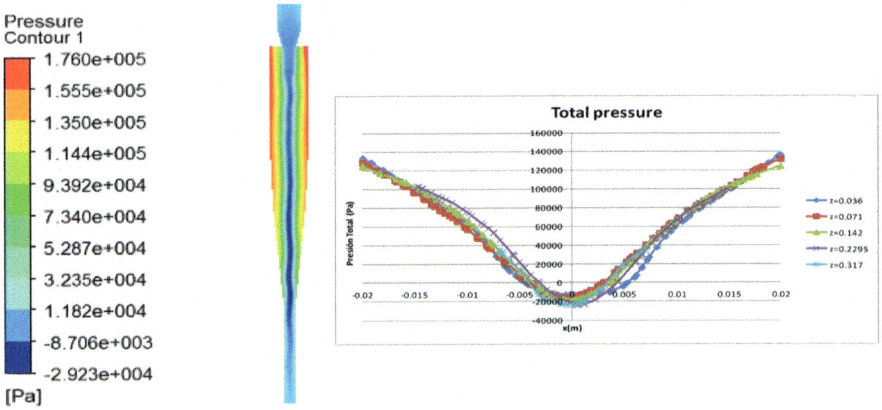

Fig. 6.22 Pressure contour inside a hydrocyclone [1]

the hydrocyclone was almost entirely filled with a packed bed of particles, yielding a rope discharge, with nearly all the process water reporting to the overflow (Fig. 6.23).

Evans et al. [15] presented experimental results and CFD using LES in a 45 mm hydrocyclone with and without an inserted a solid rod. See Fig. 6.24 (Fig. 6.25).

Lim et al. [21], using LES CFD computation, calculated pressure and velocity contours in a 45 mm hydrocyclone, see Fig. 6.26.

Karimi et al. [18] also used experiments and LES to model the air core. Figure 6.27 shows the development of the air core.

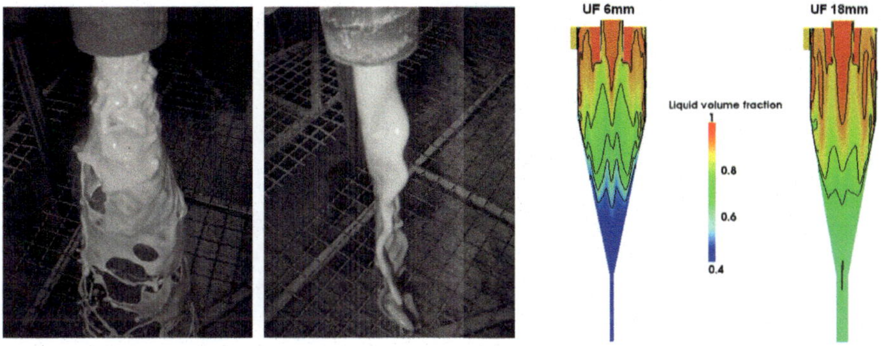

a) 10% solids feed b) 30% solids feed c) Water volume fraction

Fig. 6.23 Influence of feed concentration and spigot diameter on the over- and underflow

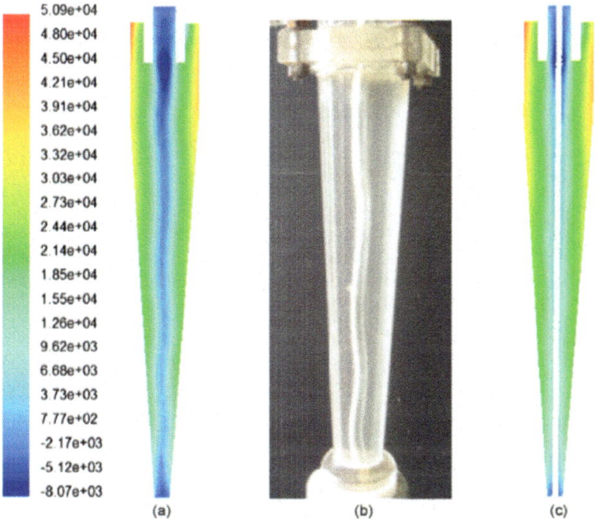

a) Operating with air core b) Experimental c) With 4-mm inserted rod

Fig. 6.24 Pressure contour on a 45 mm hydrocylone [15]

6.3.5 TFM Simulation

Ghodrat et al. [17] presented a numerical method based on TFM, a model in which both fluid (water and air) and solid phases are treated as interpenetrating continua. Particles of each size or density are represented by one phase. They studied hydrocyclones with different conical configurations with high a feed solids concentration: 30% solids by volume. The numerical results show that the cyclone performance is

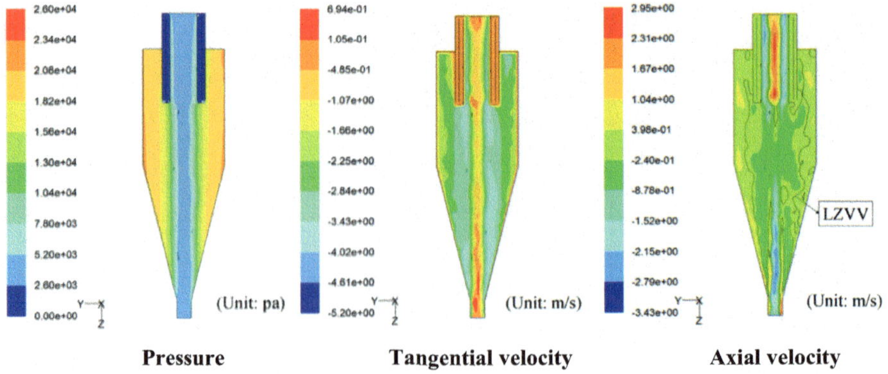

Fig. 6.25 Simulated contours [6]

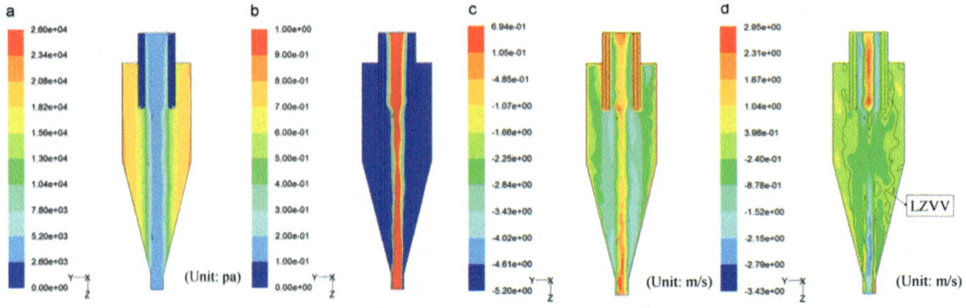

Fig. 6.26 Contour plots of **a** pressure, **b** volume fraction of air, **c** tangential velocity and **d** vertical velocity obtained from CFD simulations. The dark line in (**d**) represents the locus of zero vertical velocity [21]

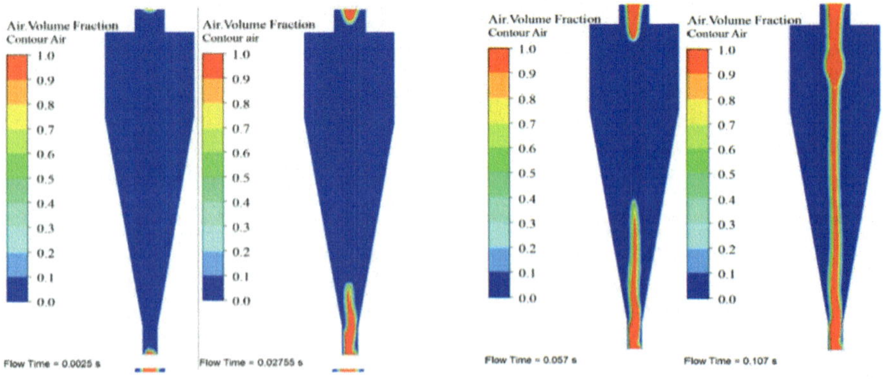

Fig. 6.27 Air core development from 0 to 0.107 s [18]

sensitive to both the length and shape of the conical section, as well as the feed solids concentration (Figs. 6.28 and 6.29).

(a) **Comparison of different turbulent models**

A hydrocyclone generally operates under highly turbulent flow conditions. In such a turbulent flow, eddies of multiple-length scales ranging from as large characteristic length scale of the system to as small as Kolmogorov-length scale, exist. Researchers

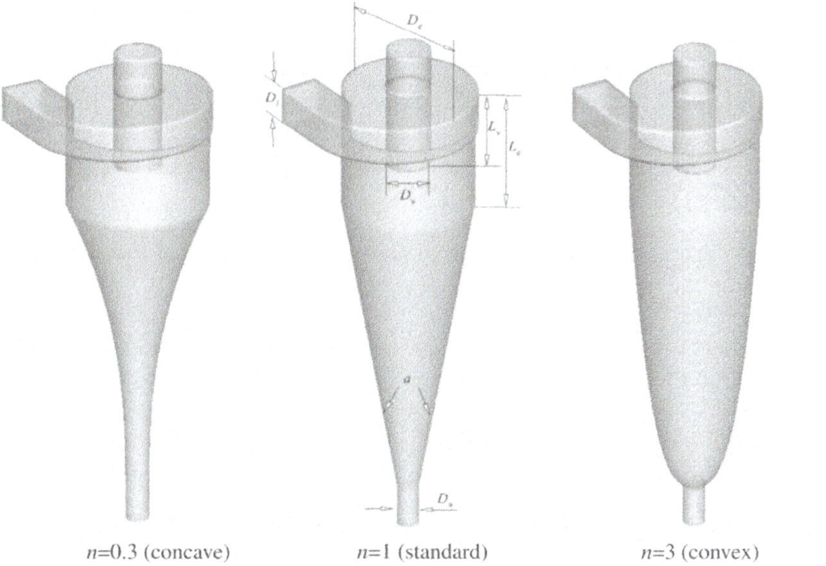

Fig. 6.28 Hydrocyclones with different cone shapes

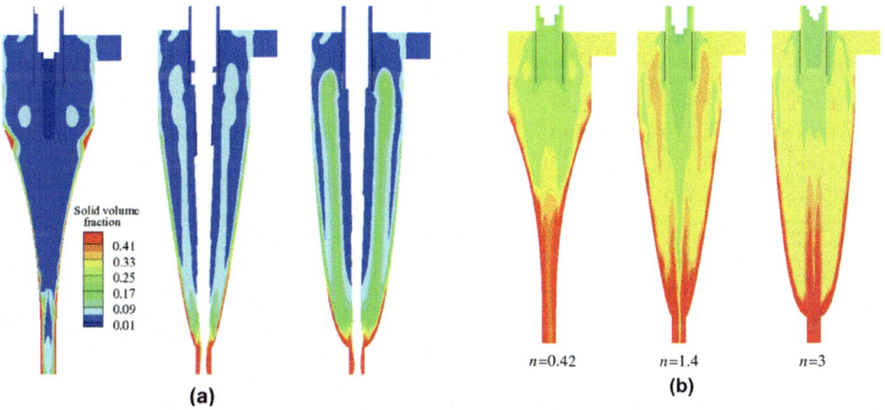

Fig. 6.29 Distributions of solid volume fraction in the hydrocyclones with different conical shapes at feed solids concentrations of **a** 30% solids and **b** 4.1% solids

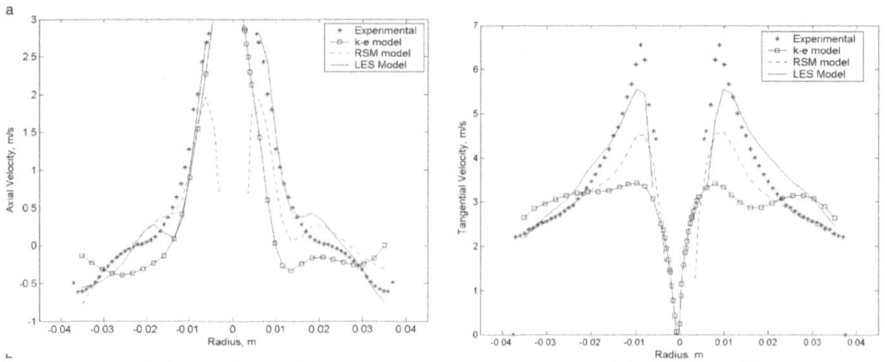

Axial velocity profile. Tangential velocity profile.

Fig. 6.30 Comparison of the three turbulence models—$k - \varepsilon$, RMS and LSM—at the same hydrocyclone level [11]

are making considerable efforts to understand the impacts of the various turbulent models on the performance of hydrocyclones.

As a result of recent progress, the LES technique has been used for hydrocyclone simulation [11]. It seems that the more computationally expensive LES provides the best solution for capturing time-dependent vortex oscillations and non-equilibrium turbulence (Fig. 6.30).

As result of recent works, the LES model is gaining popularity in comparison to the other models presented. However, Slack et al. [28] demonstrated that the DSM model performs remarkably well in comparison to LES without such prohibitively high computational costs.

Analysis of different turbulence models was carried out by Stephens and Mohanrangam [29] for the turbulent flow inside a 75 mm diameter hydrocyclone. Four different turbulence models were tested, three of the Reynolds stress class: ANSY-CFX release 11.0, shear stress transport model (SST), shear stress transport model (SST-CC) with a curvature correction model and the Reynolds stress model with a penalty of solving an additional five equations for varying stresses. Experimental results of [22] were used to compare the four numerical findings for the different turbulence models tested, with SST-CC giving the best prediction.

Ghodrat et al. [17] used TFM to present their work and a comparison of other computational methods with experimental data of Hsieh (1988) (Fig. 6.31).

Karimi et al. [18] presented a comparison of LES and RSM for the tangential velocity of a hydrocyclone to experiments and simulation of Hsieh (1988) (Fig. 6.32).

Future work should investigate an appropriate way to model turbulence flow within a hydrocyclone. Large-Eddy simulation (LES) is suggested, as this approach solves

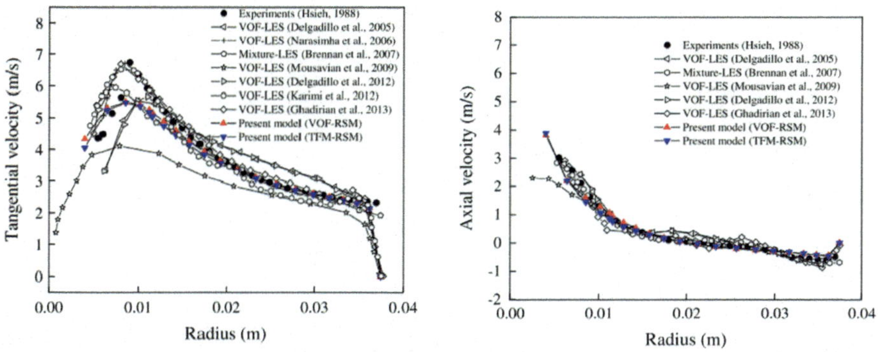

Fig. 6.31 Comparison of the measured (Hsieh and Rajamani 1988) and predicted tangential and axial velocities in a 75 mm hydrocyclone operated without solids feed. The horizontal plane for the measurement is located 60 mm from the top wall

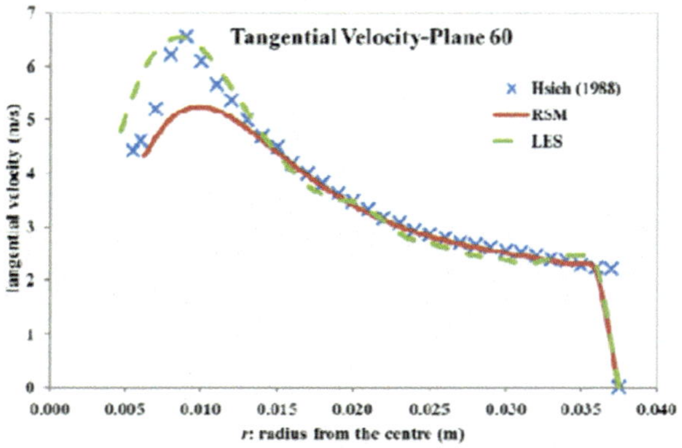

Fig. 6.32 Air core development from 0 to 0.107 s [18]

for the largest-scale motions of the flow while approximating or modeling only small-scale motions. It can be considered a kind of compromise between the Reynolds-averaged Navier–Stokes equations (RANS) approach and the direct numerical simulation (DNS) approach. The LES approach, although still very costly computationally, is becoming a more viable option for the solution of transitional and turbulent flows.

6.4 Additional Experimental Measurements

Brennan et al. [4] measured water velocities inside a 150 mm hydrocyclone using an Aeroprobe system. Figure 6.33 shows the component of the system (Fig. 6.34).

LES Fluent predictions were typically 14% higher than the measured with Aeroprobe.

Measurement system Krebs DF6 150 mm hydrocyclone.

Fig. 6.33 Five-hole pressure Aeroprobe system to measure velocity in a hydrocyclone

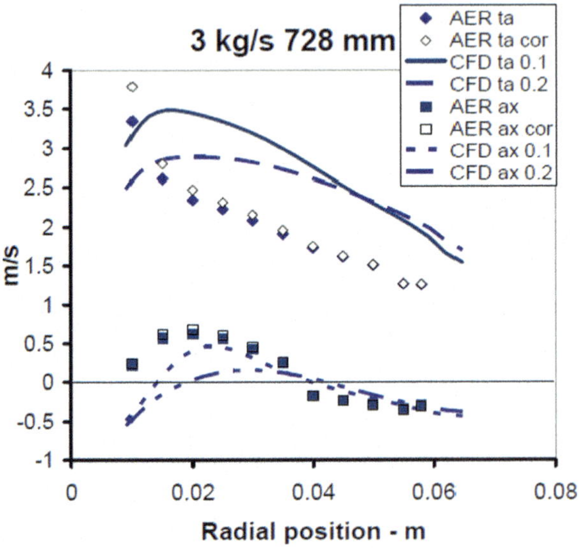

Fig. 6.34 Comparison of experimental tangential velocity measured with Aeroprobe and prediction with CFD

Vakamalla et al. (2017) setup consisting of an electrical resistance tomography and a high speed camera to measure the air core size at different operating conditions of a 75 mm diameter hydrocyclone. See Fig. 6.35. A comparison with CFD measurement is shown in Fig. 6.36.

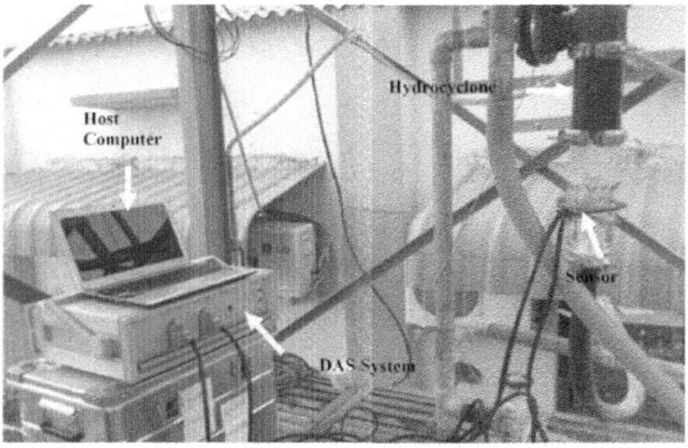

Fig. 6.35 Setup of an Electrical Conductivity Tomograhpy AND A High speed camara system to measure the air-core size (Vakamlla 2014)

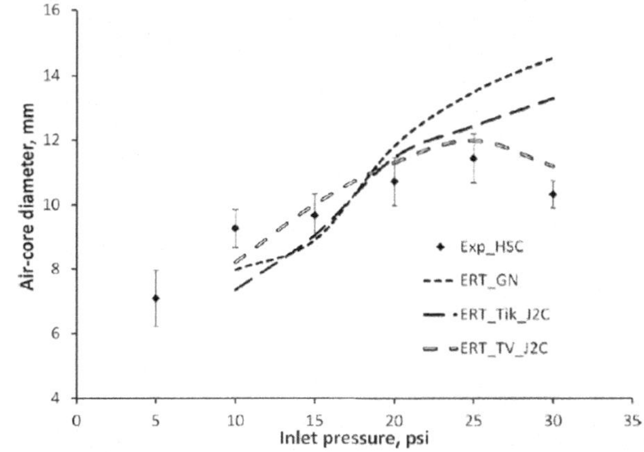

Fig. 6.36 Comparison of air-core diameter by CFD against a hifh speed camera and electrical Resistance tomography of a 76 mm hydrocyclone operating at different pressure

Ke [19] used an experimental setup consisting of a Particle Image Velocimetry (PIV) system, a CCD camera and ND-Yag and Optical Devices, to study the fluid flow in a hydrocyclone. The following figures show the systems used (Figs. 6.37 and 6.38).

In this section we have seen newly developed methods of measuring the velocity profiles in hydrocyclones that may be added to the classic ones seen earlier in this chapter and in Chap. 5. Most of these methods are used by the same researchers to compare them with their CFD simulations and they haven't been used for simulations by other researchers. In this field, the classical experimental data from Kellsal (1952, 1963), Knowls (1973), Pericleus (1986), Dadvison (1988) and Rajamani and co-workers Rajamani et al. (1988, 1991) continue to be the most reliable and used by most investigators.

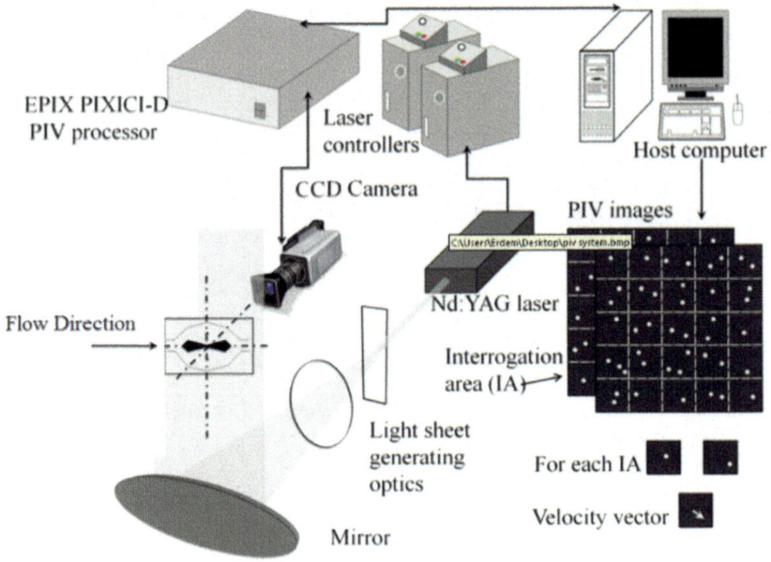

Fig. 6.37 PIV system

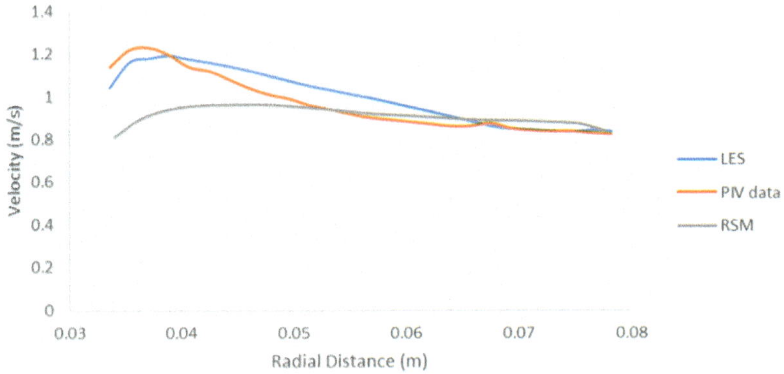

Fig. 6.38 Comparison of the CFD and PIV measurements

References

1. Aguilera GF, Arismendy JE (2014) Development of a hydrodynamic model of a hydrocyclone including the simulation of air-core effect, using CFD. Holos 30(3):184–192
2. Bhaskar KU, Murthy YR, Raju MR, Tivari S, Srivastava JK, Ramakrishnan N (2007) CFD simulation and experimental validation studies on hydrocyclone. Miner Eng 20:60–71
3. Bloor MIG, Ingham DB (1984) A theoretical investigation of the fluid mechanics of the hydrocyclone. Filtr Sep, July/August, pp 266–269
4. Brenann MS, Fry M, Vaarashimha M, Holtham PN, Water velocity measurements inside a hydrocyclone using an Aeroprobe and comparison with CFD predictions.

5. Chih YH, Wu SJ, Wu RM (2011) Particles separation and tracks in a hydrocyclone. Tamkang J Sc Eng 14(1):65–70
6. Chuan EW, Chen Y, Wu R (2010) Experimental and computational studies of multiphase hydrodynamics in a hydrocyclone separator system. Chem Eng Sci 65(24):6415–6424
7. Cui B, Wei D, Gao W, Liu W, Feng Y (2014) Numerical and experimental studies of flow field in hydrocyclone with air core. Trans Nonferrous Met Soc China, 242642−2649
8. Cullivan JC, Williams RA, Dyakowski T, Cross CR (2004) New understanding of a hydrocyclone flow field and separation mechanism from computational fluid dynamics. Min Eng 17(5):651–660
9. Davailles A, Climent E, Bourgeois F (2012) Fundamental understanding of swirling flow pattern in hydrocyclones. Sep Purif Technol 92:152–160
10. Davidson MR (1988) Similarity solutions for the flow in hydrocyclones. Chem Eng Sci 43(7):1499–1505
11. Delgadillo JA, Rajamani RK (2005) A comparative study of three turbulence-closure models for the hydrocyclone problem. Int J Mineral Process 77:217–230
12. Delgadillo JS, Al KM (2012) CFD simulation of a hydrocyclone in the absence of an air core. J Min Metall Sect B-Metall 48(2):197–296
13. Doby MJ, Kraipech W, Nowakowski AF (2005) Numerical prediction of outlet velocity patterns in solid–liquid separators. Chem Eng J 111(2–3):173–180
14. Doby MJ, Nowakowski AF, Dyakowski T (2008) Understanding air core formation in hydrocyclones by studying pressure distribution as a function of viscosity. Int J Miner Process 86(1–4):18–25
15. Evans WK, Suksangpanomrung A, Novakowski FN (2008) The simulation of the flow within a hydrocyclone operating with an air core and with an inserted metal rod. Chem Eng J 143(1–3):51–61
16. Gao SL, Wei D, Liu W, Ma L, Lu T, Zhang R (2011) CFD numerical simulation of flow velocity characteristics of hydrocyclone. Trans Nonferrous Met Soc China 21:2783–2789
17. Ghodrat M, Qi Z, Kuang SB, Ji L, Yu B (2016) Computational investigation of the effect of particle density on the multiphase flows and performance of hydrocyclone. Min Eng 90:55–69
18. Karimi M, Akdogan G, Bradshaw SM, Mainza A (2012) Numerical modelling of air core in hydrocyclones. In: Ninth international conference on CFD in the minerals and process industries CSIRO, Melbourne, Australia, 10–12 Dec 2012
19. Ke R (2016) Experimental and CFD investigations of the fluid flow inside a hydrocyclone separator with an air core. Master of Science Thesis, Department of aerospace Engineering, Case Western Reserve University
20. Lilley DG, Chigier NA (1971) Heat and Mass Transfer in Recirculating Flows, Combust. Flame, 16, 177, 7
21. Lim EWC, Chen Y-R, Wang C-H, Wu R-M (2010) Experimental and computational studies of multiphase hydrodynamics in a hydrocyclone separator system. Chem Eng Sci 65:6415–6424
22. Monredon TC, Hsie KT, Rajamani RK (1992) Fluid flow model of the hydrocyclones: an investigation of device dimensions. Int J Miner Process 35:65–83
23. Narashimha M, Spripriya R, Banerjee PK (2005) CFD modelling of hydrocyclone—prediction of cut size. Int J Miner Process 75:53–68
24. Nowakowski AF, Dyakowski T (2003) Investigation of swirling flow structure in hydrocyclones. Chem Eng Res Des 81(A):862–873
25. Nowakowski AF, Cullivan JC, Williams EA, Dyakowski T (2004) Application of CFD to modeling the flow in hydrocyclones. Is this a realizable option or still a research challenge? Min Eng 17(5):661–669
26. Olson TJ, Van Ommen R (2004) Optimizing hydrocyclone design using advanced CFD model. Miner Eng 17:713–720
27. Pericleous KA, Rhodes N (1986) The hydrocyclone classifier—a numerical approach. Int J Min Process 17:23–43
28. Slack MD, Prasad RO, Bakker A, Boysan F (2000) Advances in cyclone modelling using unstructured grids. Trans IchemE 78:1098–1104

29. Stephens D, Mohanarangam K (2009) Turbulence model analysis of flow inside a hydrocyclone. In: Conference: seventh international conference on CFD in the minerals and process industries, December 2009
30. Zhu G, Liow JL, Neely A (2012) Computational study of the flow characteristics and separation efficiency in a mini-hydrocyclone. Chem Eng Res Des 90:2135–2147

Chapter 7
Phenomenological Hydrocyclone Models

Abstract The structural and operational simplicity of a hydrocyclone hides the high complexity of its classification mechanism. Efforts to describe hydrocyclone behavior have led to two types of models: empirical and phenomenological. Empirical models are based on experimental observation of hydrocyclone performance and consist of the statistical correlation of numerous experimental data. In some cases, simple conceptual models are used to interpret the results, while in others, the device is considered a black box. The best-known empirical models are those of Lynch and Rao [19–21], Plitt [25, 26], Nageswararao [23] and Sepúlveda [32]. Experience has shown that these models have serious limitations. Their principle defect is that they must be used only in a small area of the conditions for which their parameters have been determined [22] (Svarovsky [33]). They do not permit the reasons for the equipment behavior to be understood and, therefore, design optimization cannot be based on them.

Nomenclature

a	Short-circuit of particles to the underflow
a_1	Maximum value in the leakage curve
a_2	Slope of the leakage curve
a_3	Particle size for which $f(x) = a_1/2$
b_o	Leakage to the overflow $= Q_\ell / Q_F$
$f(x)$	Short circuit of particles to the overflow
$c(x)$	Classification value for a particle of size x
c_f	Adjustable parameter in Eq. (7.2)
d_F	Feed diameter
d_c	Hydrocyclone diameter
d_o	Overflow diameter
d_u	Underflow diameter
d_n	Diameter of the air core
d_v	Exterior diameter of the vortex finder tube
d_{v*}	Dimensionless exterior diameter of the vortex finder tube

F. Concha A. and J. L. Bouso A., *Fluid Mechanics Fundamentals of Hydrocyclones and its Applications in the Mining Industry*, Fluid Mechanics and Its Applications 126,
https://doi.org/10.1007/978-3-030-67913-2_7

$\boldsymbol{e}_r$	Radial unit vector
$f(z)$	Empirical correction for the radial velocity
$\boldsymbol{g}$	Acceleration of gravity vector
g	Acceleration of gravity
k	Parameter in the tangential velocity
k_p	Adjustable parameter in the capacity Eq. (7.23)
l	Distance from the apex to the bottom of the vortex finder
l_{pc}	Length of the conical section of the hydrocyclone
l_{fpc}	Length of the cylindrical section of the hydrocyclone
r	Radial cylindrical coordinate
$r*$	Dimensionless radial coordinate
r_c*	Dimensionless coordinate given by Eq. (7.5)
$s(x)$	Selectivity value of a particle of size x, fraction sent to the underflow
t	Time
$\boldsymbol{u}$	Relative solid–fluid velocity vector
v_r	Radial fluid velocity
$v\phi$	Tangential fluid velocity
v_z	Axial fluid velocity
v_o	Fluid velocity of the exterior flow at the edge of the boundary
$\boldsymbol{v}_p$	Particle velocity vector
v_{pr}	Radial particle velocity
$v_p\phi$	Tangential particle velocity
v_{pz}	Axial particle velocity
x	Particle size
x_o	Distance along the lateral wall from the bottom of the vortex finder to the apex
x_{50}	Characteristic separation size, particle size where c(x) = 0.50
$x*$	Parameter in Eq. (7.22)
z	Axial cylindrical coordinate
z_o	Axial coordinate of the apex
A_F	Feed cross-sectional area
B	Parameter in Eq. (7.1)
$\boldsymbol{F}$	Resultant force vector applied to the particles
L	Characteristic Reynolds number (see Eq. 7.5)
Q_F	Feed volume flow
α	Angle of the hydrocyclone cone
θ	Angle of feed flow into the hydrocyclone
β	Adjustable parameter in Eq. (7.7)
γ	Incomplete gamma function
δ_m	Thickness of the boundary layer
δ_o	Parameter in the boundary layer
υ	Kinematic suspension viscosity
μ	Suspension viscosity
ρ	Suspension density
ρ_s	Particle density
ρ_f	Fluid density

Phenomenological models, on the other hand, are based on the study of the physical phenomena occurring within a hydrocyclone when it classifies particles. The first such studies were conducted in the early 1950s and have progressed very slowly due to the formidable task of solving sets of non-linear differential equations that describe the flow in the device [12].

The tangential velocity must be modeled as a function of the radial coordinate in the hydrocyclone in order to ascertain the centrifugal force generated in the device by the vortex motion. This force is responsible for the distribution of the particles by size along the radius. Based on this idea, Rietema [27] proposed the theory of the radius of stability, according to which the equilibrium between the centrifugal and radial drag and buoyancy force on a particle is obtained at different radii of the hydrocyclone (radius of stability) for different particle sizes. Particles whose position is smaller than the radius of stability will leave through the overflow while all the other particles will leave the device through the underflow. This theory considers the radial balance of forces only and assumes that the radius of stability is always reached by all particles.

7.1 Equilibrium Orbit Model

It is known that the axial and tangential velocities in a hydrocyclone are on the order of m/s while the radial velocity is one order of magnitude smaller; therefore, the velocity of a small particle in the radial direction can be considered in Stokes regime. Consider the fluid flow in the hydrocyclone with a radial velocity v_r. If a particle of radius r is dragged through a fluid by the centrifugal force with a velocity u_r in the radial direction of the hydrocyclone, the drag force F_D exerted by the fluid on the particle is:

$$F_D = -6\pi \mu r (u_r - v_r) \tag{7.1}$$

while the centrifugal force is:

$$F_C = (\rho_s - \rho_f)\frac{v_\theta^2}{r}\left(\frac{4}{3}\pi r^3\right)$$

where ρ_s and ρ_f are the solid and the fluid velocity, respectively, and v_θ is the tangential velocity of the particle. The balance of these forces gives:

$$r = -\frac{9}{2}\frac{(u_r - v_r)}{(\rho_s - \rho_f)v_\theta^2} \tag{7.2}$$

The particle with a 50% chance of passing to the overflow will be called d_{50} and we will assume that its position corresponds to the locus of zero particle velocity; thus:

$$d_{50} = \frac{9}{2} \frac{\mu}{\left(\rho_s - \rho_f\right)} \frac{v_r}{v_\theta^2} \tag{7.3}$$

Equation (7.3) is called the equilibrium orbit hypothesis.

7.2 Flow Pattern in a Hydrocyclone

The first study of the tangential velocity distribution from a fluid mechanics point of view was done by [27] The author assumed turbulent axis-symmetrical flow. He also assumed that the radial velocity is constant, and that the tangential velocity depends on the radial coordinate only. He obtained a tangential velocity distribution, in terms of a modified Reynolds number, with a shape like that in Kelsall's experimental observations. The work of Rietema was generalized by Upshadrata et al. [34], assuming that the suspension is a non-Newtonian fluid of the pseudo-plastic type and that the radial velocity is proportional to r, which is close to the actual distribution. The result was obtained in terms of a modified Reynolds number and gives Rietema's solution as a special case when the fluid is Newtonian.

Bloor and Ingham [1] went one step further by obtaining a solution for the three components of the velocity in the hydrocyclone. These authors assumed that the flow field in the equipment can be divided into a boundary layer along the walls and an interior inviscid flow. They further assumed, using Kelsall's result, that the flow is axis-symmetric, and that the tangential velocity is a function of the radial coordinate only. They postulated a specific vorticity function, obtaining a solution for the velocity distribution valid in the interior inviscid flow. In another paper the same year, Bloor and Ingham [2] discussed a primitive classification model, establishing the geometrical locus of the stability orbits based on the three-dimensional velocity distribution and on the momentum balance for the particles, neglecting all gravitational effects. In 1978 Brayshaw generalized Bloor and Ingham's method by introducing different constitutive equations for the vorticity function, thereby obtaining better classification curves.

Davidson [15] also modified Bloor and Ingham's method, searching for similar solutions of v_ϕ and v_z, but maintaining all other assumptions, which led to improved classification results. In 1975 Bloor and Ingham studied the flow in the boundary layer near the top wall of the hydrocyclone, explaining the leakage of water to the overflow. One year later [5] studied the boundary layer flow at the walls of the conical section, assuming the velocity distributions and obtaining the boundary layer thickness and the flow within the boundary layer.

A different approach was used by two Russian researchers, Mikhaylov and Romenskiy [24], who solved the Navier–Stokes equations using experimental information (not given in the paper) on the turbulent shear stresses in the hydrocyclone. They assumed a tangential velocity distribution to fit experimental information and solve the continuity and linear momentum equations for the other two velocity components. The results are excellent not only for the velocity distribution, but also for the pressure.

In a series of papers, Shubert and Neesse [28–31] studied the effect of the turbulence in a hydrocyclone, measuring the turbulence with electrodes. They established that the dispersion coefficient is a function of the hydrocyclone diameter and the tangential velocity and calculated a classification curve based on the turbulent dispersion of the particles.

Hsieh and Rajamani [18] solved the complete Navier–Stokes equations for the turbulent flow in a hydrocyclone using the method of finite differences. In a first step they solved the equations for a Newtonian fluid and compared the resultant velocity distributions with experimental information obtained with a laser Doppler velocimeter in a 75-mm glass hydrocyclone. The effect of the viscosity on the flow was studied using a mixture of glycerin in water with different compositions. The turbulence was modeled with the Prandtl mixing length theory and the air core was modeled empirically. The obtained solution was very good except near the air core. The main problem with this approach is the excessive time spent on the calculations, on the order of several hours with a large mainframe computer. In a second part of their work the authors combined the numerical solution for the velocity distribution with a momentum balance to obtain the particle trajectories and, from them, the classification curve. It is interesting to note that none of the calculated trajectories have the S shape considered typical or normal for a hydrocyclone in the mineral processing field. To better understand this section, see Chap. 3.

It is important to point out that all the research done to date on hydrocyclone hydrodynamics has contributed to a better understanding of the performance of the device, but none has produced a sufficiently complete model that surpasses the effectiveness of the empirical models for simulation purposes (see Chap. 5). In 1992, Barrientos and Concha developed a classification model for a hydrocyclone based on the work of Bloor and Ingham. Unlike Bloor and Ingham, they considered an underflow stream in their model and calculated the flow partition including a model for the air core. All fluid dynamics and metallurgical parameters were calculated. The model could simulate the behavior of a conical hydrocyclone using a small personal computer, providing the same information as the aforementioned empirical models. For more details see Chap. 5.

7.3 Phenomenological Model

Concha and Barrientos [12] and Barrientos and Concha [8] presented a phenomenological model of a hydrocyclone based on the experimental observations of Hsieh and Rajamani [17, 18] and the works of Bloor and Ingham [4, 5].

They expressed that the tangential fluid entry produces a three-dimensional vortex in the interior of the hydrocyclone. The high tangential and axial velocities, which have values of several meters per second, and the fact that all velocities must vanish at the hydrocyclone walls, enable the flow near the walls of the equipment to be regarded as a viscous boundary layer and an inviscid flow far from it [4, 5]. On the other hand, the high velocities within the device produce low pressures near the axis, sucking air from the outside through the apex and forming an air core.

They divided the equipment into six zones, as shown in Fig. 7.1.

Zone I

Zone I is the upper cylindrical part of the hydrocyclone, from the top of the device to the bottom of the vortex finder tube. Due to the presence of a toroidal vortex in this zone, it can be assumed to be well-mixed and therefore homogeneous, with the same composition as the feed to the hydrocyclone.

Zone II

Zone II is the lower cylindrical part of the hydrocyclone. It has most of the properties of the conical section below and serves as a transition region between the homogeneous Zone I and the conical Zone III. We assume the existence of a sink at the interface of Zones I and II, close to the vortex finder tube, and more precisely from point p, where the axial velocity is zero, to the border of the vortex finder tube

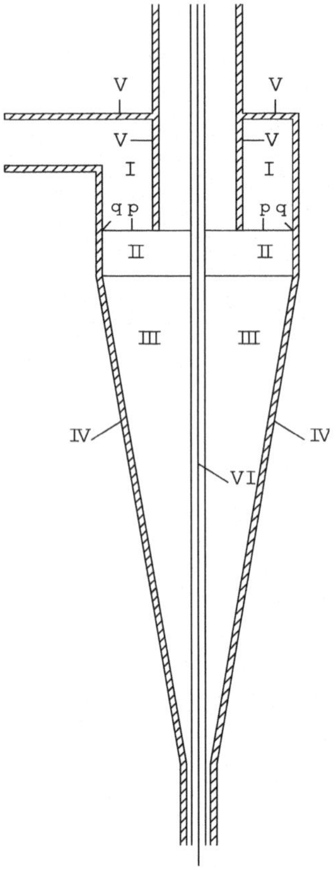

Fig. 7.1 Zones in a conical hydrocyclone

(see Fig. 7.1), such that all particles reaching this region from Zone II will flow out through the vortex. In Zone II we assume that Bloor and Ingham's solution [1, 2] for a viscid flow is valid with an empirical correction $f(z)$ for the radial velocity. With this correction the radial velocity $v_r(r, z)$ is set to zero at the I/II interface, giving the following equations for the radial axial and tangential velocity:

$$v_r = -Bz^{-2}r^{3/2}(1 - f(z)) \tag{7.4}$$

$$v_z = \frac{1}{2}Br^{-1/2}\left(3\alpha - \frac{5r}{z + l_{pc}}\right) \tag{7.5}$$

$$v_\phi = \frac{2k}{d_c}\frac{\gamma\left(\frac{4}{5}, \frac{2}{5}Lr^{*5/2}\right)}{r^*\gamma\left(\frac{4}{5}, \frac{2}{5}L\right)} \tag{7.6}$$

The tangential velocity v_ϕ expressed by Eq. (7.6) gives an excellent approximation of the available experimental information of [16]. Nevertheless, in order to make the calculations more rapidly, an empirical fitting to the prediction from the incomplete gamma function γ was made, by the combination of a free and a forced vortex with a common boundary at:

$$r^* = 0.367 - 0.00111L \tag{7.7}$$

In the above equations B, k, L and r^* are parameters that depend on the operational conditions in the device and are expressed by [12]:

$$k = -\frac{c_f Q_a}{2A_a}, \quad L = \frac{\rho B(d_c/2)}{\mu}, \quad r^* = \frac{2r}{d_c} \tag{7.8}$$

$$B = \frac{\varepsilon Q_F}{2\pi\alpha\left[\left(\frac{d_c}{2}\right)^{5/2} - \left(\frac{d_r}{2}\right)^{5/2}\right]\left[\frac{2}{\alpha\ell_{pc}+d_d} - 1\right]} \tag{7.9}$$

where c_f in (7.8) is a number that takes into account the fact that not all the flow enters the hydrocyclone tangentially, d_c is the hydrocyclone diameter, A_f is the cross section of the feed opening, Q_f is the feed pulp volume flow rate, z_0 is the coordinate specifying the position of the apex, l_{pc} is the length of the conical section of the hydrocyclone, d_O and d_U are the internal diameters of the vortex finder and apex tubes, respectively, and ε is a parameter.

The correction f_z for the radial velocity is based on the experimental observations of [16]:

$$f(z) = \left(\frac{z - z_o - l_{pc}}{l - l_{pc}}\right)^4 \tag{7.10}$$

where ℓ is the distance from the apex to the bottom part of the vortex-finder tube.

Zone III

Zone III is the region in the conical section of the hydrocyclone of inviscid flow outside the boundary layer. We assume that in this region Bloor and Ingham's solution for the velocity profiles is valid [1–3, 6, 7]. Point q in the boundary between Zones I and II denotes the border of the lateral wall boundary layer. The velocity distribution in this region III is the same given for Zone II, except that here $f(z) = 0$.

Zone IV

Zone IV is the boundary layer on the lateral wall of the hydrocyclone. The boundary layer starts at the feed and develops its maximum thickness when reaching the interface of Zones I and II. This thickness is relatively constant over the rest of the wall but diminishes when approaching the apex [5]. The maximum thickness of this boundary layer is given by:

$$\delta_m = \frac{\delta_o}{\left(v_o x_o / \nu\right)^{1/2}}, \quad \delta_o = \beta x_o \tag{7.11}$$

where x_o is the distance along the lateral wall from the bottom of the vortex finder to the apex; v_0 is the exterior flow velocity, parallel to the lateral wall at the I/II interface just outside the boundary layer; ν is the fluid kinematical viscosity and $\delta_0 = \beta x_0$, where β is a number. Bloor and Ingham's model predicted $\beta = 1.88$, but we use this parameter for upscaling purposes. By integrating the axial velocity within the boundary layer, the total flow in the boundary layer may be calculated.

Zone V

Zone V corresponds to the boundary layer over the top wall of the hydrocyclone. Bloor and Ingham [4] calculated the flow in this layer that descends through the outer wall of the vortex finder tube and discharges directly into the overflow. They called this flow leakage and gave the following value for the volume flow within the boundary layer:

$$Q_\ell = \frac{\pi d_c |k| Q^*\left(d_v / d_c\right)}{\mathrm{Re}^{1/5}} \tag{7.12}$$

with the Reynolds number given by $Re = \frac{(|k|/d_c) d_c}{U}$ and where Q_ℓ is the leakage volume flow; $|k|$ is the parameter, with dimensions of [L²/T], given in (7.5); d_c is the hydrocyclone diameter and the dimensionless flow rate is $Q^*\left(d_v^*\right) = 0.153 - 0.0897 d_v^* - 0.0621 d_v^{*2}$, where $d_v^* = d_v / d_c$ and d_v is the external diameter of the vortex finder tube.

Zone VI

Zone VI is the air core, which is assumed to have a cylindrical shape. The air core is modeled according to the work of Concha et al. (1994):

$$d_n = \frac{0.303\sigma d_r}{2703\mu\left(1 + \frac{0.678}{\Delta p^{2.61}}\right)\left(\frac{d_d}{d_r}\right)^{-0.20} - 0.138} \tag{7.13}$$

where d_n is the air core diameter; d_O and d_U are the vortex and apex diameters, respectively; Δp is the pressure drop and μ and σ are the suspension viscosity and surface tension at the overflow concentration.

7.4 Mathematical Model

Concha et al. [14], developed a hydrocyclone simulator, called HYDROSIM, to predict the operation of a conical hydrocyclone from the characteristics of feed material, the geometry of the equipment and the operating conditions. The program was based on the phenomenological model of the hydrocyclone, described in the previous section, which solves the equations describing the fluid velocity field within the hydrocyclone as the particle trajectories inside the classifier. From these equations it is possible to determine the short circuit, the classification function and the size distribution of the overflow and underflow. The simulation model is based on the following assumptions:

- Zone I is fully mixed and contains a homogeneous suspension, such that all particles in this zone have the same opportunity to enter the boundary layer or pass to Zone II. Therefore we selected the hydrocyclone feed at the interface between Zones II and III between points p and q. (See Fig. 7.1).
- Particles entering the lateral boundary layer will never leave it because there is no force capable of forcing them out; therefore, they are discharged from the hydrocyclone through the apex. From Zones II and III particles will enter the boundary layer by classification and join the particles coming from the feed. Therefore, we postulate that the flow in the lateral boundary layer transfers a certain number of particles directly from the feed (Zone I) to the underflow without classification. These particles are the *short circuit* to the underflow and must have the composition of the well-mixed Zone I (the feed). During the transit of the short-circuited stream other particles join by classification, as we have already said, and are discharged together with the non-classified particles.
- Early works accepted the hypothesis that a well-designed and operated hydrocyclone suffers no leakage of particles to the overflow. The basis for this belief is that selectivity curves normally reach the value of 100% at a certain particle size and, therefore, particles of this size and larger do not leak to the overflow. We think that there is leakage in all cases, and we assume that the flow in the upper boundary layer constitutes a direct transfer of particles from the feed to the overflow. But not all particles entering the boundary layer leave through the overflow since some particles, especially coarser particles during their movement in the boundary layer from the feed to the overflow, have the chance to escape

the boundary layer due to the effect of the drag of the fluid flow and the effect of gravity.

- Classification occurs in Zones II and III and is obtained by solving the momentum equations for individual particles and tracing their trajectories.
- Integrating the axial velocity component at the apex, between the boundary layer and the air core, and adding the boundary layer flow yields the underflow volume rate and the underflow concentration.
- The air core is the main cause of roping phenomena. For a fixed geometry, the air core diameter depends on the operating conditions and fluid properties. In some conditions the air core blocks most of the apex, leading to roping.

From experiences with empirical models we know that five equations are needed to simulate the operation of a hydrocyclone: (1) the classification function; (2) the selectivity function, (3) the short circuit to the underflow, (4) the characteristic separation size and (5) the capacity-pressure drop relationship. We will see in the following section that a sixth equation is needed to allow for the leakage of particles to the overflow.

7.4.1 Classification Function

The classification function $C(x)$ represents the fraction of particles of size x in the well-mixed zone of the hydrocyclone (Zone I) that reports to the underflow by classification. In our model this definition requires that the particles undergo classification following trajectories that pasess through Zones II and III. By calculating the trajectories of enough particles, initially evenly distributed at the I/II interface, it is possible to estimate the classification function. To calculate it the momentum equations for the individual particles are solved and their trajectories in Zones II and III must be traced. If the relative solid–fluid flow occurs in the Stokes regime, that is, if the particle Reynolds number $\mathrm{Re} = \rho|\boldsymbol{u}|x/\mu << 1$, where x is the particle diameter, ρ and μ are the suspension density and viscosity, respectively, and $\boldsymbol{u}$ is the solid–fluid relative velocity, Euler's equation reduces to:

$$3\pi\mu x\boldsymbol{u} + \Delta\rho(\pi/6)x^3\left(\boldsymbol{g} + \left(v_\phi/r\right)\boldsymbol{e}_r\right) \tag{7.14}$$

where $\Delta\rho = \rho_s - \rho_f$ is the solid-suspension density difference, $\boldsymbol{g}$ is the acceleration of gravity vector, v_ϕ is the tangential velocity component and $\boldsymbol{e}_r$ is the radial unit vector. Thus:

$$\boldsymbol{u} = -\frac{\Delta\rho x^2}{18\mu}\left(\boldsymbol{g} - \frac{v_\phi^2}{r}\boldsymbol{e}_\mathrm{r}\right) \tag{7.15}$$

where the particle velocity components are:

$$v_{pr} = v_r + \frac{\Delta\rho x^2}{18\mu}\frac{v_\phi^2}{r} \tag{7.16}$$

$$v_{p\phi} = v_\phi \tag{7.17}$$

$$v_{pz} = v_z + \frac{\Delta\rho x^2 g}{18\mu} \tag{7.18}$$

In these equations v_r, $v\phi$ and v_z are given by Eqs. (7.1) to (7.7) and $\boldsymbol{g} = -g\boldsymbol{k}$, where $\boldsymbol{k}$ is the unit vector pointing in the upward direction. The particle trajectories are obtained by integration of Eq. (7.15).

When the particle Reynolds number for the solid–fluid motion is greater than $\mathrm{Re}_p > 1$, it is necessary to use a more general equation for the hydrodynamic force. In this case we use the concept of drag coefficient and Eq. (7.14) must be replaced by:

$$-3\rho|u|\mathrm{C_d}(\mathrm{Re})\boldsymbol{u} + 4\Delta\rho x\left(\boldsymbol{g}+\frac{v_\phi^2}{r}\boldsymbol{e}_r\right) \tag{7.19}$$

where $|\boldsymbol{u}|$ is the magnitude of the relative solid–fluid velocity and C_D is the drag coefficient, which can be calculated by Concha and Almendra's equation [10] or by Concha and Christiansen's equation [11] if the particles are not spherical. The relative solid–fluid velocity components are now:

$$r \text{ direction} \quad |\boldsymbol{v}_p - \boldsymbol{v}|C_d(\mathrm{Re})(v_{pr} - v_r) = -\frac{4}{3}\frac{\Delta\rho v_\phi^2}{\rho r} \tag{7.20}$$

$$z \text{ direction} \quad |\boldsymbol{v}_p - \boldsymbol{v}|C_d(\mathrm{Re})(v_{pz} - v_z) = -\frac{4}{3}\frac{\Delta\rho g}{\rho r} \tag{7.21}$$

$$\phi \text{ direction} \quad v_{p\phi} = v_\phi \tag{7.22}$$

where

$$|\boldsymbol{v}_\mathrm{p} - \boldsymbol{v}| = \left\{(v_{pr} - v_r)^2 - (v_{pz} - v_z)^2\right\}^{1/2} \tag{7.23}$$

is the magnitude of the relative particle–fluid velocity. The particle trajectories are obtained by integration of Eqs. (7.20) and (7.21).

The plot of the classification values versus size is the *classification function*, which can be parameterized through the *sharpness index* (*S.I.*) and the characteristic separation size x_{50}, defined by:

$$S.I. = x_{75}/x_{25} \tag{7.24}$$

where x_{25},x_{50} and x_{75} are defined by $C(25) = 0.25$, $C(50) = 0.50$ and $C(75) = 0.75$. An ideal classification yields an $S.I. = 1$.

7.4.2 Short Circuit to the Under Flow

By integrating the velocity field, at the I/II interface $r = R$, from the wall to the border of the boundary layer δ_m.

$$a = \frac{\left\{\left(\frac{d_c}{2}\right)^{5/2} - \left(\frac{d_c}{2} - \delta_m\right)^{5/2}\right\} - \left\{z_o + \ell_{pc}\right\}\left\{\left(\frac{d_c}{2}\right)^{3/2} - \left(\frac{d_c}{2} - \delta_m\right)^{3/2}\right\}}{\alpha\left\{\left(\frac{d_c}{2}\right)^{5/2} - \left(\frac{3\alpha\left(z_o+\ell_{pc}\right)}{5}\right)^{5/2}\right\} - \alpha\left\{z_o + \ell_{pc}\right\}\left\{\left(\frac{d_c}{2}\right)^{3/2} - \left(\frac{3\alpha\left(z_o+\ell_{pc}\right)}{5}\right)^{3/2}\right\}} \tag{7.25}$$

Based on experimental observations, Barrientos and Concha [8] showed that, when a hydrocyclone is working properly, the air core has a diameter close to that of the apex and, in general, only a small amount of water passes from the classification zone (III) to the underflow. This implies that almost all the water appearing at the underflow comes from the boundary layer flow. According to this observation, our hypothesis on the origin of the short circuit produces exactly the observed result that the short circuit is proportional to the water split.

7.4.3 Leakage to the Overflow

Barrientos et al. [9] obtained the leakage of particles to the overflow by calculating the trajectories of particles of different sizes within the boundary layer. The particles that crossed the boundary layer edge before reaching the vortex finder wall returned to the well-mixed Zone I without going to the overflow. All the rest constituted the leakage. It was shown that the fraction of particles that leak to the overflow can be expressed in the form:

$$b(x, x^*) = b_\ell f(x, x^*) \tag{7.26}$$

where $b_\ell = Q_\ell / Q_F$ is the fraction of the total volume flow represented by the leakage Q_ℓ (given by Eqs. (7.9)), Q_F is the feed volume flow rate of pulp into the hydrocyclone and x^* is the greatest size that leaks to the overflow through the upper boundary layer. By definition $f(x)$ is given by [9]:

$$f(x) = \frac{a_1}{1 + \exp(a_2(x - a_3))} \tag{7.27}$$

where a_1 is the maximum value in the leakage curve, a_2 is the slope of the curve and a_3 is the particle size for which the value of $f(x) = a_1/2$. These parameters are a function of the geometry of the hydrocyclone:

$$a_1 = 230.7(1 - \exp(-7.420(d_v/d_c))) - 180.3 \tag{7.28}$$

$$a_2 = 0.476\exp(-4.251(d_v/d_c)) + 0.226(1 - \exp(-0.0114\alpha)) - 0.0251 \tag{7.29}$$

$$a_3 = 1.371\exp(8.090(d_v/d_c)) - 207.9\exp(-0.0283\alpha) + 284.7 \tag{7.30}$$

where d_v, d_c and α are the external diameter of the vortex finder tube, the hydrocyclone diameter and the semi-angle of the hydrocyclone body.

7.4.4 Selectivity Function

In our model, we assume that as soon as the pulp enters the hydrocyclone it splits into three fractions: a fraction associated with the upper boundary layer that comprises the leakage to the overflow, a fraction related to the lateral boundary layer that constitutes the short circuit to the underflow and a fraction entering the well-mixed zone. The material in this well-mixed zone undergoes the classification action, splitting into a fine product that, after mixing with the leakage, emerges as the overflow, and a coarse product that, after joining the short circuit, becomes the underflow. Taking these factors into account, the selectivity function can be written in terms of the classification function in the form given by [9].

Knowing the classification function $C(x)$, the short circuit "a" and the leakage $b(x/x^*)$, the selectivity function $S(x)$ can be calculated with equation

$$S(x) = a + (1 - a - b(x/x^*))C(x) \tag{7.31}$$

7.4.5 Separation Size

The characteristic separation size x_{50} is determined by interpolation of the values in the classification curve.

7.4.6 Capacity-Pressure Drop Relationship

The fact that an inviscid model is used for the description of the velocity field in the hydrocyclone prevents the theoretical determination of the pressure drop in the equipment. Therefore, we use a modification of Plitt's empirical equation for the capacity [25], which could predict the capacity of our pilot circuit within 1% error:

$$Q_F = k_p \Delta p^{0.56} d_c^{0.21} d_F^{0.53} \ell^{0.16} \left(d_u^2 + d_o^2\right)^{0.49} \exp(-0.0031\varphi_F) \tag{7.32}$$

where Δp is the pressure drop, Q_F is the inlet volume flow rate, φ_F is the volume fraction of solids in the feed, d_c is the hydrocyclone diameter, ℓ is the distance between the bottom of the vortex finder and the apex and the subscripts d_u and d_o correspond to the underflow and overflow diameters, respectively. The parameter k_p must be determined experimentally.

7.5 HYDROSIM Simulator

Program description

A simulation program, based on the phenomenological model presented above, solves the equations of the fluid velocity field within the hydrocyclone, describing the particle trajectories inside the classifier, which predicts the operation of a conical hydrocyclone considering the geometric characteristics of the device and its operating conditions. The selectivity and classification functions are determined from the particle trajectories. The program is organized into four modules:

(a) Geometric data entry
(b) Experimental data entry
(c) Parameter evaluation
(d) Simulation.

The geometric and operating variables necessary for parameter estimation are: $d_c, d_o, d_v, d_u, \ell_{pci}, \ell_{pc}, \ell_v, \alpha, k_p, \theta, d_F, Q_F(\text{or } \Delta p), \varphi_F, Q_u, \varphi_u, \rho_s, \rho_f$ and $S(x)$.

Simulation menu

The following steps should be followed for hydrocyclone simulations:

(a) Geometric data entry
(b) Experimental parameter entry
(c) Parameter evaluation
(d) Simulation
(e) End.

(a) Geometric data entry

The hydrocyclone geometry should be entered in this file. If no data are entered, the geometry of a 150 mm diameter hydrocyclone is used by default. Data may be entered by files or keyboard.

File name	Default30.geo
Hydrocyclone diameter	15 cm
Cylindrical length	25 cm
Feed diameter	5.5 cm
Internal vortex finder diameter	4.5 cm
External vortex finder diameter	9.6 cm
Length of the vortex finder	19 cm
Length of the conical section	50 cm
Internal apex diameter	3.0 cm

(b) Experimental data entry

An industrial or pilot plant test must be performed to obtain data for further simulation. In this file data on feed flow and concentration, overflow and underflow should be entered, in addition to the selectivity function.

Username Arturo Barrientos	Experimental data	Particle size [μm]	Selectivity function [%]
Institution: University of Concepcion	Q_F: 0.00533 [m/s]	300.0	100
Test identification: Test 1	ϕ_F: 0.139 [–]	212.1	100
File name: Exp30.dat	Q_a: 0.00162 [m^3/s]	150.0	98.56
	ϕ_a: 0.558 [–]	106.1	96.71
	P: 10 [psi]	75.0	92.96
	ρ_s: 2.65 [t/m^3]	53.0	63.93
		37.5	40.63
		26.5	18.93
		18.8	16.31
		13.3	14.08
OK		9.4	15.29

(c) Parameter evaluation

For the geometric and experimental data, the selectivity and classification functions must be determined. Numerous particle trajectories are calculated to obtain the selectivity function and the hydrocyclone parameters are adjusted to the experimental selectivity values. The adjusted parameters are:

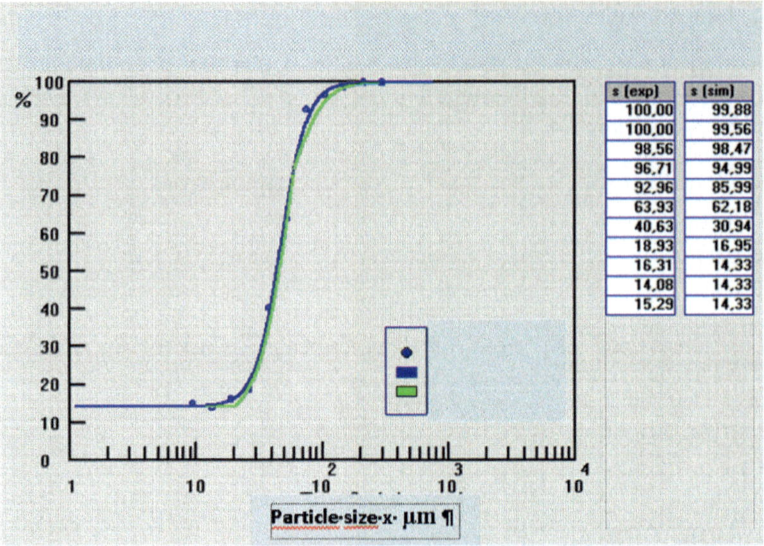

Fig. 7.2 Hydrocyclone parameters adjusted to the selectivity function; the blue points are experimental, the blue line is experimental selectivity and the green line is the adjusted selectivity function

a	b	x_{50}	CF	ε	β	K_p	K_b	H
14.33%	11.67%	43.37	0.80	0.7124	1.7227	$1.85 \times 10^{-0.3}$	3.311	2.00

Results are shown in Fig. 7.2.

(d) **Simulation**

The simulation menu includes

(a) Determination of flow pattern
(b) Determination of streamlines
(c) Determination of particle trajectories
(d) Classification results
(e) End.

With the material particle size distribution and other parameters determined in the previous section, the characteristics of the flows in the hydrocyclone are simulated.

Flow pattern (Figs. 7.3, 7.4 and 7.5).

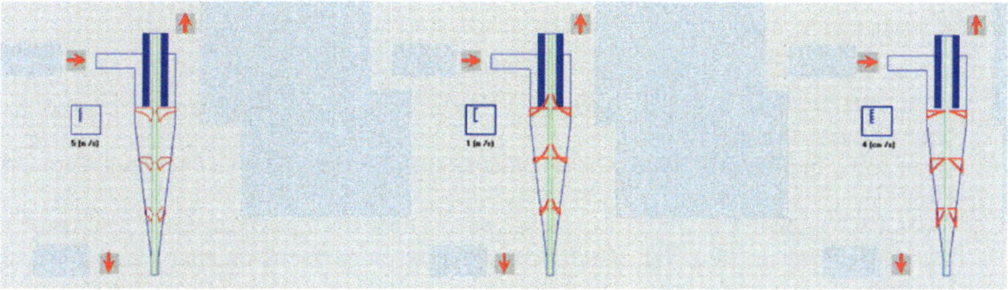

Tangential velocity Axial velocity Radial
velocity

Fig. 7.3 Velocity patterns in the simulated hydrocyclone

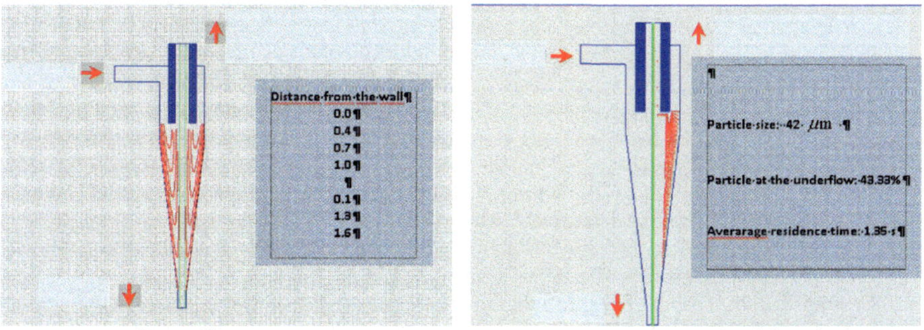

Streamlines Particle traject

Fig. 7.4 Simulation of the streamlines and particle trajectories of the simulated hydrocyclone

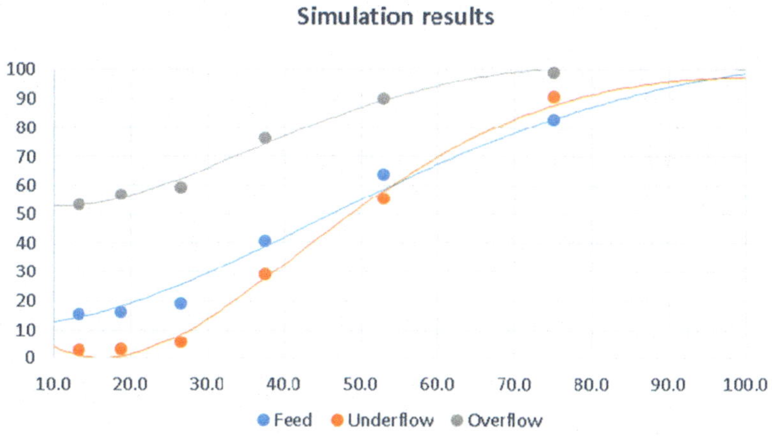

Fig. 7.5 Particle size distribution results for a 150-mm-diameter hydrocyclone

References

1. Bloor MIG, Ingham DB (1973) Theoretical investigation of the flow in a conical hydrocyclone. Trans Inst Chem Eng 51:36
2. Bloor MIG, Ingham DB (1973) On the efficiency of the industrial cyclone. Trans Inst Chem Eng 51:173
3. Bloor MIG, Ingham DB (1975) Turbulent spin in a hydrocyclone. Trans Inst Chem Eng 53:1
4. Bloor MIG, Ingham DB (1975) The leakage effect in the industrial cyclones. Trans Inst Chem Eng 53:7
5. Bloor MIG, Ingham DB (1976) Boundary layer flow on the side walls of conical hydrocyclones. Trans Inst Chem Eng 54:276
6. Bloor MIG, Ingham DB (1983) Theoretical aspects of hydrocyclone flow. In: Wakeman RJ (ed) Progress in filtration and separation, vol 3. Elsevier Sci Publ Co., New York, pp 57–148
7. Bloor MIG, Ingham DB (1987) The flow in industrial cyclones. Journal of Fluid Mech 178:507
8. Barrientos A, Concha F (1992) Phenomenological model of classification in conventional hydrocyclones. In: Kawatra K (ed) Comminution: theory and practice. AIME, New York, Chapter 21, pp 288–305
9. Barrientos A, Etcheverry J, Concha F (1995) The leakage of particles through the top boundary layer and its effect on the classification in a conventional hydrocyclone. Int J Miner Process, submitted
10. Concha, F. and Almendra, E.R., Settling velocities of particulate systems. 1. Settling velocities of individual spherical particles, Int. J. Miner. Process. 5 349–367. 1979.
11. Concha F, Christiansen A (1986) Settling velocities of particulate systems. 5. Settling velocities of suspensions of particles of arbitrary shape. Int J Miner Process 5:309–322
12. Concha F, Barrientos A (1991) Monografía "Modelación Fenomenológica de Hidrociclones". Proyecto PNUD/CHI/ 88/011, Agosto 1991
13. Concha F, Barrientos A, Montero J, Sampaio R (1994) Air core and roping in hydrocyclones. In: Proccedings of the 8th European Symposium of Conminution, Stockholm, Suecia, vol II, pp 814–823
14. Concha F, Barrientos A, Muñoz L, Bustamante O, Castro O (19969) A phenomenological model of hydrocyclone. Hydrocyclone 96, Mechanical Engineering Pub., pp 63–82
15. Davidson MR (1988) Similarity solutions for flow solutions for flow in hydrocyclones. Chem Eng Sci 43(7):1–13
16. Hsieh KT (1988) Phenomenological model of hydrocyclone. Ph.D. Thesis, University of Utaa
17. Hsieh KT, Rajamani K (1988) Phenomenological model of hydrocyclone: model development and verification for single phase flow. Int J Mineral Process 22:223
18. Hsieh KT, Rajamani K (1988b) Phenomenological model of hydrocyclone: numerical solution and verification of particle movement. In: Forssberg E (ed) Proceedings XVI IMPC, Amsterdam, p 377–388
19. Lynch AJ, Rao TC (1968) The operating characteristics of hydrocyclone classifiers. Ind J Tech 6:106–116
20. Lynch AJ, Rao TC, Prisbrey KA (1974) The influence of hydrocyclone diameter on reduced efficiency curves. Int J Mineral Process 1:173–181
21. Lynch AJ, Rao TC (1975) Modelling and scale-up of hydrocyclone classifiers. In: Proceedings of the 11th international mineral processing congress, Cagliari, Italy, Paper 9, pp. 245–269
22. Medronho RA, Svarovsky L (1984) Tests to verify hydrocyclone scale-up procedure. In: 2nd international conference on hydrocyclones, Bath, England, pp 1–14
23. Nageswararao K (1978) Further developments in the modelling and scale-up of industrial hydrocyclones. Ph.D. Thesis, University of Queensland (JKMRC), Brisbane, Australia
24. Mikahylov PM, Romenskiy AA (1974) On the calculation of fluid dynamics in liquid cyclones. Fluid Mech Soviet Res 3:154
25. Plitt LR (1976) A mathematical model of the hydrocyclone classifier. CIM Bull. December 114

26. Plitt LR, Flintoff BC, Stuffco (1978) Roping in hydrocyclones. In: Wood P (ed) Proceedings of the 3rd international conference on hydrocyclones, Oxford, BHRA. Elsevier, Amsterdam, paper A-3, p 21–34
27. Rietema K (1961) Performance and design of hydrocyclones, Part I to IV. Chem Eng Sci 15:298–325
28. Schubert H (1983) Wet classification and wet screening of fine particles. Part Sci Technol 1:393–407
29. Schubert H (1985) A hydrocyclone separation model in consideration of the turbulent multi-phase flow. Part Sci Technol 3:1–13
30. Schubert H, Nesse T (1973) The role of turbulence in wet classification. In: The Inst. Mining Met., 10th Int. Mineral. Processing Congress, Cannes paper 29, pp 5–27
31. Schubert H, Nesse T (1980) A hydrocyclone separation model in consideration of the turbulent multi-phase flow. In: International conference on hydrocyclones, BHRA, Fluid engineering, pp 23–36
32. Sepúlveda J (2002) Cyclosim
33. Svarovsky L (1984) Hydrocyclones, HoltRinehard and Wiston, London p 198
34. Upshadrata KR, Ketcham VJ, Miller JD (1984) Tangential velocity profiles for pseudoplastic power-law fluids in the hydrocyclone; a theoretical derivation. Int J Mineral Process 20:309–318

Chapter 8
Hydrocyclone Design

Abstract It is common to call "design" the election of the sizes of the parts of hydrocyclones chosen from those provided by the manufacturers. It is really a geometric selection for which empirical formulas exist (see Chap. 9). In this chapter we will develop a design for hydrocyclones as a mixed-flow turbo machine including different inlet, vortex and apex geometry choices producing the greatest centrifugal energy from the angular momentum of the entrance to obtain the maximum fluid flow with minimum head loss when performing classification.

8.1 Hydrocyclone as a Turbo Machine

The characteristics of the velocity components in a hydrocyclone lead it to be classified as a mixed-flow turbo machine. In a turbo machine, there are two classification criteria: design and similarity criteria. The design criteria consider the following variables: flow rate, elevation height and rotation velocity. In general, characteristic curves are defined for different velocities of the same turbo machine. In the case of the hydrocyclone, this is a matter of different inlet and vortex and apex geometry choices, increasing the versatility of the device. These choices are equivalent to different impellers in a turbo machine. The similarity criteria in a hydrocyclone seek similarity in the apex-to-vortex-diameter ratio.

The difference between a turbo machine and a hydrocyclone is that, in the case of the former, the piece interchanging energy is the impeller, which gives or receives energy, producing work, while in a hydrocyclone the goal is not work, but rather the production of the greatest centrifugal energy from the angular momentum of the entrance to obtain the maximum fluid flow with minimum head loss when performing classification.

F. Concha A. and J. L. Bouso A., *Fluid Mechanics Fundamentals of Hydrocyclones and its Applications in the Mining Industry*, Fluid Mechanics and Its Applications 126,
https://doi.org/10.1007/978-3-030-67913-2_8

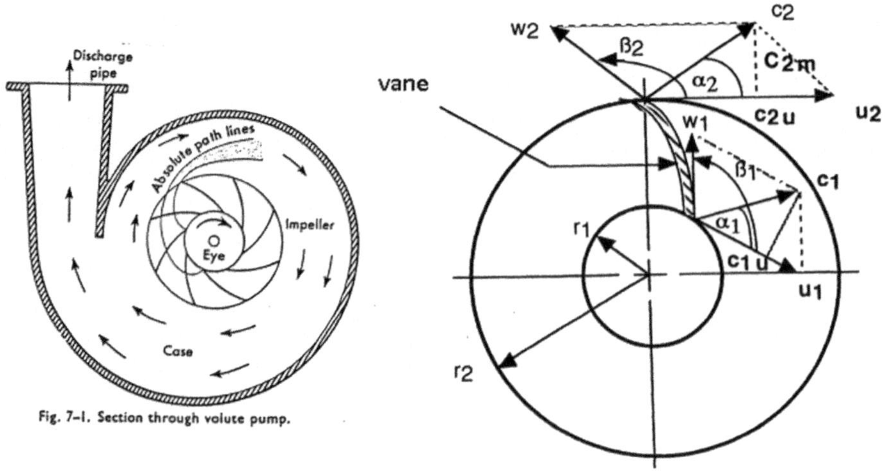

Fig. 8.1 Fluid velocity in the impeller of a pump

8.1.1 Turbo Machine Design Criteria

Euler's equation is the basic equation for turbo machines. It expresses the energy exchange in the impeller in turbo machines such as pumps, turbines, turbochargers, steam turbines and gas turbines. Consider Fig. 8.1, which shows an impeller with velocities w_1 and w_2, and u_1 and u_2 [13].

The angular momentum Ω and power P of the device are given by:

$$\begin{aligned} \Omega &= \rho g Q H \\ P \equiv \Omega &= \rho Q (u_2 r_2 - u_1 r_1) \end{aligned} \tag{8.1}$$

Since the power is equal to the angular momentum,

$$\begin{aligned} \rho Q (u_2 r_2 - u_1 r_1) &= \rho g Q H \\ H &= \frac{u_2 r_2 - u_1 r_1}{\rho} \end{aligned} \tag{8.2}$$

where H is the total hydraulic column height and r_1 and r_2 are the diameters of the impeller and the interior of the turbo machine.

8.1.2 Hydrocyclone Inlet Design

There are three main inlet geometries for tangential hydrocyclone inlets: (a) without involute, (b) with 180° involute and (c) with 360° involute. Figure 8.2 shows the three

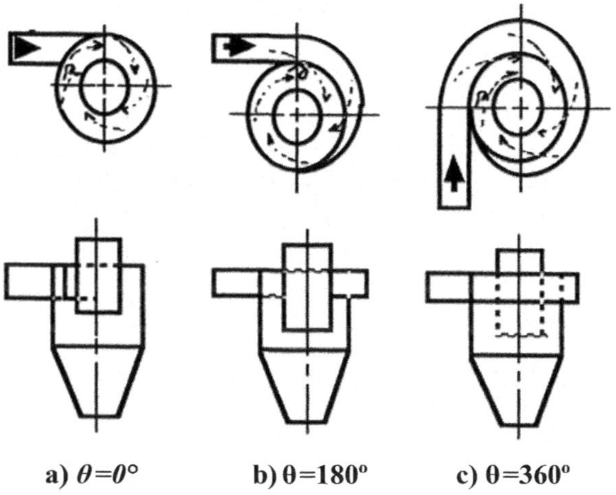

Fig. 8.2 Hydrocyclones with involutes at different angles

types of hydrocyclone inlets. The difference in performance of these three types is reflected in their head loss, which decreases as the angle increases.

The pressure coefficient C_p relates the pressure drop Δp, the flow rate Q and the fluid density ρ, and is defined as the pressure forces divided by inertial forces, that is:

$$C_p = \frac{\Delta p}{(1/2)\rho v_F^2} \tag{8.3}$$

The flow rate Q_F for the three types of fluid inlet is given by:

$$Q_F = A\sqrt{\frac{2}{\rho C_p}}\sqrt{\Delta p} \tag{8.4}$$

According to Soto [13], for the three inlet design alternatives, the pressure coefficients C_P are: (a) 50.0 k, (b) 43.40 k and (c) 36.80 k, where k is a constant. The decreasing pressure coefficients show that the flow rate for the same pressure drop increases with the involute angle θ (Fig. 8.3).

The change in involute angle θ from zero (Krebs hydrocyclone) to 360 increased the feed flowrate,for the same inlet pressure in 29% for the conical and the flatt botom hydrocyclone respectively (Fig. 8.4).

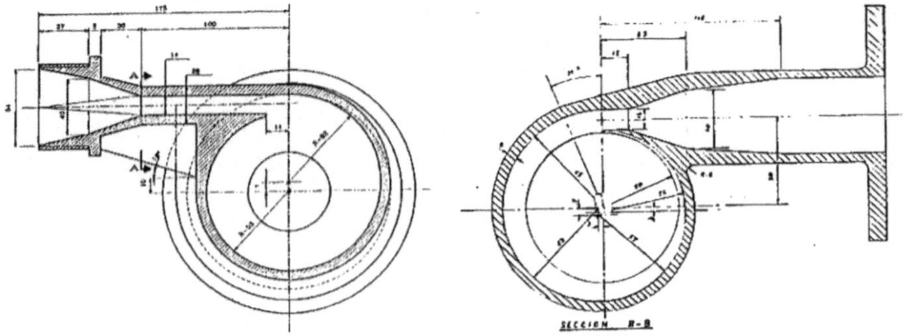

a) Involute θ=0°. b) Involute θ=360°.

Fig. 8.3 Conical and flat-bottom hydrocyclones, 100 mm diameter [13]

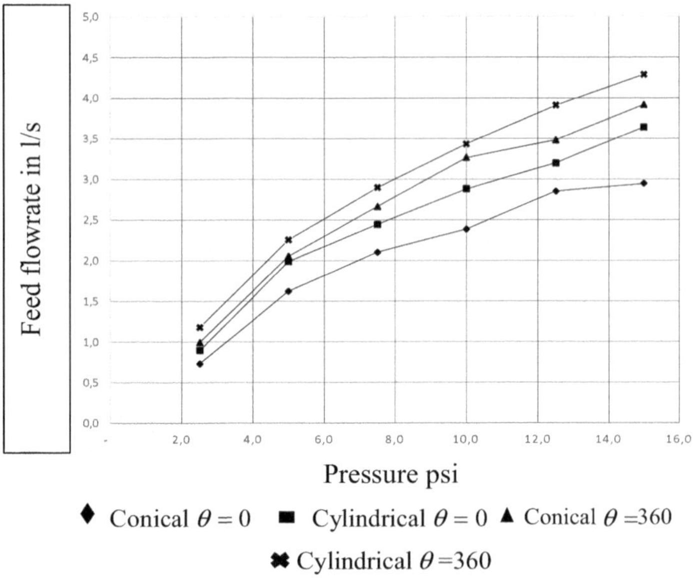

Fig. 8.4 Capacity versus pressure for conical and flatt bottom hydrocyclones for θ equal zero and 360 [13]

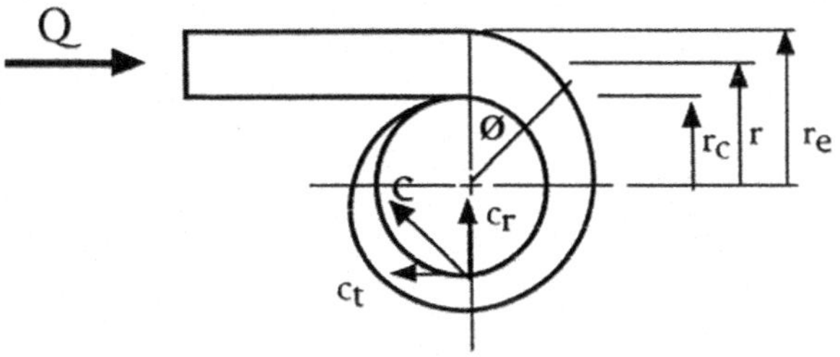

Fig. 8.5 Vortex chamber

8.1.3 Pressure–volume Flow in a Hydrocyclone as a Turbo Machine

Figure 8.5 shows a hydrocyclone with an involute inlet as the inverse of a turbo machine; that is, the inlet of the former is the outlet of the latter.

A hydrocyclone does not generate power; therefore, Eq. (8.2) leads to

$$\rho Q(v_{\theta 2} r_2 - v_\theta r) = 0$$
$$v_\theta = v_{\theta 2}\frac{r_2}{r} \tag{8.5}$$

where v_θ and $v_{\theta 2}$ are respectively the tangential velocities in the entry of the hydrocyclone and the inlet of the involute. Equation (8.5) shows the velocity in the hydrocyclone of inlet radius r_1 in terms of the velocity and involute entry radius r_2.

For a vortex chamber, such as that shown in Fig. 8.5, the tangential velocity is given by:

$$v_\theta = \frac{Q_F}{H \times r \times \ln(r_e/r_c)} \tag{8.6}$$

For a given geometry, the capacity of a hydrocyclone is given by:

$$Q_F = m\pi\left(r_o^2 + r_u^2\right)\sqrt{2gH} \tag{8.7}$$

where m is the flow coefficient, and H is the total head at the feed inlet, r_o and r_u are the radius of the overflow and the underflow, respectively, and ρ is the fluid density. Since $H = p/\rho g$,

$$Q_F = m\pi\left(r_o^2 + r_u^2\right)\sqrt{2P/\rho} \tag{8.8}$$

From a mechanical point of view, the hydrocyclone performance for given values of flowrate Q and pressure P is represented by the value of the fluid coefficient m.

$$m = \frac{1}{\sqrt{\frac{2}{\rho}}\pi r_u^2\left(1 + \left(\frac{r_o}{r_u}\right)\right)} \frac{Q}{\sqrt{P}} \tag{8.9}$$

$$K(\rho, r_u, r_o) = \frac{1}{\sqrt{\frac{2}{\rho}}\pi r_u^2\left(1 + \left(\frac{r_o}{r_u}\right)\right)},$$

we have:

$$m = K(\rho, r_u, r_o)\frac{Q}{\sqrt{P}} \tag{8.10}$$

8.2 Industrial Application of the Design

Weir Minerals patented [2] its new Cavex hydrocyclone entry for a hydrocyclone based on the work of Soto [13], Mora [9], Concha and Castro [4], Véjar [11] and Bustamante and Mora (1996) and tested it at La Coipa Mine and the Mantos de Oro concentrator in Chile [3].

The laminar spiral inlet geometry design of the Cavex hydrocyclone see Fig. 8.6, provides a natural flow path with no sharp edges or square corners allowing the feed stream to blend smoothly with rotating slurry inside the chamber and reducing significantly the turbulence intensity throughout the cyclone increases the wear life of the liners.

With less turbulence, it reduces wear and extends the lifespan of the product. In addition, the minimal flow resistance allows Cavex hydrocyclones to process a much higher slurry volume than typical hydrocyclones with equivalent conditions

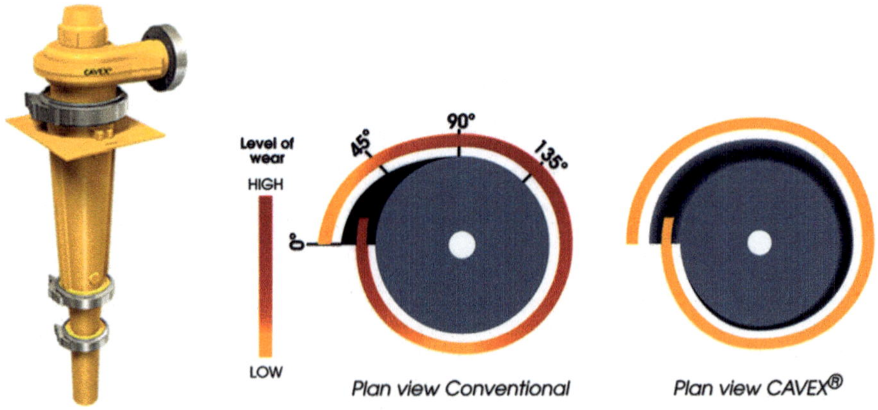

Fig. 8.6 Cavex hydrocyclone showing the spiral inlet geometry

and producing a significantly finer product than the standard hydrocyclone. The Cavex hydrocyclones is equipped with ceramic spigots reducing the bypass of fines in comparison with Krebs gMax, see Fig. 8.6.

Olson and Ommen [10] designed a new spiral inlet for a 250 mm Krebs hydrocyclone using 3D CFD, see Fig. 8.7. The goal was to eliminate or spread this wear pattern over the entire inlet.

Figures 8.8 shows a reduction in wear witween the standard and new inlet design and Fig. 8.9 shows effective wear patern for both design (Fig. 8.10).

The performance of the standard and the gMAX inlet designs are shown in Fig. 8.11 The simulation was performed with a 89 mm vortex finder and 70 kPa pressure drop. The modeled performance of the new inlet design presented much smaller recovery to the underflow.

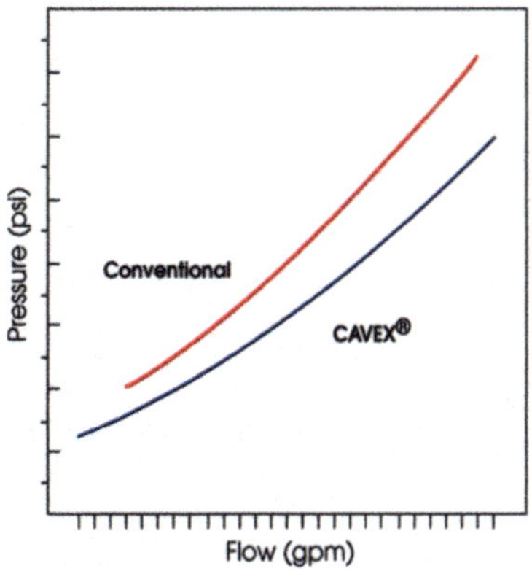

(a) Capacity versus inlet pressure

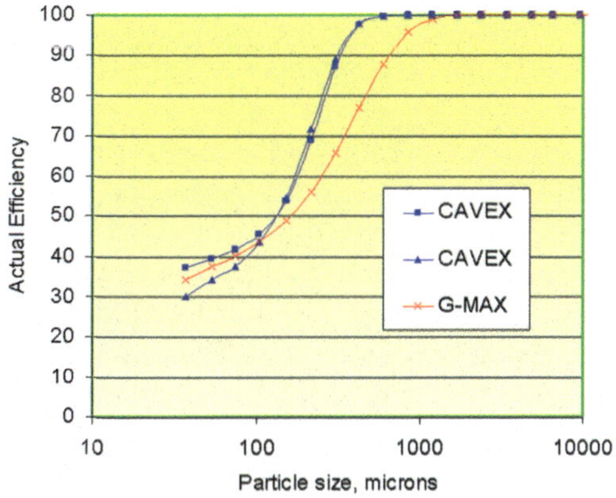

(b) Cavex versus gMax.

Fig. 8.7 Capacity versus pressure and selectivity of the new design at Mantos Blancos [3]

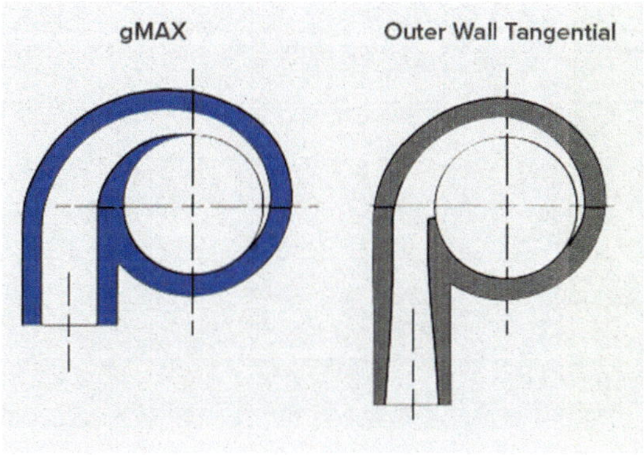

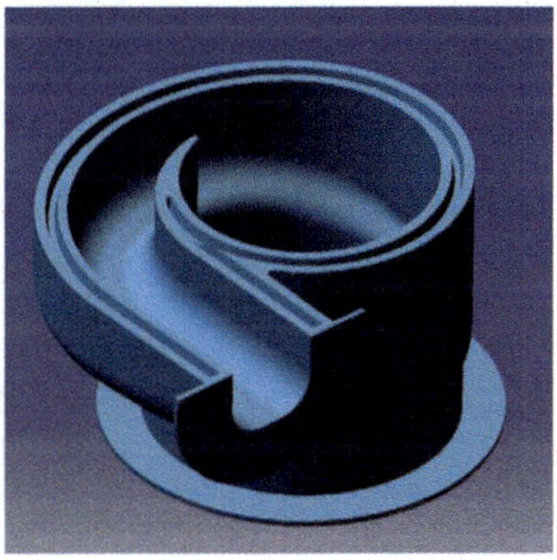

Fig. 8.8 A 3D model of inlet for a 250 mm Krebs gMAX [7, 10].

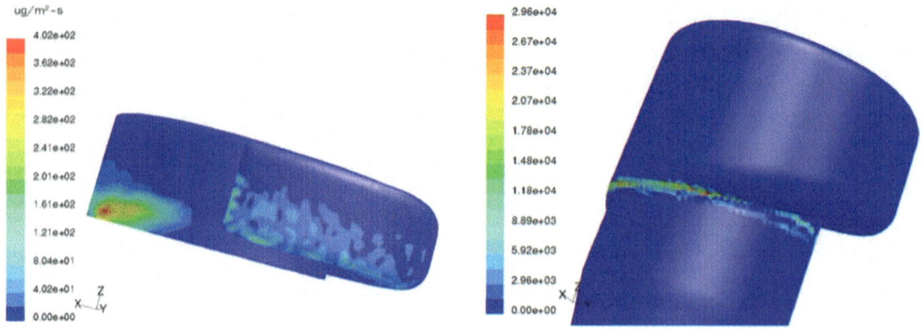

a) CFD erosion for the standard inlet b) CFD erosion for the new inlet design

Fig. 8.9 Erosion in the top part of the hydrocyclone

Fig. 8.10 The resulting wear patterns showed a dramatic reduction in wear on the gMAX inlet head liner as a result of its improved geometry

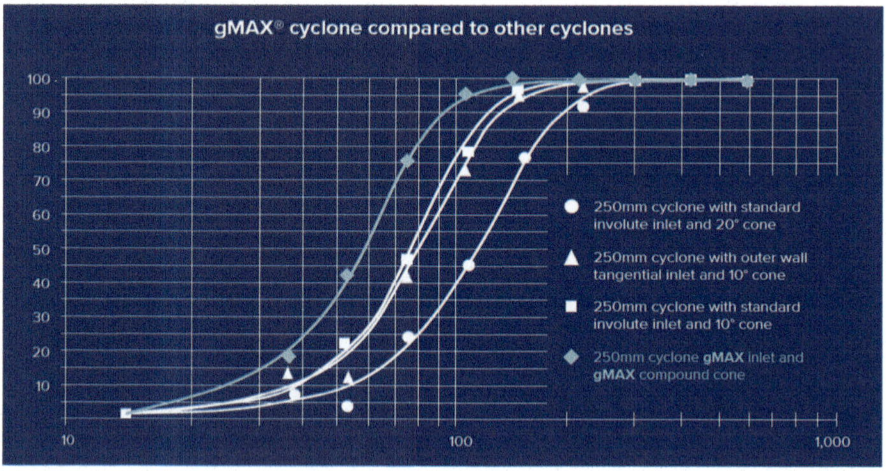

Fig. 8.11 Comparison of classification curves for gMAX hydrocyclone and other hydrocyclones

In 1961 Professor Helmut Trawinski as Technical Manager of Dorr Oliver G.m.b.H. design Dorr Oliver process equipment such as hydrocyclones, hydro classifiers, centrifuges and thickeners. Leaving this company in 1961, and a short stay in Sharples, a centrifuge business, joint the German firm Amberger Kaolin Werke, AKW where he began the design and construction of new hydrocyclones constructed in polyurethane, a new elastomer recently developed by the German firm Bayer, with the name of "Vulkollan." He designed new hydrocyclones with new geometries, highlighting mainly the tangential entrance in 180° and rectangular inlet section to promote laminar flow. Until relatively few years ago the hydrocyclone feed entry section in almost all manufacturers was circular, while today all new hydrocyclones available on the market have rectangular sections following the guidelines set by AKW (Fig. 8.12).

Davailles et al. focused on hydrocyclone design and operation under high solids concentrations, up to 50% solids. They designed and operated a 100 mm hydrocyclone with a tangential inlet pipe decreasing in size at the entrance of the hydrocyclone (Fig. 8.13).

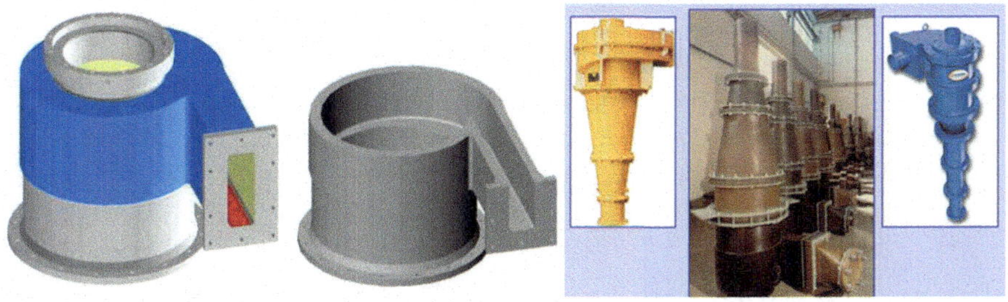

Fig. 8.12 Hydrocyclones with involute feed part and inlet rectangular section

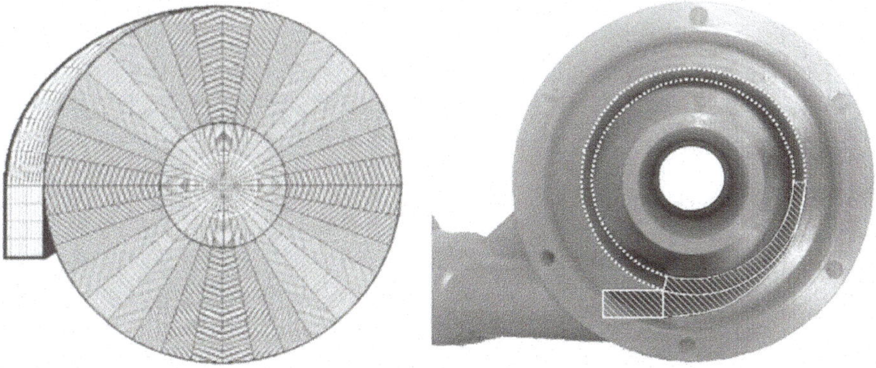

(a) Cross-section view of the mesh (b) Tangential inlet pipe with sudden reduction

Fig. 8.13 Neyrtec's HC100, 100 mm diameter hydrocyclone injection design

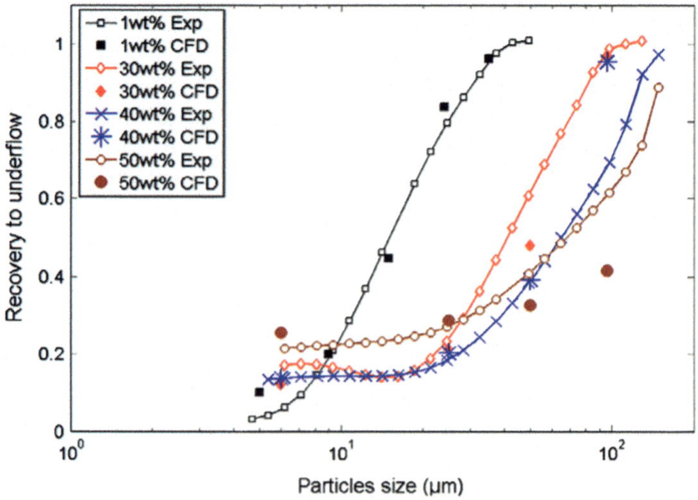

Fig. 8.14 Selection functions for various feed solids concentrations

By varying feed solids concentrations and apex diameters, several distinct operating regimes were produced inside the hydrocyclone (Fig. 8.14).

Dense regions can form inside the hydrocyclone, mainly toward the bottom part of the conical section, which dissipate energy through viscous effects.

8.3 Vortex Finder Length and Apex-to-Vortex Ratio

Fernandez et al. studied the effect of the vortex finder tube length on the efficiency of classification in a 10-cm hydrocyclone for a cut size of 10 μm (Fig. 8.15).

They found that a value of one tenth of the total hydrocyclone length was the most efficient and that the least efficient configurations were those with a vortex finder bottom near the juncture between the cylindrical and the conical part, where high turbulence may occur. This is in accordance with predictions.

Apex-to-Vortex Ratio

Kilavuz and Gülsoy (2011) analyzed the effect of the vortex-to-apex-diameter ratio on the cut size d_{50} and bypass for small hydrocyclones d_o/d_u. Figure 8.1 shows that increasing d_o/d_u increases the cut size for a given inlet pressure and increases cut size if the pressure is increased and Fig. 8.11b indicates the effect on the solid bypass and the water recovery (Fig. 8.16).

They found that the bypass of solids was significantly higher than either the recovery of water to the underflow or volumetric recovery of feed slurry to the underflow. These results contrast with the models usually used in small diameter hydrocyclones.

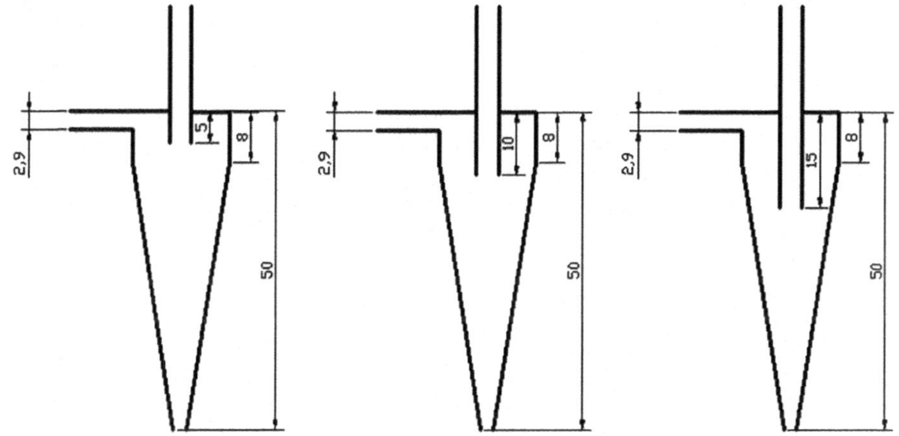

Fig. 8.15 Three vortex finder lengths in a 10 cm hydrocyclone (dimensions given in cm)

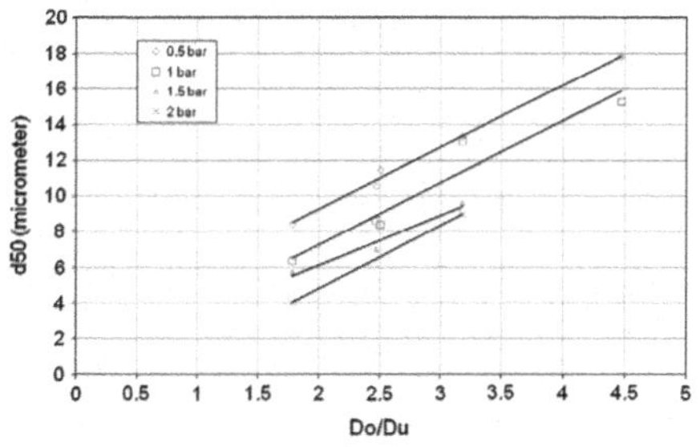

(a) Effect on the cut size

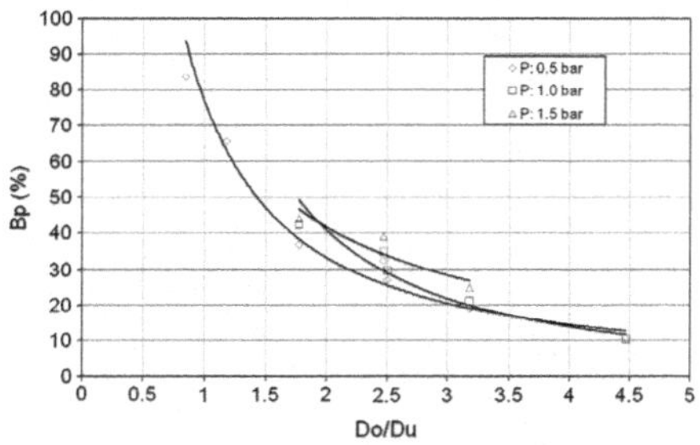

(b) Effect on the solid bypass

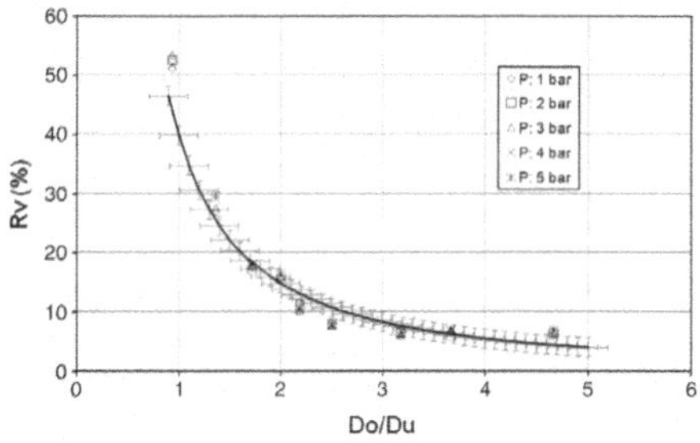

(c) Effect on pulp recovery

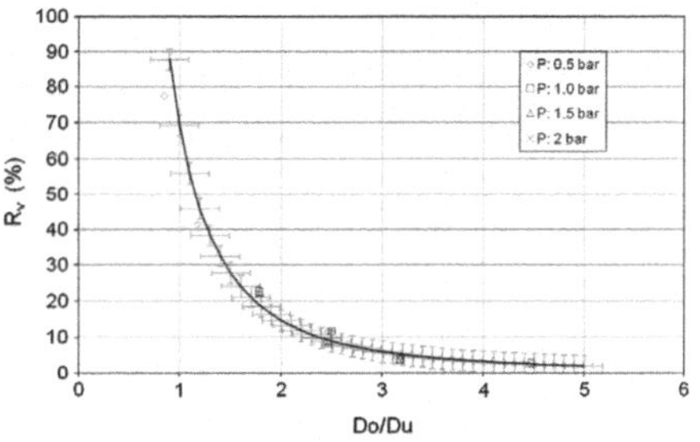

(d) Effect on water recovery

Fig. 8.16 Effect of the vortex-to-apex ratio on the performance of a 50.8 mm hydrocyclone

Desai and Praveen studied six mm diameter hydrocyclone models with combinations of 20° and 26° cone angles and 40, 50 and 65 mm apex diameters that were designed, manufactured and tested for performance evaluation. The models were designed for the removal of particulate matter 48 μm and larger from irrigation water. The authors observed that the cone angle and the apex diameter were the most important design of these hydrocyclones.

As the apex diameter increases, the underflow also increases, carrying more particulate matter to the collection basket. If the apex is greater than the vortex, the hydrocyclone will drain itself through the underflow and perform no separation function. However, if the underflow apex is too small, larger solid particulate matter will accompany smaller particulate matter through the overflow.

8.4 Cylinder Length and Cone Angle

Using computational fluid dynamics (CFD) analysis, guided then design of Krebs g-MAX hydro cyclone with sharply angled upper cones followed by longer and smaller angled lower cones. This combination maximizes tangential velocity in the upper part of the cyclone. It also provides a long residence time in the critical separation zones of the cyclone's lower section. The result is a substantially finer separation, with fewer fines in the underflow and less coarse material bypass to overflow [7] (Figs. 8.17 and 8.18).

Desai and Praveen found that, the increase in cone angle from 20° and 26° increases the inlet pressure necessary to maintain a flow rate of 25 m^3/h with a cut size of 48 μm, for a 198 mm diameter hydrocyclone used for irrigation.

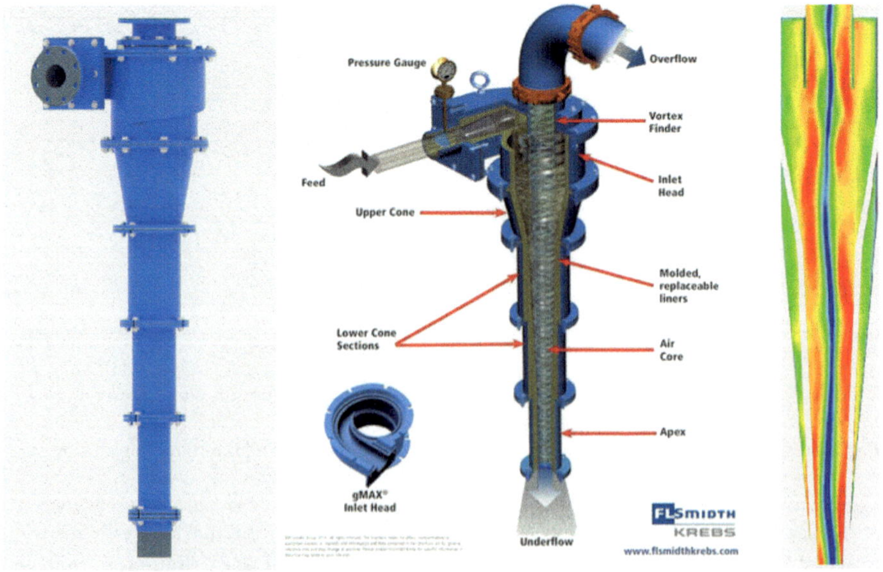

Fig. 8.17 CFD showed the convenience of changing the shape of the hydrocyclone cones [7]

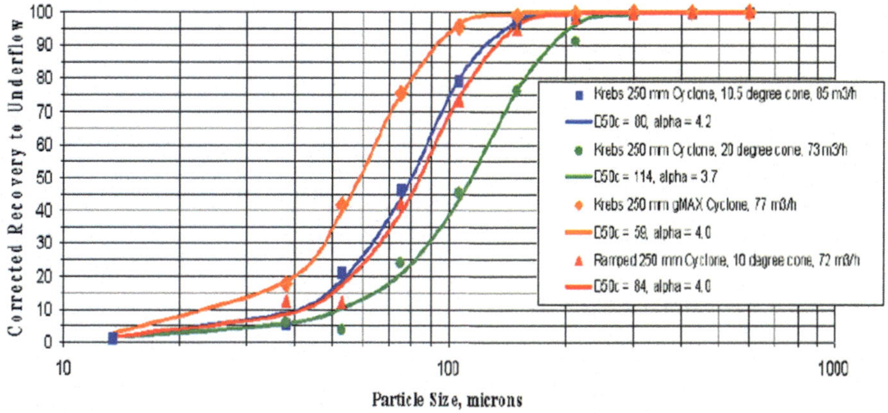

Fig. 8.18 Comparison of the Krebs gMAX classification curve with other Krebs hydrocyclones [7]

Ghodrat et al. presented a numerical study of hydrocyclones with high feed solids concentrations, up to 30% solids by volume, for different conical configurations. The numerical results show that the cyclone performance is sensitive to both the length and shape of the conical section, as well as the feed solids concentration. The separation efficiency of the hydrocyclones decreases as feed solids concentration increases when the conical section length is long enough, while when the conical section length is too short, the separation efficiency initially drops drastically and then varies within a relatively narrow range (see Fig. 8.19).

Chica et al. [6] addressed the phenomena in a hydrocyclone from an energy perspective. From a mechanical energy balance the main results with Newtonian suspensions showed that, although energy dissipation in the discharge of a hydrocyclone is very small compared with the dissipation in the hydrocyclone body and therefore does not control the behavior of the hydrocyclone, it provides sufficient

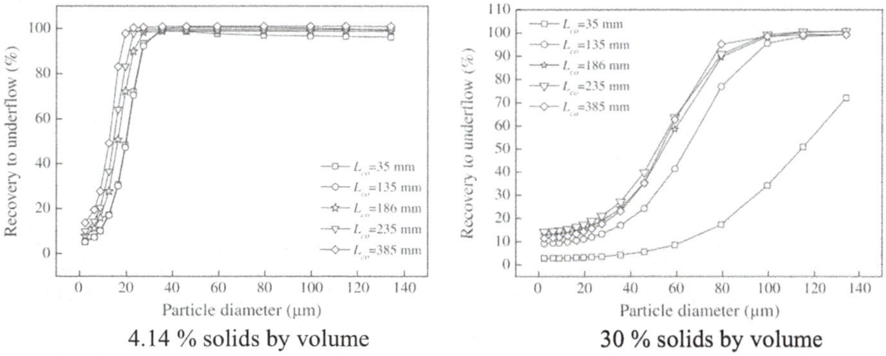

Fig. 8.19 Effect of conical section length on selectivity at different solids concentrations

information to describe the discharge shape in terms of dimensionless variables, such as the geometric relationship among the apex and vortex, the flow partition, the air core diameter and the discharge angle (Fig. 8.20).

The normalized air core, that is, the air-core-to-apex-diameter ratio d_{air}/d_a, depends strongly on the apex-vortex ratio, d_a/d_v. For values of $d_a/d_v < 0.3$ the discharge is roping regardless of the operating condition. Similarly, for $d_a/d_v > 0.8$ the discharge is spray. In the interval of $0.34 > d_a/d_v > 0.80$, the underflow shape depends on the apex-vortex and the underflow-overflow ratios. Figure 8.21 shows that for $0.5 > Q_u/Q_o$ we always have rope and for $Q_u/Q_o > 0.5$ we always have spray, regardless of the apex energy dissipation (Fig. 8.22).

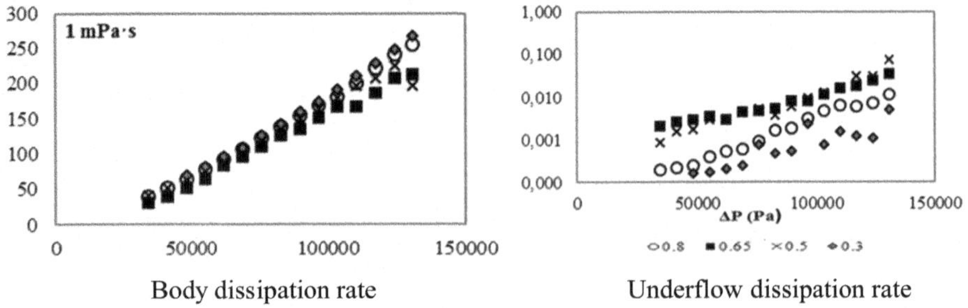

Fig. 8.20 Energy dissipation in the hydrocyclone body and discharge against the inlet pressure for a hydrocyclone 100 mm in diameter and 750 mm in length

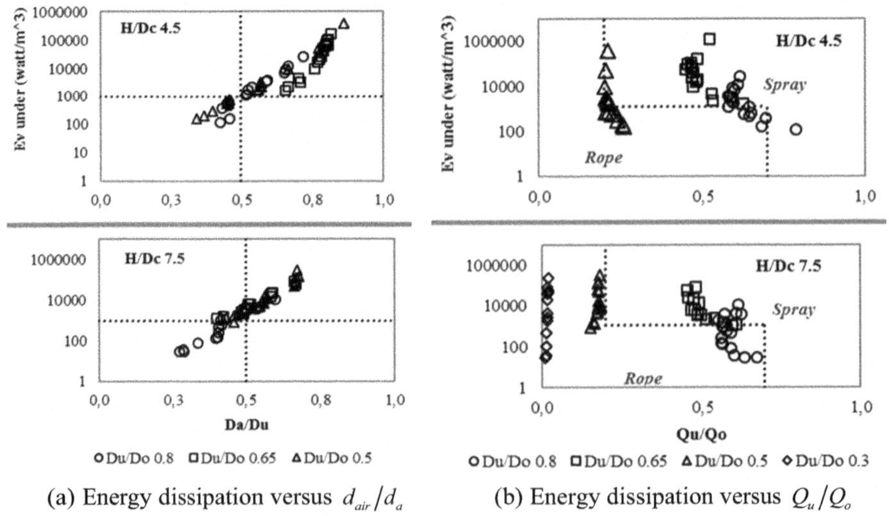

Fig. 8.21 Energy dissipation related to the ratios between air core diameter and apex diameter, d_{air}/d_v, and the flow split ratios, Q_u/Q_o

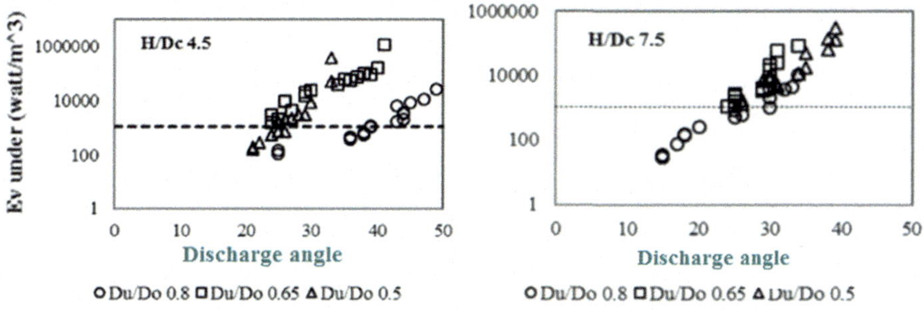

Fig. 8.22 Relationship between the discharge angle and underflow dissipation energy for two H/d_c

In 2016 Ghodrat et al. used a two-fluid numerical model to present the reduced classification functions for pulps with particles of different densities in a 1000 mm hydrocyclone at two solids concentrations (Fig. 8.23).

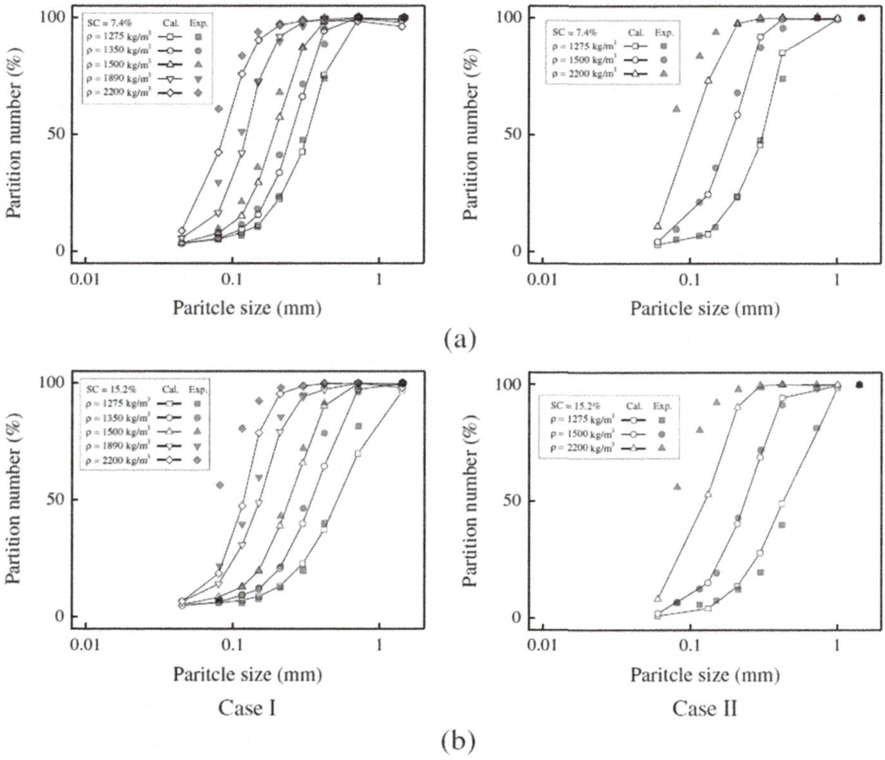

Fig. 8.23 Comparison of the simulated and measured partition curves of particles in 1000 mm hydrocyclones at solids concentrations of **a** 7.4% and **b** 15.2%

8.5 Apex Design

Using CFD, Lanyue et al. [8] studied the short-circuit of a conical and four cylindrical hydrocyclones with five different apex configurations, namely A to D, in Fig. 8.24.

The result of this study showed that the conical hydrocyclone (A) has the highest short-circuit to the underflow while al cylindrical hydrocyclones have about halve that value. With respect to the short-circuit to the overflow the tendency is that the conical hydrocyclone has the minimum value and the traditional flat bottom hydrocyclone (D) has the highest and the lowest sharpness of separation (Fig. 8.25).

To validate the CFD simulations, he used Hsieh and Rajamani measurements using water as the fluid with a hydrocyclone with the same geometry and operating conditions as Hsieh and Rajamani. The results shown in Fig. 8.26 show an error of less than 5% in the tangential velocity.

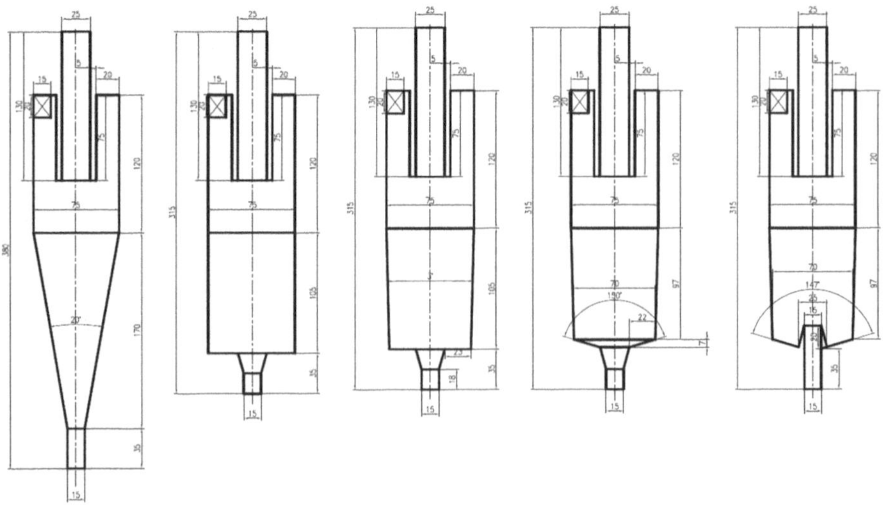

(a) type A (b) type B (c) type C (d) type D (e) type E

Fig. 8.24 Several apex design

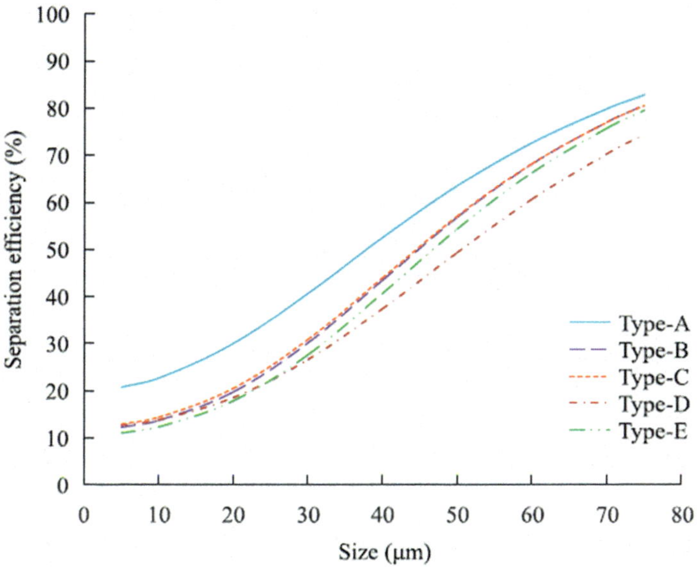

Fig. 8.25 Separation efficiency of the hydrocyclones with 5 apex design

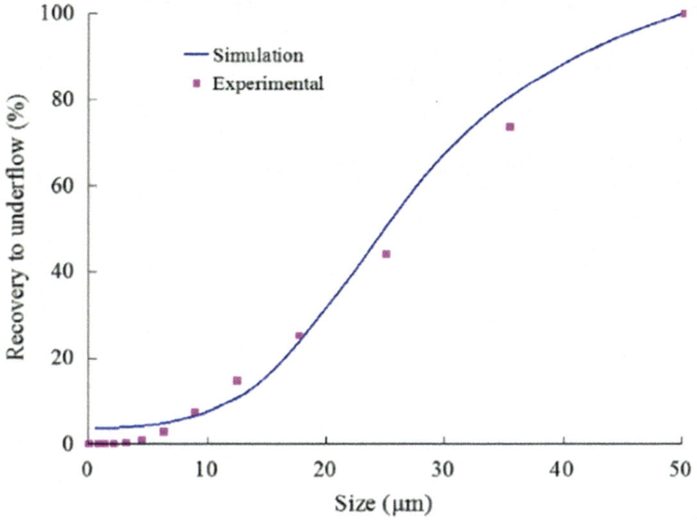

Fig. 8.26 Comparison of the measured and simulated classification curves

References

1. Álvarez C (1995) Alternative apex design for Krebs 150 mm hydrocyclone. Engineering thesis, University of Concepción (in Spanish)
2. Castro O (2002) Cyclone separator having an inlet head US Patent 8104622 B2
3. Castro O (2003) Application of a new generation of Hydrocyclones at La Coipa: a Placer Dome Gold Plant in Chile. In: 1st international symposium on hydrocyclones, Hydrocyclones 03, Mount Nelson Hotel, Cape Town, South Africa, 24–26 Sept 2003
4. Concha F, Castro O (1995) FONDEF Project MI-08, Design of Classification equipment for the national industry. Cooperative Research U. de Concepción - CIMM, Financing of FONDEF-Codelco
5. Concha F, Barrientos A, Muñoz A, Castro A (1996) Phenomenological model of a hydrocyclone. In: Hydrocyclone-96 Conference, Cambridge UK, 2–4 April 1996
6. Chica L, Bustamante O, Barrientos A (2013) Underflow mechanical energy dissipation in hydrocyclones: new modeling approach, Dyna, Medellín, Colombia, N° 181, pp 136–145. ISSN 0012-7353 (in Spanish)
7. Krebs 04–204-gMax brochure. (PDF)
8. Lanyue J, Liu P, Zhang Y, Yang X, Wang H, Gui X (2019) Design layer structure for improving the particle separation performance of a hydrocyclone. Powder Technol 350:1–14
9. Mora H (1994) Similarity in hydrocyclone. Engineering thesis, University of Concepción (in Spanish)
10. Olson TJ, Ommen RV (2004) Optimizing hydrocyclone design using advanced CFD model. Miner Eng 17:713–720
11. Véjar JA (1996) Design of the vortex finder of an industrial hydrocyclone. Engineering thesis, University of Concepción (in Spanish)
12. Yang J, Sun G, Zhan M (2015) Prediction of the maximum-efficiency inlet velocity in cyclones. Powder Technol 286:124–131
13. Soto C (1995) Design of the inlet of hydrocyclones. In: Design of new classifier equipment for the mining industry. Engineering thesis, University of Concepción (in Spanish)
14. Trawinski, H. 1961; 1963
15. Zapata MA (1996) Design of the inlet of an industrial hydrocyclone. Engineering thesis, University of Concepción (in Spanish)

Chapter 9
Hydrocyclone Selection

Abstract There are two types of empirical models: so-called *design* or *selecting* models and *simulation* models. The first is used to select commercial hydrocyclones that fulfill given capacity and classification conditions and the second, based on experimental work, is used to determine the capacity, split size, short circuit and classification function parameters that permit the simulation of a hydrocyclone or various hydrocyclones.

9.1 Arterburn Hydrocyclone Selection Model

The best-known hydrocyclone design model is *Arterburn's* model, also known as the *Krebs* model. Arterburn [1] developed an empirical model that estimates the size and number of hydrocyclones required for a given comminution-classification operation. The method was developed for the normalized and standardized geometry of the Krebs hydrocyclone based on the company's experience with pilot and industrial operations. They defined the *normal hydrocyclone* geometry as follows:

$$
\begin{aligned}
&\text{Normal geometry:}\\
&\qquad A_F = 0.5d_c^2\\
&\qquad d_O = 0.35d_c\\
&\qquad L = d_c\\
&\qquad 10^\circ < \alpha < 20^\circ\\
&\text{Normal operating conditions:}\\
&\qquad \text{Fluid: water at } 20\,^\circ\text{C}\\
&\qquad \text{Solid spherical particles density } \rho_s = 2.65\,\text{g}/\text{cm}^3\\
&\qquad \text{Solid feed volume fraction } \varphi < 0.01\\
&\qquad \text{Pressure drop } \Delta p = 70 \text{ kPa (10 psi)}
\end{aligned}
\tag{9.1}
$$

F. Concha A. and J. L. Bouso A., *Fluid Mechanics Fundamentals of Hydrocyclones and its Applications in the Mining Industry*, Fluid Mechanics and Its Applications 126,
https://doi.org/10.1007/978-3-030-67913-2_9

where A_F is the inlet area in cm^2, d_c is the hydrocyclone diameter, d_O and d_U are the vortex and apex diameters and L is the length of the cylindrical part of the hydrocyclone, all in cm, and α is the angle of the conical portion.

Arterburn proposed an empirical equation to relate the hydrocyclone diameter, necessary to obtain a normalized d_{50}^n size (particle size under normal operating conditions), to the actual cut size x_s by introducing corrections to account for the actual operating conditions. Finally, he proposed an equation that gives the capacity of a hydrocyclone according to its diameter and operating conditions. The design to meet the target consists of seven stages: (1) selection of the d_{50} from the desired cut size x_s, (2) calculation of the normalized d_{50}, (3) calculation of the hydrocyclone diameter, (4) selection of the apex and vortex diameters, (5) calculation of the hydrocyclone capacity, (6) calculation of the necessary number of hydrocyclones and (7) simulation of the overflow particle size distribution.

(1) **Estimation of d_{50}^n.**

Based on Arterburn's experimental data, the following graph and equation gives the relationship between d_{50}^n and x_s (Fig. 9.1):

(2) **Estimation of actual d_{50} from the normalized d_{50}^n.**

Since the actual material properties and operating conditions are different from normalized conditions, d_{50} must be calculated from the normalized d_{50}^n:

$$d_{50} = x_s \times (15.102 - 3.161\ln(F_{80}(x_s)))$$

$$d_{50}^n / d_{50} = f_1(w) \times f(\Delta p)_2 \times f_3(\Delta\rho)$$

$$f_1(\varphi) = \left(\frac{100}{100 - 1.9\varphi_F}\right)^{1.43}, \quad \varphi_F \text{ in \% solids}$$

$$f(\Delta p)_2 = (22.55\Delta p)^{-0.28}, \quad \Delta p \text{ in psi}$$

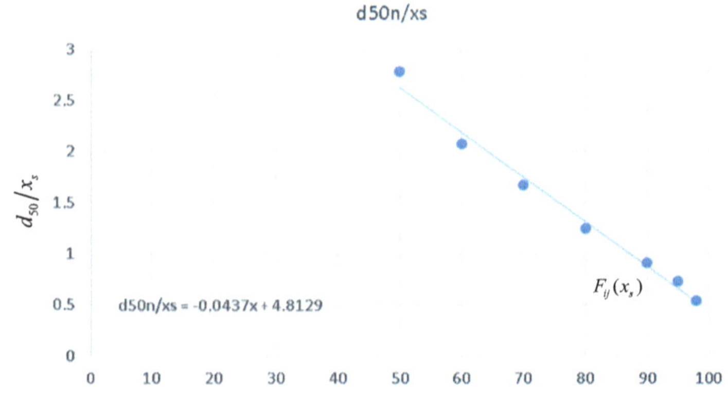

Fig. 9.1 $F_{ij}(x_s)$Relationship between d_{50}^n and x_s according to Arterburn's experimental data

$$f_3(\Delta\rho) = \left(\frac{1.65}{\rho_s - \rho_f}\right)^{0.50}, \quad \rho_i \text{ in ton}/\text{m}^3 \tag{9.2}$$

(3) **Hydrocyclone diameter**

The hydrocyclone diameter is a function of the normalized d_{50}^n

$$d_c = 2.06\left(d_{50}^n\right)^{1.516} \tag{9.3}$$

where d_{50}^n is given in microns, while d_c is in centimeters.

(4) **Hydrocyclone capacity**

The hydrocyclone capacity, expressed in m^3/s of inlet pulp, is given in terms of the hydrocyclone diameter and inlet pressure:

$$Q_c = 2.3 \times 10^{-6} \times (d_c)^{2.0351} \times (\Delta p)^{0.5} \quad \Delta\text{p in psi}$$

where d_c is given in cm and Δp in psi.

(5) **Determination of the required number of hydrocyclones**

The number of hydrocyclones necessary to meet the desired production depends on the total pulp feed flow rate and the capacity of each hydrocyclone. Arterburn comments that the calculated capacity of each hydrocyclone is based on water; therefore, it is conservative.

$$N_c = \frac{Q_F}{Q_c} \tag{9.4}$$

(6) **Selection of the underflow concentration**

Arterburn suggests that the pulp in discharge should have a concentration $0.50 < \varphi_u < 0.53$ for a grinding circuit and $0.40 < \varphi_u < 0.45$ for a regrinding circuit. Mular and Jull [9.18] established a restriction on the concentration of the underflow to avoid roping:

$$\varphi_U < 0.49 + 0.54\varphi_o \tag{9.5}$$

(7) **Selection of apex and vortex diameters**

The apex diameter is a function of the overflow flow rate:

$$d_U = 2.62 Q_u^{0.447}, \; d_O = 0.35 d_c \tag{9.6}$$

[2] presented an equation defining the range of d_U/d_O where roping can occur:

$$0.34 < d_U/d_O < 0.5 \tag{9.7}$$

Example of hydrocyclone design according to Arterburn

Consider the grinding-classification circuit shown in Fig. 9.2. Select the size and number of hydrocyclones to meet the requirement of the feed coming from a rod mill at 350 t/h (1), 300% circulating load and operating conditions given in Table 9.1 (Tables 9.2 and 9.3).

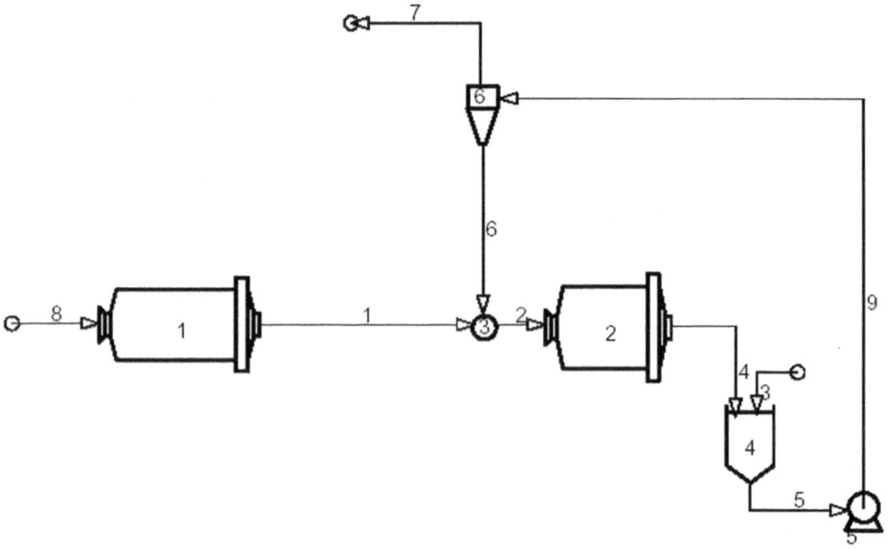

Fig. 9.2 Grinding-classification circuit

Table 9.1 Alternatives

Direct circuit	Designation; Units	Alternatives			
Data		1	2	3	4
Feed flowrate	F_f t/h	350	350	350	350
Cut size	x_s mm	150	150	150	150
	F(xs) %	80	80	80	80
Solid density	Ps kg/m^3	2650	2650	2650	2650
Liquid density	Pf kg/m^3	1000	1000	1000	1000
Assumed underflow concentration	w_D % solid by weight	70	70	70	70
Desired overflow concentration	w_R % solid by weight	30	30	30	30
Assumed feed pressure	D_p psi	5	10	15	20
Circulating load	cc %	300	300	300	300
Restriction	$0.34 < dD/dR < 0.50$				

Table 9.2 Available hydrocyclones

Available hydrocyclones		Sizes
Krebs	[in]	D-4, D-6, D-10, D15, D20, D-26, D-30, D-50
Dorrclone FR Heavy duty	[in]	4, 6, 12, 18, 24, 36, 48
Warman	[in]	C-4, C-6, C-10, C15, C20, C-26
Weir Cavex	[mm]	40, 100, 150, 250, 400, 500
ERAL	[mm]	10, 20, 40, 50, 75, 100, 150, 200, 250, 325, 400, 500, 625, 750, 1000, 1250
Mozley	[mm]	10, 22, 44, 60, 75, 100, 125, 150, 250

9.2 Hydrocyclone Simulation Models

The simulation of a hydrocyclone entails the calculation of the separation size, classification function, short circuit and particle size distribution of the under and overflow.

When the hydrocyclones of an operating plant are simulated, the number of hydrocyclones and their dimensions, as well as the feed to the plant, are specified. The variables in this case are: mill capacity F_1 (1 and 8 in Fig. 9.2) and the particle size distribution of the feed. The simulation gives the recirculation load and the product size distribution, the pressure drop Δp and flow through the apex and vortex of the hydrocyclone. The results of these simulations generally show that the magnitude of the changes that can be achieved by maintaining a pressure drop in the range of $\Delta p_{\max}/\Delta p_{\min} \approx 4$ and a concentration in the correct range is restricted.

The best-known hydrocyclone simulators are the Lynch and Rao and Plitt simulators and their variations developed by different researchers. These simulation models consist of four empirical equations that express the (a) capacity, (b) separation size, (c) water balance and (d) classification function.

Experimental data

Experimental data were obtained from a mass balance around a hydrocyclone with a given geometry to be used with all simulation models. From the mass balance and the particle size distribution the short circuit "a" and the parameters d_{50} , d_{25} and d_{75} were obtained; they are given in Table 9.4 (Table 9.5).

$$x_n - \frac{x_n - x_{n+1}}{c_n - c_{n+1}} \times (0.50 - x_n),\ x_{n+1} + \frac{x_n - x_{n+1}}{c_n - c_{n+1}}(c_n - 0.5).$$

Table 9.3 Mass balance around the hydrocyclone

Mass balance							
Feed							
Solid flowrate	F_f	t/h	O + U	1,050.0	1,050.0	1,050.0	1,050.0
Mass concentration	w_F	% en peso	FF/(D/xD + R/xR)	48.5	48.5	48.5	48.5
Volume concentration	j_F	% en volumen	100pfwF /(ps(100-wF) + pfwF)	26.2	26.2	26.2	26.2
Pulp density	ρ_F	kg/m^3	100rsrf /(rs(100-wA) + rfwA)	1,432.1	1,432.1	1,432.1	1,432.1
Water flowrate	Q_{fw}	t/h	QRAg + QDAg	1,116.7	1,116.7	1,116.7	1,116.7
Mass flowrate	F*	t/h	A + QAA	2,166.7	2,166.7	2,166.7	2,166.7
Volume flowrate	Q^*_f	m^3/s	1000A/pF/3600	0.420	0.4	0.4	0.4
Overflow							
Solid flowrate	F_o	t/h	O	350.0	350.0	350.0	350.0
Mass concentration	w_o	% en peso	Wo	30.0	30.0	30.0	30.0
Volume concentration	jo	% en volumen	100p f_{wo} /(p $_s$(100-w_o) + $\rho_f$wo)	13.9	13.9	13.9	13.9
Pulp density	ρ_o	kg/m^3	100$\rho_s$pf /(ρ_s(100-wo) + $\rho_f w_o$)	1,229.7	1,229.7	1,229.7	1,229.7
Water flowrate	Q_{OW}	t/h	R/w_o(100-w_o)	816.7	816.7	816.7	816.7
Mass flowrate	o*	t/h	A + Qow	1,166.7	1,166.7	1,166.7	1,166.7
Volume flowrate	Q^*_o	m^3/s	1000Xo/pR/3600	0.3	0.3	0.3	0.3
Underflow							
Solid flowrate	u	t/h	((CC-100)/100)xo	700.0	700.0	700.0	700.0
Mass concentration	w_u	% en peso	wu	70.0	70.0	70.0	70.0
Volume concentration	j_u	% en volume	j = ρ_fw/(ρ_s(100-w) + ρ_fw)	46.8	46.8	46.8	46.8
Pulp density	ρ	kg/m^3	100$\rho_s$pf /(ρ_s(100wu) + pfwu)	1772.6	1772.6	1772.6	1772.6
Water flowrate	Q_{uW}	t/h	D/w_U(100-w_U)	300.0	300.0	300.0	300.0
Mass flowrate	Q^*_u	t/h	D + Q_{uw}	1000.0	1000.0	1000.0	1000.0

(continued)

Table 9.3 (continued)

Mass balance

Volume flowrate	Q^*_u	m^3/s	$Q^*_u = u^*/pu$	0.157	0.157	0.157	0.157

Results

Separation size	dso Mm	d50 = Xs(15.102–3.161Ln(F (Xs))	187.0	187.0	187.0	187.0
Factor 1	fl(jA)	$(53/(53\text{-}9f))^{143}$	2.650	2.650	2.650	2.650
Factor 2	f2(Dp)	$3.27(6.897Ap)^{-028}$	1.213	0.999	0.892	0.823
Facto 3	f3(Ap)	$(1650/(ps\text{-}1000))^{05}$	0.932	0.932	0.932	0.932
Actual particle size	d^n50 Mm	d^n50 = Xs/(fl(jF)Xf2(Ap)Xf3(Ap))	62.4	75.8	84.9	92.0
Hydrocyclones						
Hydrocyclone Diameter	dc cm	$dc = 0.206(d^n50)^{1515}$	108.1	145.0	172.2	194.6
Hydrocyclone Diameter	dc in	$dc = 0.0811(d^n50)^{1515}$	43	57	68	77
Capacity of one hydrocyclone	Q^*_{Fc} m^3/s	$Q_{fc} = 2.3 \times 10^{-6}(d_c)^{20351}\ (Ap)^{05}$	0.0708	0.4073	0.5781	0.7410
Number of hydrocyclones	N	Q*Fc	5.93	1.03	0.73	0.57
Number of hydrocyclones	N (Integer + 1)	Q*F/Q*Fc	6	2	1	1
Underflow diameter	dy cm	$d_U = 2.62(1000xQ_U)^{0447}$	11.3	18.4	25.1	25.1
Overflow diameter	d_O cm	$d_O = 0.35d_c$	37.83	50.76	60.28	68.11
Roping is formed if		0.34 < dy/do < 0.50	0.30	0.36	0.42	0.37

Hydrocyclone selection

Diameter	cm	dc	150.00
Diameter	in	dc	59
Normalized cut size		d^n50	0.4073
Cut size		d50	1.03
Feed diameter	cm	dF	37.5
Underflow diameter	cm	du	20
Overflow diameter	cm	do	52.5
Numer of hydrocyclones	–	N	2.00

Table 9.4 Particle size analysis, selectivity, classification and short circuit

	mm		Feed			Underflow					Overflow					
Mesh	Size	Average	t/h	fF(x)	Ff(x)	Rf(x)	t/h	f„(x)	Fu(x)	Ru(x)	t/h	fO(x)	Fo(x)	Ro(x)	S(x)	C(x)
28	600	505			100.00				100.00				100.00			
35	425	357	2.12	3.49	96.51	3.49	2.12	8.60	91.40	8.60	0.00	0.00	100.00	0.00	1.00	1.00
48	300	252	1.68	2.76	93.74	6.26	1.68	6.80	84.60	15.40			100.00	0.00	1.00	1.00
65	212	178	3.03	4.99	88.76	11.24	2.81	11.40	73.20	26.80	0.22	0.60	99.40	0.60	0.93	0.91
100	150	126	5.90	9.72	79.03	20.97	3.92	15.90	57.30	42.70	1.98	5.50	93.90	6.10	0.66	0.58
150	106	89	10.27	16.91	62.13	37.87	4.86	19.70	37.60	62.40	5.37	14.90	79.00	21.00	0.47	0.34
200	75	63	12.41	20.44	41.68	58.32	3.63	14.70	22.90	77.10	8.79	24.40	54.60	45.40	0.29	0.11
270	53	45	8.75	14.41	27.27	72.73	2.27	9.20	13.70	86.30	6.48	18.00	36.60	63.40	0.26	0.07
270	38	19	16.56	27.27		100.00	3.38	13.70		100.00	13.18	36.60		100.00	0.20	0.00
	10	3				100.00				100.00				100.00		
	1	TOTAL	60.72	100.00			24.67	100.00			36.01	100.00				0.20
			wf =	48.60			w	65.50			w	41.30				
d25								78.84								
d50								113.88								
d75								153.02								

Table 9.5 Hydrocyclone material and operational information

	Units	Feed	Underfow	Overflow		Design	Units	Data	Units	Data
Solid mass flow	t/h	60.72	24.67	36.01	Solids density		g/cm^3	2.65		
Solids percent	%w/w	48.60	65.50	41.30	Liquid densoty		g/cm^3	1.00		
Volume fraction	–	0.26	0.42	0.21	Hydrocyclone diameter	do	in	20.00	mm	628.00
Pulp Density	g/cm3	1.43	1.69	1.35	Length of the vortex finder	h	in	13.40	mm	340.36
Water mass flow	t/h	64.22	12.99	51.18	Hydrocyclone feed diameter	dp	in	5.00	mm	127.00
Pulp mass flow	t/h	124.94	37.66	87.19	Hydrocyclone apex diameter	du	in	2.50	mm	63.50
Pulp volume flow	m^3/h	87.13	22.30	64.77	Hydrocyclone vortex diameter	do	in	5.00	mm	127.00
Volumen pulpa	l/min	1,452.19	371.73	1,079.50	Length of the vortex finder	h	in	13.40	mm	340.36
Volumen agua	l/min	1,070.30	216.57	853.02	Operating pressure	P	psi	8.00	Pa	55,158
Cone angle	20									

9.2.1 Lynch and Rao Simulation Model

The Lynch and Rao simulation model [5] requires a set of experiments to calculate parameters K_1, K_2 and K_3 and the sharpness index $S.I.$, which is dependent on the hydrocyclone size and type and the material to be treated. The required data are: (a) apex and vortex diameter, (b) feed volume flow, (c) feed pressure, (d) weight fraction of solids in the feed, (e) overflow mass rate, (f) solids concentration by weight in the overflow, (g) overflow sieve analysis and (h) discharge particle size analysis.

(a) **Capacity-pressure relationship**

$$Q[\ell/\mathrm{min}] = K_1 d_v \Delta p^{0.5} (1 - w_F)^{0.125} \tag{9.8}$$

where Q and w_F are the volume flow rate and feed concentration solids percentage, d_O is the vortex diameter in [cm], Δp [psi] is the pressure drop and K_1 is a constant.

(b) **Mass flowrate of water in the overflow**

$$R_w[\mathrm{t/h}] = K_2 + 1.07(100 - w_F) - 3.95 d_U \tag{9.9}$$

where d_a is in [cm].

(c) **Separation size**

$$\log d_{50}(\mu m) = K_3 + \frac{d_O}{2.6} + \frac{d_U}{3.5} + \frac{\Delta p}{10.7} - \frac{R_w}{52} \tag{9.10}$$

where the diameters are in [cm].

(d) **Classification function**

$$c(x) = \frac{\exp[\lambda(x/d_{50}) - 1]}{\exp[\lambda(x/d_{50}) + \exp(\lambda) - 2]} \tag{9.11}$$

where λ is given in the implicit equation for the sharpness index, $S.I.$

$$S.I. = \frac{\ln[\{\exp(\lambda) + 2\}/3]}{\ln[3\exp(\lambda) - 2]} \tag{9.12}$$

Table 9.6 Results

Parameters		Particle size	Exper	Simulation
K1	24.7	357.07	1.00	1.00
K2	21.27	252.19	1.00	0.98
K3	-0.777	178.33	0.91	0.87
a	0.2	126.1	0.58	0.59
d25	78.84	89.16	0.34	0.32
d50	113.88	63.05	0.11	0.17
d75	135.02	44.88	0.07	0.09
A	3.82	19.49	0.00	0.03

Parameter determination for Lynch and Rao's equations

The first step in this procedure is to conduct an experiment on a hydrocyclone with a given pulp of a known flow rate, particle size distribution and solid concentration, calculating the parameters K_1 , K_2 and K_3 and the values of a, d_{25}, d_{50} and λ.

Parameter determination

See Table 9.6.

Simulation

See Fig. 9.3.

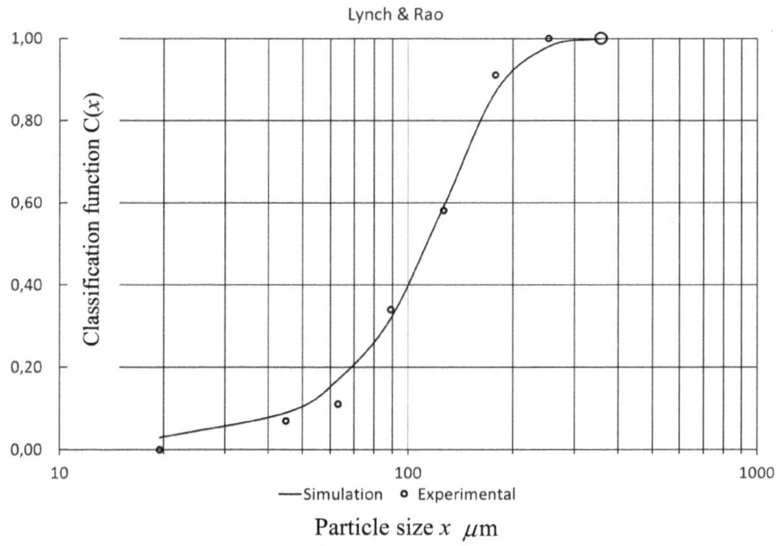

Fig. 9.3 Lynch and Rao's experimental and simulated classification functions

9.2.2 Plitt Simulation Model

In Plitt's equation [8], all lengths are measured in [cm] and the pressure drop Δp in [psi]; h is the length of the vortex finder, d_F is feed tube diameter, φ_F and φ_U are the volume fractions of solids in the feed and underflow, respectively, and S is pulp split, that is, the ratio of the mass flow of pulp in the discharge to that in the overflow. In this model a set of experimental results is required to calculate parameters K_1 to K_4. Plitt's empirical simulation model, revised by [3], is summarized below, with the following units used: Size x and d_{50} in [μm], d_i and h in [cm], ΔP in [kPa], Q_i in [ℓ/min], ρ_i in [g/cm^3], φ_i in [volume fraction] and w_i in [% solids].

(a) **Separation size**

$$d_{50}\,(\mu\text{m}) = K_1 \frac{39.7 d_c^{0.46} d_F^{0.60} d_O^{1.21} \eta^{0.5} \exp(6.3\varphi_F)}{d_O^{0.71} h^{0.38} Q_F^{0.45} \left(\rho_s - \rho_f\right)^{0.5}} \tag{9.13}$$

(b) **Slurry split S_s and short circuit a**

$$S = K_2 \frac{18.62_s^{0.24} \left(d_U/d_O\right)^{3.31} h^{0.54} \left(d_U^2 + d_O^2\right)^{0.36} \exp(5.4\varphi_F)}{d_c^{1.11} (\Delta P)^{0.24}} \tag{9.14}$$

The volume of slurry recovery R_v and the solids recovery R_c to the underflow are

$$R_v = \frac{S}{(1+S)}, \quad R_s = \sum_x s(x) f_F(x) \tag{9.15}$$

and the short circuit is

$$a = \frac{R_v - R_s \varphi_F}{1 - \varphi_F} \tag{9.16}$$

(c) **Pressure-capacity relationship**

$$\Delta p\ (\text{kPa}) = K_3 \frac{1.88 Q_F^{*1.78} \exp(5.5\varphi_F)}{d_c^{0.37} d_F^{0.94} h^{0.28} \left(d_U^2 + d_O^2\right)^{0.87}} \tag{9.17}$$

(d) **Classification function $c(x)$**

$$c(x) = 1 - \exp(-0.693(x/d_{50})^{\lambda}$$
$$\lambda = \frac{1.5725}{\ln(S.I)}$$
$$S.I. = F_4 \times 1.94 \times \exp(-1.58R_v)\left(\frac{d_c^2 h}{Q_F^*}\right)^{0.15}, \quad Q_F^* \text{ in } [\ell/\text{min}] \qquad (9.18)$$

Parameter determination of Plitt's equation

The summary of the results are given below.

Rv	0.256			
Rs	0.406			
a	0.203			
K1	1.309			
K2	0.758			
K3	0.039			
F4	1.506			
1	2.371			
SI	0.52			
x	S(x)	S(x)	C(x)	C(x)
	Exp	Sim	Exp	Sim
357.1	1.00	1.00	1.00	1.00
252.2	1.00	0.99	1.00	0.99
178.3	0.93	0.89	0.91	0.87
126.1	0.66	0.67	0.58	0.59
89.2	0.47	0.46	0.34	0.32
63.0	0.29	0.33	0.11	0.16
44.9	0.26	0.26	0.07	0.07
19.5	0.20	0.21	0.00	0.01

Simulation

See Fig. 9.4.

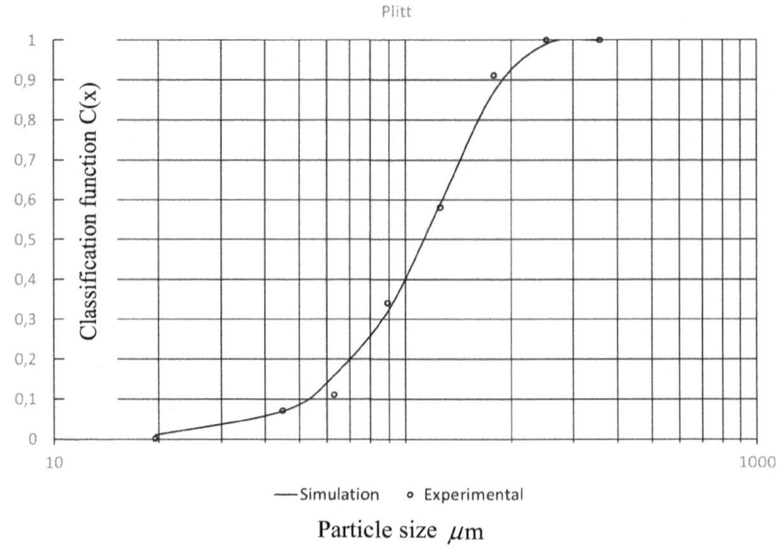

Fig. 9.4 Plitt's experimental and simulated classification functions

9.2.3 CYCLOSIM Hydrocyclone Simulator

The Molycop Tools, developed by Sepúlveda [9], presented a new hydrocyclone simulation based on the work of Lynch and Rao [4], which was incorporated into their Cyclosim-Single, Simula V2 Static Concentration Plant Simulator (Table 9.7).

(a) **Pressure-capacity relationship**

$$\Delta P(\text{psi}) = \alpha_1 \frac{Q_F^{*1.45} \exp\left(-7.63\varphi_F + 10.79\varphi_F^2\right)}{d_c^{0.20} L^{.0.15} d_F^{0.51} d_O^{1.65} d_U^{0.53}} \tag{9.19}$$

where P is the pressure, $Q_F^*[\text{m}^3/\text{h}]$ is the feed pulp volume flow rate and φ_F its volume fraction concentration, d_c [in] is the hydrocyclone diameter and d_F, d_O and d_U are the diameters of the entry, vortex and apex, respectively, measured in inches and density is in ρ_s $[\text{g}/\text{cm}^3]$.

(b) **Separation size**

$$d_{50}\,(\text{mm}) = \alpha_2 \frac{d_c^{0.44} d_F^{0.58} d_O^{1.91} \exp(11.12\varphi_F)}{d_U^{0.80} L^{.37} Q_F^{0.34} (\rho_s - 1)^{0.50}} \tag{9.20}$$

Table 9.7 Parameter determination

x	S(x)	S(x)	C(x)	C(x)
	Exp	Sim	Exp	Sim
357.1	1	1	1	1
252.2	1	0.99	1	0.99
178.3	0.93	0.89	0.91	0.87
126.1	0.66	0.67	0.58	0.59
89.2	0.47	0.46	0.34	0.32
63	0.29	0.33	0.11	0.16
44.9	0.26	0.26	0.07	0.07
19.5	0.2	0.21	–	0.01
x	S(x)	S(x)	C(x)	C(x)
	Exp	Sim	Exp	Sim
357.1	1.00	1.00	1.00	1.00
252.2	1.00	0.99	1.00	0.99
178.3	0.93	0.89	0.91	0.87
126.1	0.66	0.67	0.58	0.59
89.2	0.47	0.46	0.34	0.32
63.0	0.29	0.33	0.11	0.16
44.9	0.26	0.26	0.07	0.07
19.5	0.20	0.21	–	0.01

(c) **Short circuit**

$$a = \lambda \frac{S_s/(S_s+1) - \varphi_F R_s^c}{1 - \varphi\left[1 - \lambda(1 - R_s^c)\right]}; \text{ with } R_s^c = \sum f(x)c(x) \quad (9.21)$$

where the slurry split

$$S_s = \frac{Q_U^*}{Q_O^*} = \alpha_3 \frac{L^{0.19}(d_U/d_O)^{2.64} \exp(-4.33\varphi_F + 8.77\varphi_F)}{\Delta P^{0.54} d_c^{0.38}} \quad (9.22)$$

(c) **Reduced classification function**

$$c_r(x) = 1 - \exp\left[-0.393\left(x/d_{50}^n\right)^m\right] \quad (9.23)$$

where

$$m = \exp\left[\alpha_4 - 1.58\left(S_s/S_s + 1\right)\right]\left[d_c^2\left(L/Q_F\right)\right]^{0.15} \quad (9.24)$$

Parameter determination for the CIMM model

$$\Delta P(\mathrm{psi}) = \alpha_1 \frac{Q_F^{*1.45} \exp\left(-7.63\varphi_F + 10.79\varphi_F^2\right)}{d_c^{0.20} L^{.0.15} d_F^{0.51} d_O^{1.65} d_U^{0.53}}$$

$$d_{50}\,(\mu\mathrm{m}) = \alpha_2 \frac{d_c^{0.44} d_F^{0.58} d_v^{1.91} \exp(11.12\varphi_F)}{d_a^{0.80} L^{.37} Q_F^{0.34} (\rho_s - 1)^{0.50}}$$

$$S_s = \frac{Q_U^*}{Q_O^*} = \alpha_3 \frac{L^{0.19}\left(d_U / d_O\right)^{2.64} \exp(-4.33\varphi_F + 8.77\varphi_F)}{\Delta P^{0.54} d_c^{0.38}}$$

$$c_r(x) = 1 - \exp\left[-0.693\left(x / d_{50}^n\right)^{\lambda}\right]$$

$$\lambda = \exp\left[\alpha_4 - 1.58\left(S_s / S_s + 1\right)\right]\left[d_c^2 h / Q_F\right]^{0.15}$$

$$a = a_5 \frac{S_s / (S_s + 1) - \varphi_F R_s^c}{1 - \varphi\left[1 - \lambda(1 - R_s^c)\right]}; \quad \text{with } R_s^c = \sum f(x) c(x)$$

Simulation

See Fig. 9.5.

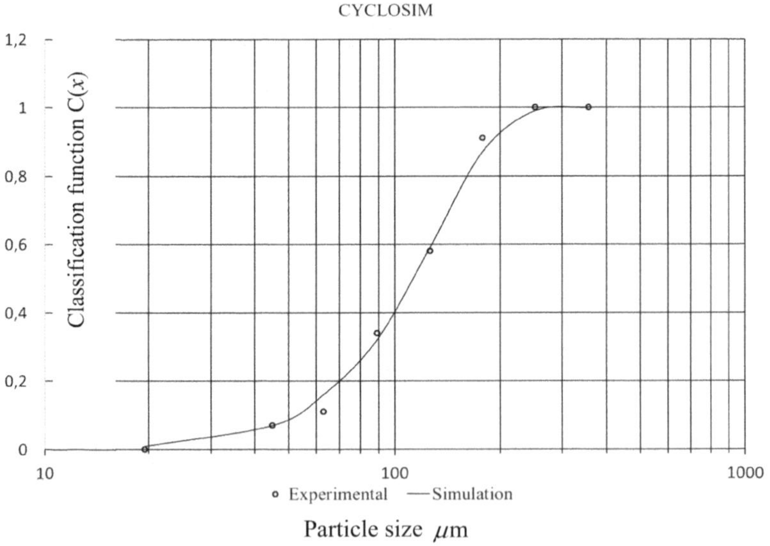

Fig. 9.5 CIMM model hydrocyclone simulator

Table 9.8 Parameter determination

Parameters		Particle size	Expe rimental	Simulation
bi	0.941	357.1	1.00	1.00
P2	15.386	252.2	1.0000	0.99
b3	7.295	178.3	0.91	0.87
a	0.2	126.1	0.58	0.59
d25	78.84	89.2	0.34	0.32
d50	113.88	63	0.11	0.16
d75	153.02	44.9	0.07	0.07
SI	0.52	19.5	0.00	0.01
1	2.37			

9.2.4 Nagewararao's Model

For Nageswarao's model [7], based on his work of 1978 [6], the factors considered relevant to describe cyclone performance were the Euler number, $Q\big/\left(d_c^2 \times (\Delta p/\rho)\right)^{0.5}$, dimensionless cut size, d_{50}/d_c, recovery of water to the underflow, R_f, and volume recovery of feed slurry to the underflow, R_v. The [6] database, complemented with data of [4], was used to determine the simulation parameters β_i (Table 9.8).

(a) **Separation size**

$$\frac{d_{50}}{d_c} = \beta_1 (d_c^{-0.65}) \left(\frac{d_O}{d_c}\right)^{0.52} \left(\frac{d_U}{d_c}\right)^{-0.50} \left(\frac{d_F}{d_c}\right)^{0.20} \left(\frac{\ell_c}{d_c}\right)^{0.20} \left(\frac{\Delta p}{\rho_s g d_c}\right)^{-0.20} \theta^{0.15} \lambda^{0.93}$$

$$\text{where } \lambda = 8.05 \times \frac{10^{1.82\varphi}}{(1-\varphi)^2} \tag{9.25}$$

(b) **Pressure-capacity relationship**

$$\frac{Q_F}{d_c^2\sqrt{\Delta p/\rho_F}} = \beta_2 d_c^{-0.10} \left(\frac{d_O}{d_c}\right)^{0.68} \left(\frac{d_F}{d_c}\right)^{0.45} \left(\frac{\ell_c}{d_c}\right)^{0.20} \theta^{-0.10} \tag{9.26}$$

where $Q\,[\ell/\text{min}]$, Δp[kPa] and d_i and ℓ_c [cm].

(c) **Recovery of water to the underflow**

$$R_f = \beta_3 (d_c)^{0.0} \left(\frac{d_O}{d_c}\right)^{-1.19} \left(\frac{d_U}{d_c}\right)^{2.40} \left(\frac{d_F}{d_c}\right)^{0.50} \left(\frac{\ell_c}{d_c}\right)^{0.20} \left(\frac{\Delta p}{\rho_s g d_c}\right)^{-0.53} \theta^{-0.24} \lambda^{0.27} \tag{9.27}$$

(d) **Classification function**

$$c(x) = 1 - \exp\left(-0.693147 \times \left(x/d_{50}\right)^{\lambda}\right) \tag{9.28}$$

Simulation

See Fig. 9.6.

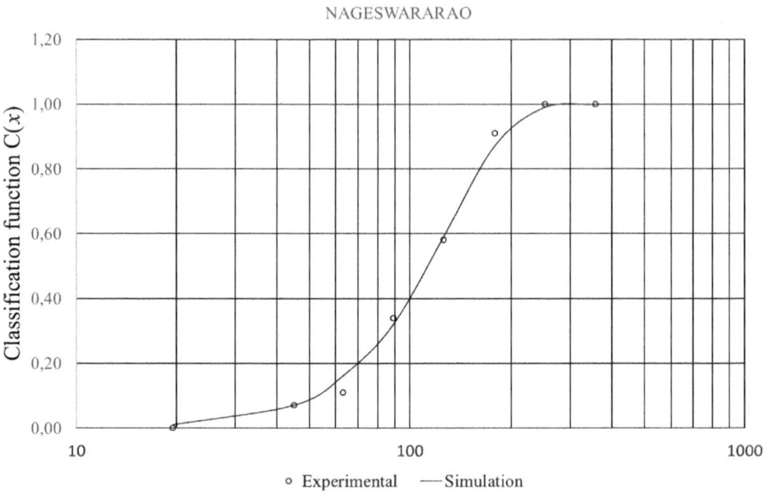

Fig. 9.6 Particle size x

References

1. Arterburn RA (1982) Chapter 32, Design and installation of comminution circuits (Mular A, Jergensen G, eds). SME-AIME, New York, NY, pp 592–607
2. Concha F, Barrientos A, Montero J, Sampaio R (1994) Air core and roping in hydrocyclones. In: Proceedings of the 8th European symposium of comminution, Stockholm, Sweden, vol II, pp 814–823
3. Flintoff BC, Plitt LT, Turak AA (1987) Cyclone modeling: a review of present technolongy. CIM Bull 80(905):39–50
4. Lynch AJ, Rao TC (1975) Modelling and scale-up of hydrocyclone classifiers. In: Proceedings of the 11th international mineral processing congress, Cagliari, Italy, paper 9, pp 245–269.
5. Medronho RA, Svarovsky L (1984) Tests to verify hydrocyclone scale-up procedure. In: 2nd international conference on hydrocyclones, Bath, England, pp 1–14
6. Nageswararao K (1978) Further developments in the modelling and scale-up of industrial hydrocyclones. Ph.D. Thesis, University of Queensland (JKMRC), Brisbane, Australia
7. Nageswararao K, Wiseman DM, Napier-Munn TJ (2004) Two empirical hydrocyclone models revisited. Min Eng 17:671–687
8. Plitt LR (1976) A mathematical model of the hydrocyclone classifier. CIM Bull, 114–123
9. Sepúlveda XX (2002) Cyclosim Hydrocyclone Simulator, Simula V2 Static concerntration Plant Simulator. Academic version of MolyCop Tools YM 2.0 SCA Metal Group. www.uc.cl/sw_educ/simula2/html/simulacion.html

Chapter 10
Hydrocyclone System Design

Abstract Once hydrocyclones are designed for given grinding-classification circuits, desliming, dewatering and other purposes, it is necessary to consider all the elements that make up the hydrocyclone system: pulp feeding tank volume and geometry, connecting pipes, pulp distributors and instrumentation. In this section we will present a general description of the necessary characteristics of the different components of the hydrocyclone system and give an example of a comminution-classification circuit in a copper concentrator.

10.1 Hydrocyclone as an Element in Grinding-Classification System Design

10.1.1 Pulp Feeding Tank

In general, the hydrocyclone feed is conveyed by a pumping system coupled to the pulp discharge from a mill. The pumping system consists of a tank that receives the pulp discharge from the grinding mill, a centrifugal pump driven by an electric motor, a transmission system of pulleys and trapezoidal belts and a piping system. In high-power drives, above 200 kW, the coupling between pump and motor is carried out by a reducer in order to adjust the pump's rotation speed.

At present and, given the considerable reduction in the costs of frequency converters, their use allows a variable-speed system, making it possible to adjust the rotation speed to that most appropriate for the pump. This also allows automated operation depending on the hydrocyclone operating pressure, the level of the pump in the tank and the control of the circulating load in the case of a closed-circuit grinding system.

The pump tank must be fitted with a series of features to ensure that it is stable and easy to maintain.

F. Concha A. and J. L. Bouso A., *Fluid Mechanics Fundamentals of Hydrocyclones and its Applications in the Mining Industry*, Fluid Mechanics and Its Applications 126,
https://doi.org/10.1007/978-3-030-67913-2_10

Cylinder-conical tank design. Prismatic tank design.

Fig. 10.1 Pumping system with centrifugal pump and electric motor

Tank geometry and volume

A cylinder-conical shape is the best design for a pulp feeding tank because it does not present inner edges with section changes that may cause unwanted sedimentation of solids. Although this design is the most appropriate, in large-volume pumping such as in closed grinding circuits, the prismatic design is more suitable (Fig. 10.1).

The most important point to be considered in the design of the box is the slope of the deposit, which in principle should not be less than 45 degrees and preferably 60 degrees when the pulps that are handled contain high-specific-weight solids that can sediment easily.

The pump-tank connection must be made by means of a large-diameter tube to promote pump suction. If possible, this duct should be provided with wear elements to allow periodic maintenance without damaging the pump casing.

A good design should consider the interior coating of the pump box, especially in the lower part of the tank, which is where abrasion is greater, as the velocity of the particles that are sucked by the pump is higher in this area. When the arrival of foreign elements is foreseen the pump box must be equipped with a grid box or mesh where the feed chutes arrive so that these foreign elements cannot reach the suction area, obstructing the pump. Ideally, the design of this protection box should allow easy removal of the deposited items.

The inlet of the feed flow into the pump box must not directly affect the suction area of the pumps, as this causes turbulence that disturbs the normal operation and, in many cases, causes cavitation phenomena that disturb the hydrocyclone operation. Until relatively recently, in critical pumping of large volumes, pump tanks were equipped with two centrifugal pumps, one in operation and the second on standby. In practice it was demonstrated that in most cases when the reserve pump was required, it was obstructed by sedimentary minerals and the pump change could not take place automatically, making the costly reserve pump system useless. It is now preferable to have a complete spare pump if the change must be made in the shortest possible time.

The pump tank must be fitted with a safety overflow conduit so that in the event of an accidental pump stoppage, the flow will not fall onto the pump and eventually the drive motor. In case of accidental pump stoppage and when handling pulps with solids in suspension that sediment easily, the pump box must be equipped with a quick-pouring device to avoid sedimentation of the solids at the bottom of the box that would block the suction of the pump, preventing its further operation. A point of extreme importance that is not normally considered during the engineering stage is the volume of the pump box, which, as is logical, must be determined depending on the flow to the pump and the particularities of the pipe that connects to the hydrocyclone.

By default, the minimum volume of the pump box should allow pumping for a minimum of 2 min without receiving new feed; based on this assumption, the standard pumping groups should consider the average flow. The determination of the minimum volume of the box should also consider the inner volume of the duct to the hydrocyclone so that the pump startup allows this duct to be filled without cavitation occurring in the pump. In general, there is a response time from the start of pumping until the arrival of fresh feed to the pumping group, which must be compensated for by the volume of the pump box.

Maintaining a minimum pulp level inside the pump box is essential to avoiding bomb cavitation, which, in addition to causing unstable hydrocyclone operation, can cause great damage to the pump itself, especially in its axis bearings. A level control is usually performed with an adjustable water addition through a valve that is controlled by a level sensor installed in the pump box.

10.1.2 Hydrocyclone Configuration

Hydrocyclone separation sizes in a grinding-classification circuit vary between 100 and 300 μm and hydrocyclone diameters depend on separation size, which in turn is related to the particle size in the feed. However, it is not always correct to use large hydrocyclones for coarse separations with large particle sizes and small diameter ones to treat fine particles.

The configuration of the cylindrical and conical zones of a hydrocyclone, its cone angle and its feeding and overflow nozzle diameters must be different for different applications. It is advisable to have the greatest possible number of part combinations to adapt to the specific conditions of a specific operation, see Fig. 10.2. Sometimes this versatility allows operation with the same hydrocyclone diameter but different configurations, for example, a 500-mm flat-bottom hydrocyclone configuration classifying at 150 μm in grinding circuits and the same hydrocyclone diameter classifying at 30 μm in long, conical configurations in regrinding circuits.

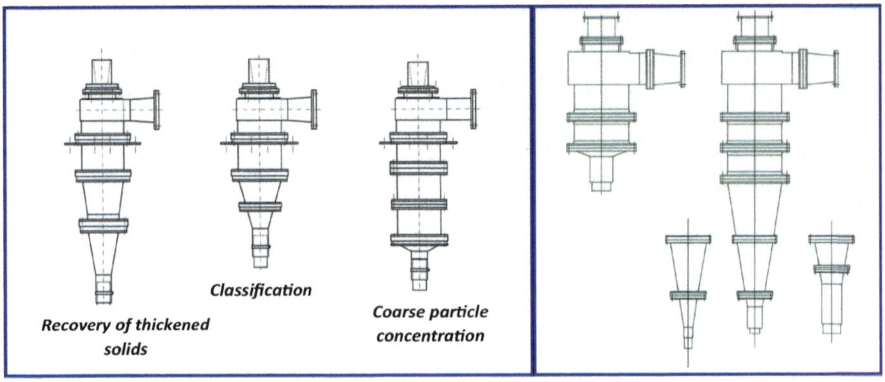

Fig. 10.2 Different hydrocyclone geometries and configurations

10.1.3 Construction Materials

Conventional construction involves steel casing and anti-abrasive rubber liners (Fig. 10.3), but reduced-weight materials such as elastomers and plastics, which reduce maintenance costs, are also used. Thus, hydrocyclones made entirely of self-supporting polyurethane, with the casing and cladding in one-piece Fig. 10.3 and others with plastic housings and interchangeable coatings made of different abrasion-resistant materials such as polyurethane, synthetic rubber and advanced ceramics have emerged, Fig. 10.4.

Parts subjected to greater wear, such as discharge nozzles, conical body bottoms, overflow nozzles and areas of the tangential inlet receiving the greatest feed impact are currently manufactured using special abrasion-resistant materials. The most frequently used materials in the coatings are advanced ceramics with light casings Fig. 10.5 and highly resilient synthetic rubbers and polyurethanes, Fig. 10.6.

10.1.4 Hydrocyclones and Connecting Pipes

As with all connecting pipelines in mineral processing plants, the connection between pipes and hydrocyclones must be designed in such a way as to achieve the least loss of pressure in them and avoid areas of possible solids sedimentation. When designing and sizing the connection piping, the significant differences between the flows circulating through the three ducts, feed underflow and overflow. Although the general principle of a limiting speed, which avoids sedimentation of the coarser solid particles, is inherent to the three conduits, one must consider the large differences in the solid concentrations and their particle sizes.

Underflows conduits must be designed with anti-abrasive liners to prevent premature wear. Due to the nature of the pulps circulating in these conduits, horizontal paths

Fig. 10.3 Steel casing and anti-abrasive rubber liners

should be avoided and minimum slopes on the order of 25° must be considered by default. The overflow currents are less problematic since in general the pulps have solids concentrations of less than 40% by weight and the particle size is lower than 500 μm. The minimum slopes by default should be on the order of 5° and the ideal materials for these pipes are HDPE and eventually polyamide, preferably treated against ultraviolet rays.

The hydrocyclone feed pipes work under pressure and with high circulating speeds, so they must have as smooth a path as possible in terms of changes in direction, curves and diameter increases or decreases. Avoidance of horizontal sections that may cause particle sedimentation when the duct is out of operation should be seriously considered. The most suitable material for these pipes is steel with a rubber lining and preferably rubber hoses that are as flexible as possible to simplify the layout. Eventually HDPE can also be used, but given high pressures, these conduits must be very thick to withstand the abrasion to which they are subjected.

In general, through the feed pipes of hydrocyclones having large diameters (greater than 300 mm) circulate heterogeneous pulps, with particles of different sizes, whereas in feed pipes of medium and small diameter hydrocyclones circulate homogeneous

Fig. 10.4 Polyurethane hydrocyclone

Fig. 10.5 Ceramics with light casings

Fig. 10.6 Highly resilient synthetic rubbers

pulps with a much narrower particle range. Regarding closed grinding circuits, regrinding circuit pulps could be considered homogeneous.

10.2 Operating Conditions

Hydrocyclones must work at a certain pressure range according to their diameter and operational requirements. Hydrocyclones of different sizes must operate at given pressures:

- between 10 and 50 mm at pressure range between 20 and 400 kPa.
- between 75 and 250 mm at pressure range between 25 and 250 kPa.
- between 300 and 500 mm at pressure range between 30 and 200 kPa.
- between 650 and 1250 mm at pressure range between 35 and 150 kPa.

The pressure range indicated for the different diameters is that considered optimal for conventional hydrocyclones, although depending on the classification requirements and the manufacturing characteristics the maximum pressures can be higher. There are special operations in which hydrocyclones operate at 15,000 kPa. The minimum pressure is established so that the internal currents of the hydrocyclone and primary and secondary whirlpools are sufficiently stable to guarantee the formation of these currents and consequently the hydrocyclone operation.

Hydrocyclone can also be fed using gravity or pumping, see Fig. 10.7.

Fig. 10.7 Closed grinding circuits with 500 m- and 400 mm-diameter hydrocyclones

Gravity feeding is infrequent and often not considered in the engineering stages despite being a very stable feeding method and logically representing considerable energy savings (Fig. 10.8).

Critical (Limiting) Transport Velocity

When pressure is applied to a pipe containing sediment, the fluid moves and, as the particles begin to suspend, the pressure decreases. Between the beginning of the movement of the particle bed and complete suspension there is a range of speeds at which the pressure of the fluid is minimum and beyond which it increases again. This speed range is called the critical transport velocity.

There are numerous correlations proposed for the critical velocity of particle suspensions from 50-mm (2-in) to 300-mm (12-in) pipes. [2] reviewed the transport velocity work of several works.

Rayo modified the Durand and Condolios (1952) expression of critical velocity and applied it to a vast number of particle sizes and size distributions and tube diameters (see Chap. 4 of this book).

In 1986, Jergensen and Samuell [4] developed a simple empirical equation with the pipe diameter, particle size, volume fraction of solids and densities of the solid and liquid as variables.

$$v_L = (0.2154\varphi + 0.1584) \times \ln(d_{50}) + (1.11\varphi + 1.19);$$

Fig. 10.8 Gravity-fed hydrocyclone batteries

$$\text{for } 5 < d_{50} < 500; \text{ in } \mu m \tag{10.1}$$

10.3 Hydrocyclones and Distributors: Design and Instrumentation

The hydrocyclone diameter is determined according to the cut or classification size. Once the hydrocyclone is selected, according to the flow needed, the number of hydrocyclones to be used in the separation is defined.

When it is necessary to use more than one hydrocyclone to meet flow needs, they are usually installed in distributors (clusters) so that the flow is distributed evenly among the hydrocyclones, with each performing the same classification work.

The best hydrocyclone distributor design is that of a chamber with a generous volume and a spherical or pseudo-spherical form, as this design allows a better balance of pressures and thus a better distribution of the pulp in all the devices.

Generally, the distributor design does not provide for a large chamber volume, impeding the perfect distribution of pulp to each hydrocyclone. By contrast, a high-volume chamber allows the inside of the chamber to have zero velocity with a more

Fig. 10.9 Distributor with 38, 100 mm diameter hydrocyclones

uniform distribution of flow to each conduit, ensuring that the various hydrocyclones operate under the same conditions (Fig. 10.9).

In addition to the central distribution chamber, the design must incorporate the collection tanks of the discharge currents (underflow) and the overflow currents of all the hydrocyclones. It is advisable that the overflow collection tank have a certain height to facilitate its exit, since the height directly influences the output speed.

In a good distributor design, all the connecting tubes to the hydrocyclones, as well as their overflow and discharge ducts, are the same, such that the loss of load in each hydrocyclone is identical. To achieve this goal, radial distribution is ideal. The first company to use this design was AKW from Germany, which called its design AKA SPIDER due to its resemblance to a spider (Fig. 10.10).

The connection of the hydrocyclone to the central chamber of the distributor is usually made through a valve that allows each one to be isolated if necessary due to maintenance needs, flow requirements or an operation pressure adjustment. These valves must not be regulating valves, but rather open-closed actuation valves, which are known as on–off valves.

In the handling of pulps with solids in suspension, two valve types are used: pinch or knife. Whichever type is used, the valve must always be open or closed to allow full passage of the pulp flow to the hydrocyclone or its total isolation. These valves usually carry remote position indicators so that the operator can know the number of hydrocyclones in operation at any time.

It is good to be able to vary the number of hydrocyclones in operation to adjust the pressure within the limits of the operation, which requires that at least 25% of the valves be automatic controlled valves, usually pneumatic. Today, due to reduced valve costs, it is usually normal for distributors, especially large ones, to have 200-mm-diameter valves (Fig. 10.11).

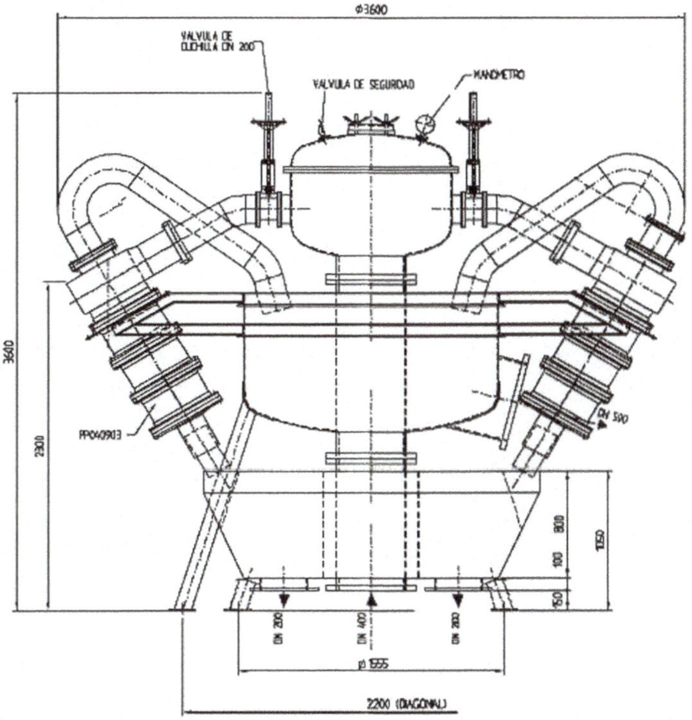

Fig. 10.10 Distributor of 4500 mm diameter hydrocyclones

Fig. 10.11 Distributor construction detail, AKA-SPIDER design

Understanding of the hydrocyclone operating pressure is fundamental. Therefore, the distributors are equipped with a pressure gauge that shows the pressure on the distributor itself and a pressure switch that emits an electrical signal proportional to the pressures so that the operator located in the central control room can know

Fig. 10.12 Closed-circuit grinding rods-balls with 500 mm diameter hydrocyclones

this pressure and eventually make decisions about operation, opening or closing hydrocyclones or acting on the speed of the feed pump to the distributor.

This mode of operation is essential in the case of closed grinding circuits, where variations in the hardness of the minerals cause continuous changes in the circulating load, requiring, for optimal operation, that the flow of the different circuit streams be adjusted. This can be achieved by varying the number of hydrocyclones in operation (Fig. 10.12).

10.4 Installation and Operation of Hydrocyclones and Batteries of Hydrocyclones

By applying the most suitable mathematical models and the best knowledge, it is possible to determine the best hydrocyclone for the desired separation size, and the number of units to be used depending on the flow to be treated (see Sect. 9.1, Chap. 9 of this book).

Another important practical criterion is that the hydrocyclone batteries should have flexibility, permitting operation with several units. It is desirable that the designed hydrocyclone battery permit opening or closing hydrocyclones to adjust the flow in magnitudes on the order of 15% to 25%, allowing the installation to adapt to the fluctuations of the feed in terms of quantity and quality. This is especially important in closed grinding circuits where changes in particle size, due to either changes in the hardness of the ore or changes in the crushing areas, have a great influence on the circulating load of the grinding-classification circuit. In operations where significant power fluctuations are envisaged, it is convenient to operate with a greater number of medium-diameter hydrocyclone units, which presents flexibility advantages compared to simply operating with the lowest possible number of large-diameter units. Operating with cyclones with smaller-than-necessary diameters often presents the disadvantage of frequent jams caused by foreign elements both in the discharge nozzle and in the feed (Fig. 10.13).

When it comes to the installation of a single hydrocyclone, little attention is usually paid to the need to control the operation over time, leading to a multitude of installations where there is no pressure measurement, the discharge and overflow ducts are not easily accessible and, in most cases, it is impossible to take samples of the different currents for analysis, see Fig. 10.14. In the case of installations with several units, other problems inherent to the distribution of the feed arise.

Distribution Chamber

The distribution chamber must have a generous volume proportional to the vortex flow in such a way as to achieve a hypothetical "zero" speed that eliminates the turbulence caused by the pipeline transport from the pump to the distributor. If possible,

Fig. 10.13 Installation of batteries of 750 mm (30 in) and 500 mm (20 in) diameter hydrocyclones in a closed grinding circuit

Fig. 10.14 Isolated hydrocyclone without access

its form should be as close to a sphere as possible to achieve an ideal equilibrium of pressures that would give rise to a uniform and homogeneous distribution to each hydrocyclone, Fig. 10.15.

The chamber should be equipped with a special manometer for pulp with a separating membrane and hydraulic damper, as well as a pressure transmitter for remote

Fig. 10.15 Semi-spherical distributor chamber

Fig. 10.16 Distributor with annular design

pressure control, in duplicate if possible. In addition, it must have a manual or automatic deaeration system, since normally all the transported air accumulates in the chamber and sometimes causes problems. Also, in compliance with safety rules, it must have a pressure-limiting valve, Figs. 10.15 and 10.16.

Overflow and Underflow Collecting Tanks

The overflow and discharge collectors of the different units must have the necessary dimensions to prevent splashing and allow the sampling of each flow. Special care should be given to the design of the overflow collection tank, the height of which must be great enough to ensure the proper output speed to evacuate the high overflow, considering the singularities of the withdrawal pipes and their level differences, Figs. 10.17 and 10.18.

The design of the discharge tank should consider the difficulty of movements of these high concentration solids flows, avoiding the formation of dunes or, instead, studying their intentional creation as a protection system against wear, Fig. 10.18.

Feed Piping

Another important point is the connection between the feed distributor and the hydrocyclones. The connection pipes must all be identical and equipped with open/closed valves in order to isolate each hydrocyclone during maintenance and maintain a given pressure. The valves should be automatic opening-closing valves, especially in the case of valves with a diameter greater than 100 mm and should never be used as a flow control system They should be provided with switches that indicate the extreme open and closed positions. In borderline cases, at least 20% of the valves should be automated to simplify operation control, Fig. 10.17.

Fig. 10.17 Distributor with deposits

Fig. 10.18 Automatic valve detail in distributor

Overflow and Underflow Piping

Like the feeding pipes, the overflow and underflow pipes must be identical for uniform distribution. Unfortunately, too many facilities do not take this into account and cause unnecessary operational problems. Both underflow and overflow hydrocyclone streams should be released freely into the atmosphere without disturbances that would cause internal turbulence. Special care should be given to the design of the overflow ducts, which must be generous in diameter (greater than that of the overflow nozzle) and have large radius curves, allowing the internal vortex to expand smoothly and progressively.

If possible, the end of the overflow duct must be located above the feeding level to avoid a siphoning effect that would cause abnormal heavy particle drags. If for any reason this is not possible, this conduit should be supplied with an aeration tube to allow the air to enter, thus avoiding the formation of the siphon. This aeration tube should be vertical and with a height proportional to the operating pressure to avoid pulp and vapor projections.

The discharge pipes must allow the coarse product to exit freely to the exterior, avoiding the possibility of obstruction by sedimentation and, a very important factor, must allow the discharge to be seen, because the form of this current clearly indicates the operating condition: spray, semi-rope or rope (Figs. 10.1 and 10.14).

Hydrocyclone Installation Angle

About 10 years ago, a trend began of installing hydrocyclones almost horizontally, within ten degrees of the horizontal line. However, the drawbacks were more significant than the advantages. Today, all the new installations maintain the criterion of placing the hydrocyclones either vertically or at a maximum of 30° relative to the vertical axis in order to build more compact batteries that occupy less space.

There is no operational disadvantage to placing the hydrocyclones in a horizontal position, and even in the upward position toward the discharge, Fig. 10.19, provided that the operating pressure is relatively high to compensate for the greater head loss. In fact, in the tailings dam wall preparation operations, hydrocyclones are placed in this position, Fig. 10.20.

The major drawbacks of installing hydrocyclones in a horizontal position, forming batteries, are that maintenance becomes more cumbersome and complicated and that the space occupied by them is much greater than that occupied by annular

Fig. 10.19 Horizontal position in tailing dams

Fig. 10.20 Hydrocyclones forming tailing walls

design distributors. The need to increase the concentrator treatment tonnage led to the discovery that changing the cyclone angle of installation resulted in a coarser classification cut, which meant reducing the task of grinding, allowing an increase in the tonnage treated by the concentration plants. Unfortunately, all these changes led to a decrease in classification efficiency, which resulted in a parallel drop in milling and flotation efficiencies. But in the case of mines with a high metal content in the ore, the changes yielded positive results, since the operation losses were compensated for by the increase in tonnage.

In this case, the logical solution was to change to larger diameter hydrocyclones, which, under normal operating conditions, will give the necessary cut size. This solution resulted in the installation of 625 mm and 750 (26 and 33 in) diameter hydrocyclones in both existing newly designed installations. Recently, large diameter hydrocyclones were redesigned in order to achieve finer cutting sizes, allowing an improvement in grinding-grading circuit efficiency. Today, these new large-diameter and large-capacity hydrocyclones allow cut sizes like those previously obtained with conventional 500 mm (20 in) diameter hydrocyclones if a properly combination of feed and overflow nozzles is chosen.

10.5 Operating Recommendations

Fortunately, despite all the empirical calculations and installation deficits, in most plants hydrocyclones operate properly thanks to the support of the operators who have no choice but to deal with the problems. If the selection has not been too bad, the collaboration of a good operators can lead to good hydrocyclone operation.

Fig. 10.21 Hydrocyclone distributor with maintenance platform

To achieve good operation, there must be easy access to the different parts of the hydrocyclone to avoid tedious and difficult maintenance and control routines; especially important is good visual control of the cyclone, see Fig. 10.21.

Large hydrocyclone batteries should have a platform. External observation of the discharge flow, either spray or rope, allows the operator expert to know the state of the cyclone. Contrary to common knowledge, a rope operation is not necessarily a symptom of a bad operation, Fig. 10.21. It is true, however, that a spray discharge allows a smoother operation, making it a desired state by less skilled operators, Fig. 10.20, independent of the platforms.

By modifying the pressure and changing the feeding, overflow and discharge nozzles, the operator can ultimately achieve the desired objective. It is very helpful to have quality hydrocyclones that allow adjustment by exchanging nozzles with different diameters, conicity changes and different lengths, along with the backing, solvency, reputation and practical experience of the manufacturer for in-situ technical service.

Underflow Discharge

External observation of the underflow, in either spray or rope, allows the operator expert to know the state of the cyclone. Contrary to common knowledge, a rope operation is not necessarily a symptom of a bad operation, Fig. 10.22. Although it is true that a spray discharge allows a safe operation, semi-rope is the preferred operation, Fig. 10.23.

A very open spray discharge is usually a sign of an oversized nozzle, which brings about a high short circuit; in operations at medium or high pressures a spray discharge can be optimal. The same can be said when working with very fine particle sizes. Consequently, at least at the beginning of the operation, the concentration of solids by sampling must be checked at the plant itself. Unless an automatic control system

Fig. 10.22 Rope discharge

Fig. 10.23 Spray discharge

is available, intermittent discharge from rope to spray denotes either a slightly low pressure or a nozzle diameter at the limit. Both factors cause sedimentation of solids in the final part of the cone, which self-destructs once the increase of internal pressure helps the expulsion, causing the intermittency.

Overflow

The overflow flow must be continuous and uniform. The discontinuity in the flow indicates a low operating pressure, causing the rupture of the internal vortex, Fig. 10.24, which can be due to extreme low operating pressure (below 20 kPa) or simply flow cuts due to the existence of air in the feed line to the hydrocyclone or a low level in the pump tank. This is easily detectable through a pressure gauge in the feed, if one exists or works, which is not unusual in most installations.

The overflow stream should completely cover the periphery of the duct, which ensures that the internal vortex continues through the outlet (Fig. 10.22). A partial flow indicates that the hydrocyclone operates at a low pressure or with insufficient flow, which can also be caused by a partial blockage of the feed nozzle (Fig. 10.25).

Vibration

An operation at a very high solids concentration sometimes leads to a slowdown of the flow, which can be appreciated by listening to the internal noise produced. Sedimentation at the apex of the cone is caused by inadequate diameter nozzles or simply by low operating pressure, which can also be perceived through sound. These phenomena, which are infrequently observable, indicate incorrect operation that should be rectified.

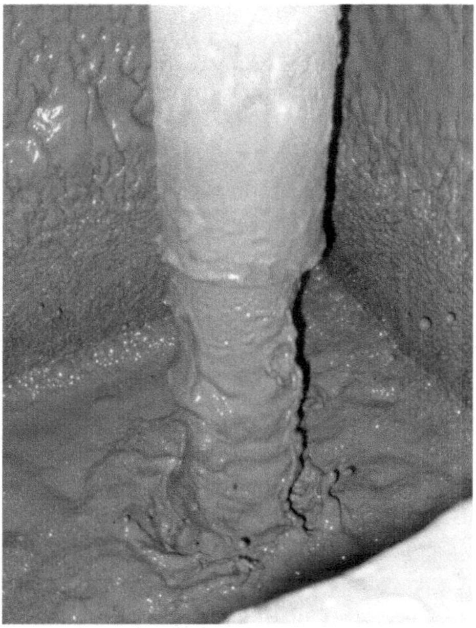

Fig. 10.24 Incorrect discontinuous overflow

Fig. 10.25 Full-tube uniform overflow

Tendencies of discharge blocking or presence of excess air (by cavitation of the pump) are converted into anomalous hydrocyclone vibrations, indicating a problem. The automatic control of the hydrocyclone discharge uses vibrating sensors to establish the limiting conditions of the operation and take corrective measures to return to the normal operating conditions.

Pressure

Hydrocyclones lack any interior mechanism and the interior flow is due to the vortex induced by the high-pressure tangential entry of the pulp. It is always essential to know the operating pressure and very useful to keep a record of it over time. In most installations, manometers are absent or, when they do exist, either do not work or are uncalibrated. Every hydrocyclone or hydrocyclone group requires a manometer, which must be of good quality and have a design suitable for working with pulps with solids in suspension. Large-diameter diaphragm manometers with a large opening to avoid obstructions and equipped with anti-vibration fluid are suitable for this work and represent a negligible expense against total investment (Fig. 10.24). Nowadays digital manometers or pressure transducers are available, allowing us to know the operating pressure at the control panel of the plant and even to carry out control or to set an alarm or warning (Figs. 10.26 and 10.27).

Sampling

It is essential to be able to control the hydrocyclone feed, which requires that a representative sample of this flow be taken with a sampling device, Fig. 10.26. It is quite difficult to control the grinding circuits with information on the overflow

Fig. 10.26 Membrane manometer in distributor

Fig. 10.27 Sampling in distributor

streams alone. New facilities are equipped with density control in the hydrocyclone feed, forming a control loop with the pump speed, water supply and tonnage feeding the mill.

Flow densimeters are invaluable devices, but require frequent calibration, contrasting with actual concentration analyzed in the laboratory. Paradoxically, most facilities lack a point to take a sample of the hydrocyclone feed, which makes it difficult to obtain the sample. When there is a single hydrocyclone, sampling is virtually impossible unless a valve was previously installed in the pump-hydrocyclone feed pipe. In operations with hydrocyclone batteries, if the central chamber lacks a sampling device, a good alternative is to slightly and briefly open the valve of one of the standby hydrocyclones, so that all the flow comes out through its discharge. If the underflow of the hydrocyclone is accessible, as it should be, a representative sample of the feed can be taken from the hydrocyclone discharge after a few minutes.

Underflow and overflow samples should be taken by means of cutters so that only a small amount of specimen is taken, but they are representative of the total flow.

References

1. AKW. Apparate+Verfahren, GmbH u. Co KG.
2. Chien AF (1994) Critical velocity of sand-fluid mixtures in horizontal pipe flow. ASME, FED-189, pp 231–246
3. Durand R, Condolios E (1952) Hydraulic transport of coal and solids material in pipes, In: Proceedings of colloquium on hydraulic transportation, London
4. Jergensen GV, Samuell EJ (1986) Slurry transportation in concentration circuits. In: Mular AL, Anderson MA (1986) Transportation of tailings design and installation of concentration and dewatering circuit. Society of Mining Engineering Inc.

Chapter 11
Instrumentation and Control

Abstract Perturbations in the ore discharge of a grinding mill affects directly the classification in hydrocyclones and deviates the grinding-classification circuit from its optimal operating point. Operational variables, including the particle size, size distribution and volume fraction of solids in the feed should be subjected to control.

In Chap. 2 we saw that a hydrocyclone has several design variables and other operational variables subject to control, including the inlet volume flow rate Q_F (or pressure p), particle size, size distribution and volume fraction of the feed, overflow and underflow. In general, hydrocyclones in the mining industry are poorly instrumented, with inlet pressure and particle size being the only measured variables. Other possible variables were suggested by Neesse et al. [6, 7], including the sediment mass stored in the conical part of the hydrocyclone, the formation of the air core and the shape of the underflow discharge. They analyzed several instruments, including optical, acoustic and weighing sensors, necessary to establish the operational state of a hydrocyclone. Another work, by Vuuren et al. [13] is on monitoring of hydrocyclones using image analysis.

It is well known that a hydrocyclone can operate with a spray, semi-rope and rope discharge, depending on several variables, but mainly depending on the particle concentration of the feed. Figure 11.1 shows the representation of these phenomena [5].

In dense flow separation, more solids are stored in the conical part of the hydrocyclone. They are partly moved toward the overflow, consequently reducing solids recovery to the underflow. The air core does not extend over the entire length of the hydrocyclone and oscillates intensively. The discharge assumes the shape of a rope and is characterized by high solids content and fewer fines, which is the advantage of this variant. The transition state between spray and rope discharge presents the advantages of both high solid recovery and high solids content in the underflow. In this case, the hydrocyclone control mechanism may be based on volume split regulation [6].

F. Concha A. and J. L. Bouso A., *Fluid Mechanics Fundamentals of Hydrocyclones and its Applications in the Mining Industry*, Fluid Mechanics and Its Applications 126,
https://doi.org/10.1007/978-3-030-67913-2_11

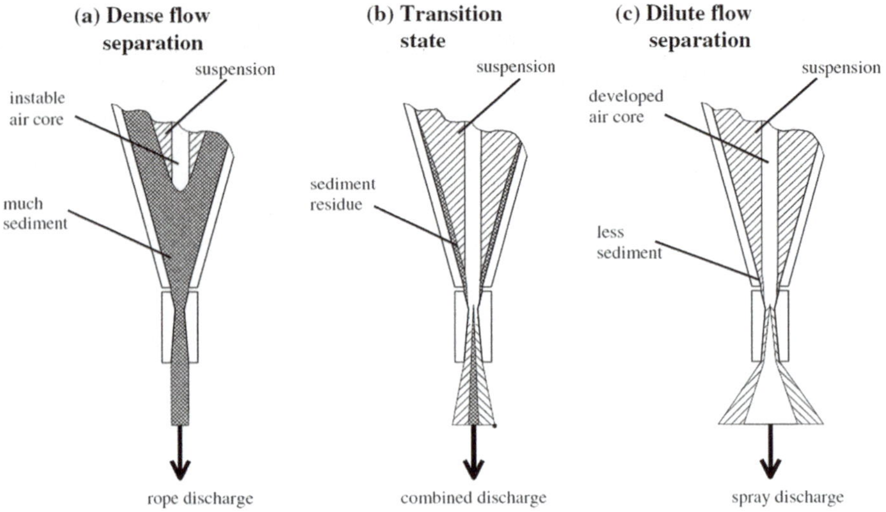

Fig. 11.1 Operating state of a hydrocyclone

11.1 Feed Concentration Measurement

If the objective of separation is to control product fines, the standard control requires instruments to measure the particle size at the overflow. Several instruments exist for this objective, the best known of which is the AUTOTEC PSI particle size analyzer. The PSI 500i is based on laser diffraction and the DFMC model uses ultrasonic scattering (Fig. 11.2).

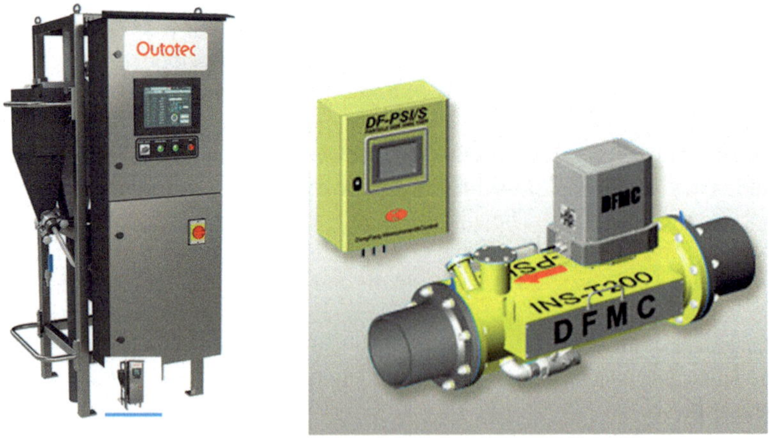

(a) 300i Particle size analyzer. (b) DFMC ultrasonic instrument.

Fig. 11.2 OUTOTEC PSI size analyzer

Fig. 11.3 PST Instrument and its installation

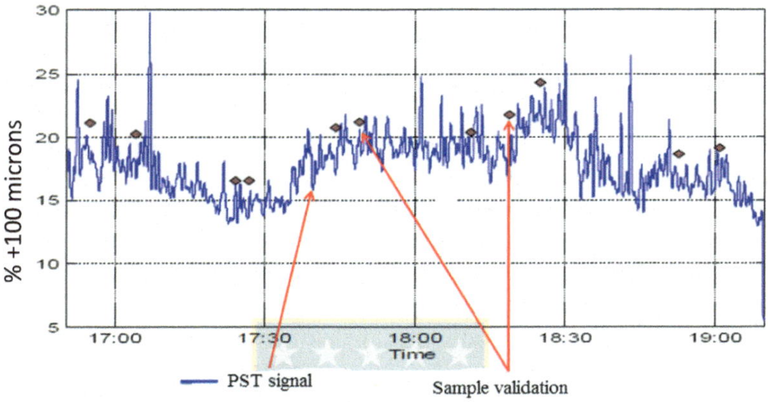

Fig. 11.4 Validation of PST with experimental sampling

Another instrument is the CYCLONEtrac PST, developed by Cidra Minerals processing [3]. The instrument is a device for the online measurement of particle size that allows the development and optimization of a particle size control system in the overflow of a hydrocyclone. Results from a copper concentrator plant using this instrument are given in the following Figs. 11.3, 11.4 and 11.5 [1].

11.2 Underflow Concentration Measurement

In a high feed particle concentration, the underflow assumes the shape of a rope and the air core is obstructed by the high solid concentration at the apex, which is stored in the lower part of the conical section of the hydrocyclone.

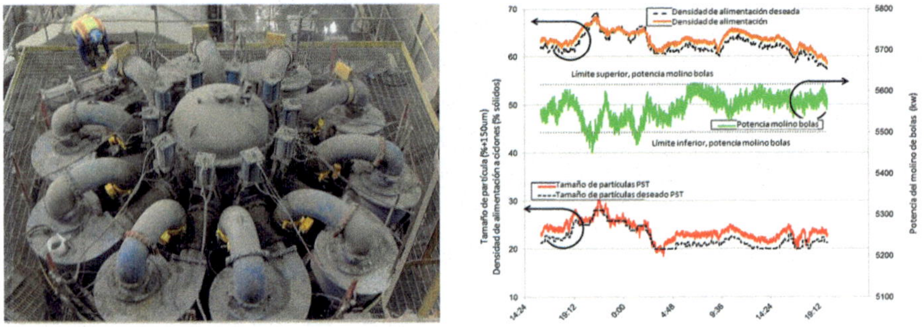

Fig. 11.5 Example of control using CYCLONEtrac PST

11.2.1 Electrical Capacitance

The development and application of electrical resistance tomography (ECT) for online auditing of an industrial hydrocyclone separation was reported by Williams et al. (1999). The work demonstrates the use of ECT to detect spraying, roping, blockage and accurate measurement of the air core size for different operational conditions based on parametric reconstruction of conductivity data in three dimensions.

Meng et al. [4] carried out solid concentration measurement in a cyclone with eight electrodes placed on the straight tube of the discharge. ECT showed as a useful technique for online measurements of solid concentration in the discharge tube of hydrocyclones which can be used for process control.

Capacity measurements provide a way to detect the degree of spray and roping of a hydrocyclone discharge. Figure 11.6 shows that pixels that are scattered at angles between 60 and 80° are spray-type discharge, and the pixels scattered at angles between 20 and 30°, correspond to roping discharge. These measurements allow one to decide how close to roping one wants to operate [6, 14].

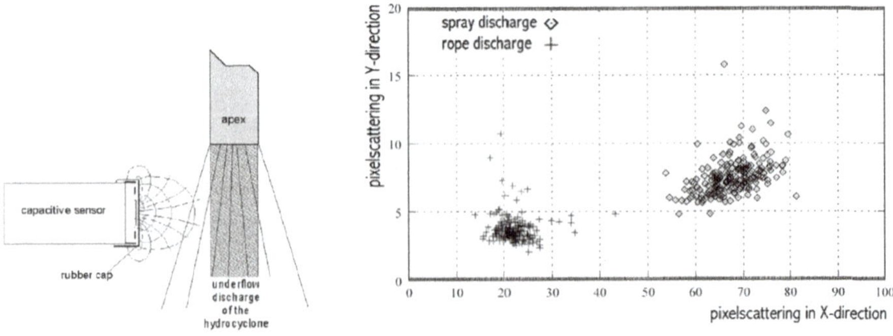

(a) Capacitive sensor. (b) Spray/rope by pattern recognition.

Fig. 11.6 Capacitive sensor for spray discharge detection

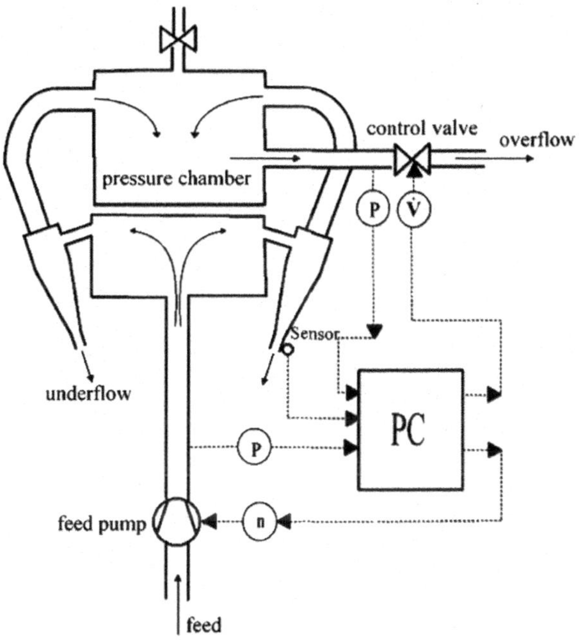

Fig. 11.7 Principle of a computer-controlled hydrocyclone battery [7]

11.2.2 Acoustic Sensor

The acoustic sensor detecting the underflow shape is installed in one of the hydrocyclones of the cluster. These signals indicating the process state are transmitted to the computer, together with the values of the pressure in the feed and overflow. The overflows of all hydrocyclones are collected in a special pressure chamber where a control valve is installed and discharged through a collecting of pipe. The system is controlled by adjusting the throttle valve and the feed pump speed. The control mechanism, actuated by stepwise throttling, intensifies the pressure inside the hydrocyclone. Consequently, the underflow discharge increases until the air core breakthrough point is reached. Then the sensor detects the spray discharge and briefly opens the control valve. Then a new control interval starts [6–8] (Fig. 11.7).

11.2.3 Electrical Conductivity Tomograhpy

Vakamalla et al. [12] setup consisting of an electrical resistance tomograph and a high speed camara to measure the air core syze at different operating conditions of a 75 mm diameter hydrocyclone. See Fig. 5.40. A comparison with CFD measuremn are shown in Fig. 5.41 (Figs. 11.8 and 11.9).

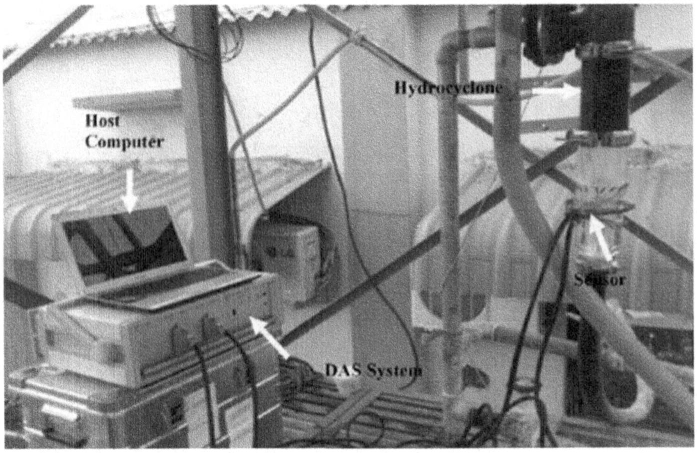

Fig. 11.8 Setup of an electrical conductivity tomograhpy system to measure the air-core size [12]

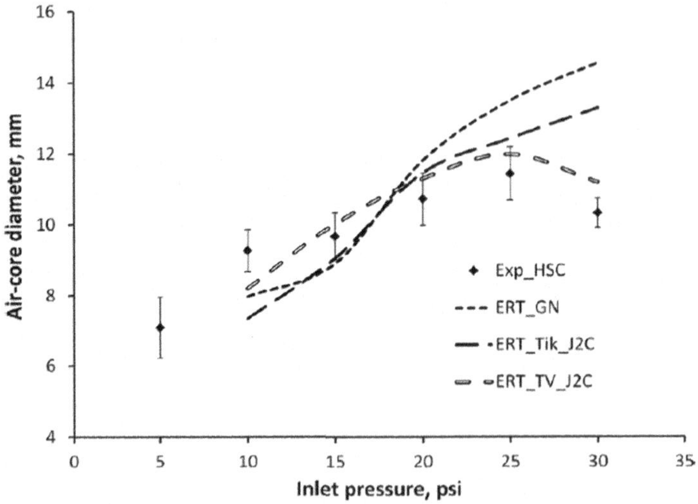

Fig. 11.9 Comparison of air-core diameter by CFD against a high-speed camera and electrical Resistance tomography of a 76 mm hydrocyclone operating at different pressure

11.2.4 Gravimetric Method

Measuring the hydrocyclone weight enables a direct, fast determination of its operating state. However, the signals are superimposed by disturbing forces coming from the flexible connections of the hydrocyclone. These forces depend on the operating conditions (pressure drop) and assemblage, necessitating special software to compensate. Therefore, this method is less recommendable.

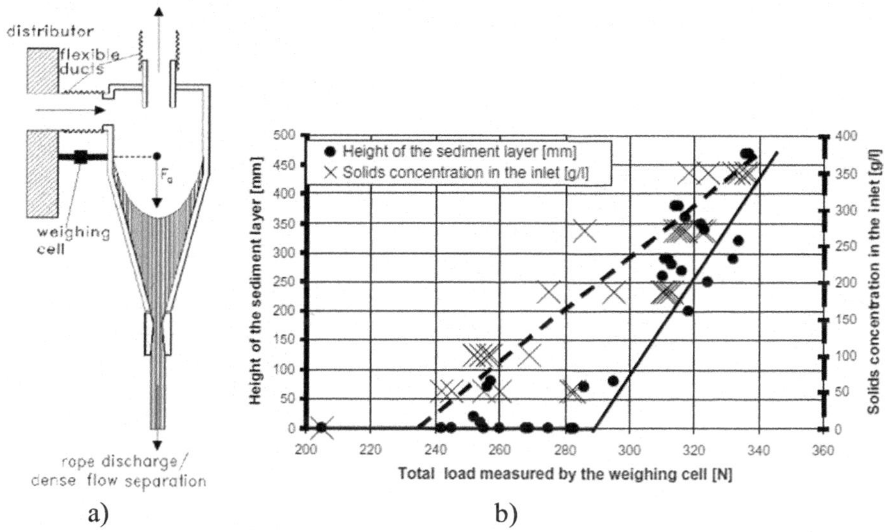

Fig. 11.10 **a** Measuring the operating state of the hydrocyclone using a weighing cell. **b** Results of the gravitational method for a 150 mm diameter hydrocyclone [6]

The hydrocyclone is attached to the distributor using an arm, where a weighing cell is also installed. All other cyclone connections on the feed side and the overflow side are flexible ducts. Figure11.7 illustrates the dependence of the measured hydrocyclone weight on feed solids concentration [6] (Fig. 11.10).

11.2.5 Image Analysis

Recorded images were transmitted from the camera to a computer, where analysis of the images was performed using the MATLABTM software package. Data analysis comprised various steps. Unlike in the [11] study, in which the entire video frame was analyzed to determine the underflow angle, only a single horizontal line was extracted from each video. By specifying only one line, the processing time was reduced significantly. The underflow width was then determined from the horizontal line. Ultimately, this is comparable to the underflow angle measurements of Petersen et al. [11], as the underflow width and angle are related to each other, provided that the camera position relative to the underflow remain completely fixed [13] (Fig. 11.11).

The results of the time series analysis show how a correlation between the underflow width variation and hydrocyclone operating conditions might be drawn; therefore, image analysis is an appropriate tool to predict the dynamic behavior of the underflow for control purposes (Fig. 11.12).

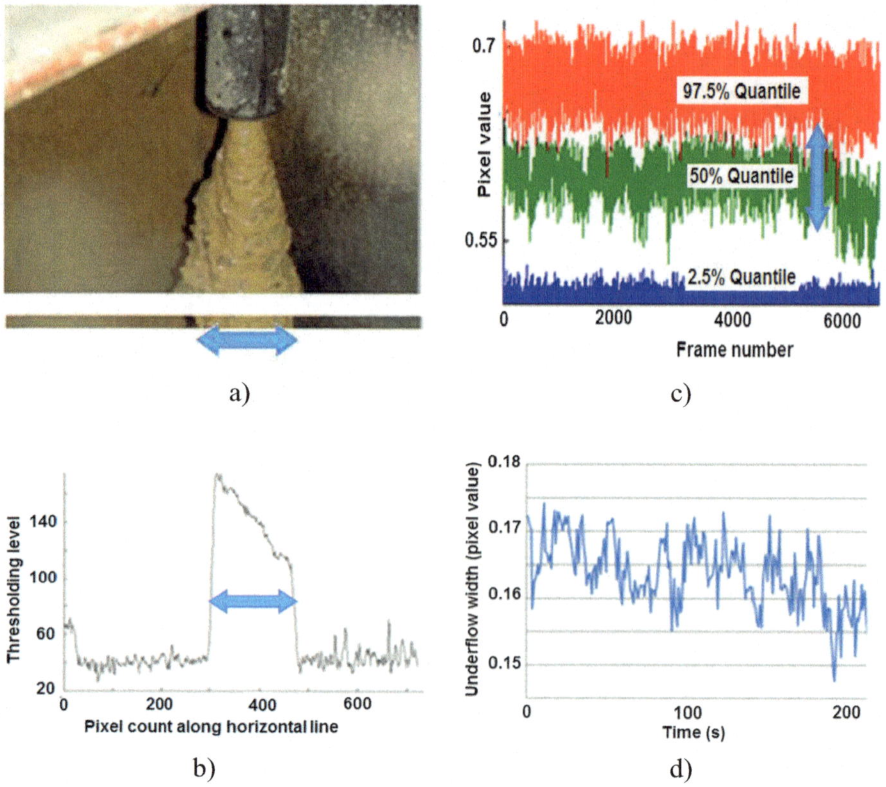

Fig. 11.11 **a** Extraction of horizontal line, **b** application of threshold limit, **c** removal of outliers using quantile distribution and **d** resultant variation of underflow width over time [13]

11.2.6 Vibration Sensing

Hydrocyclones in operation are vibrating units. Stationary vibration is generated by movements of the turbulent multiphase flow in the hydrocyclone, the eccentric and unstable movement of the air core and the discharge condition (spray or roping). Turbulent movements and particle impact on the hydrocyclone wall cause higher-frequency oscillations, while roping conditions present lower frequency.

Neesse et al. [6] measured hydrocyclone oscillation with a system consisting of an acceleration sensor, a preamplifier and computer-supported signal conditioning. The sensor mounted to the exterior, in the vicinity of the apex, indicates typical oscillation spectra for pure water, spray and rope discharge (Figs. 11.13 and 11.14).

Perez et al. [9] measured the pressure at the middle height of the cylinder, at the cone and at the apex. Pérez [10] show the waveform and fast Fourier transform (FFT) in the cylinder (top), conical (middle) and apex (down) zones, giving information on the type of discharge (spray, semi-rope or rope) (Fig. 11.15).

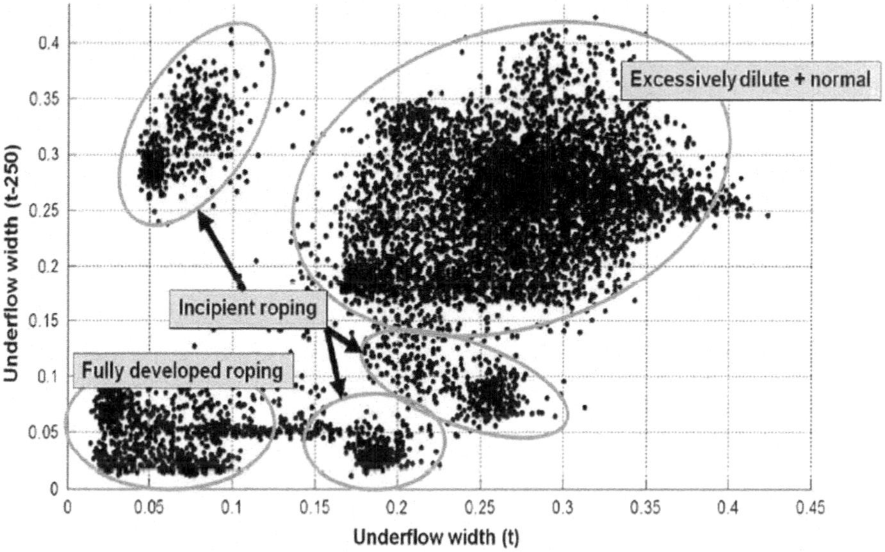

Fig. 11.12 Lag plot of underflow width variation. Gold ore [13]

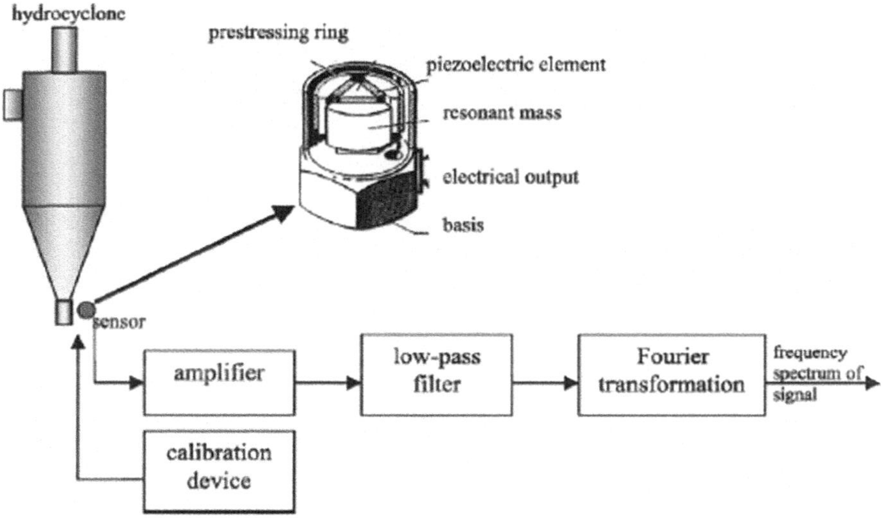

Fig. 11.13 Arrangement of the acoustic sensing system

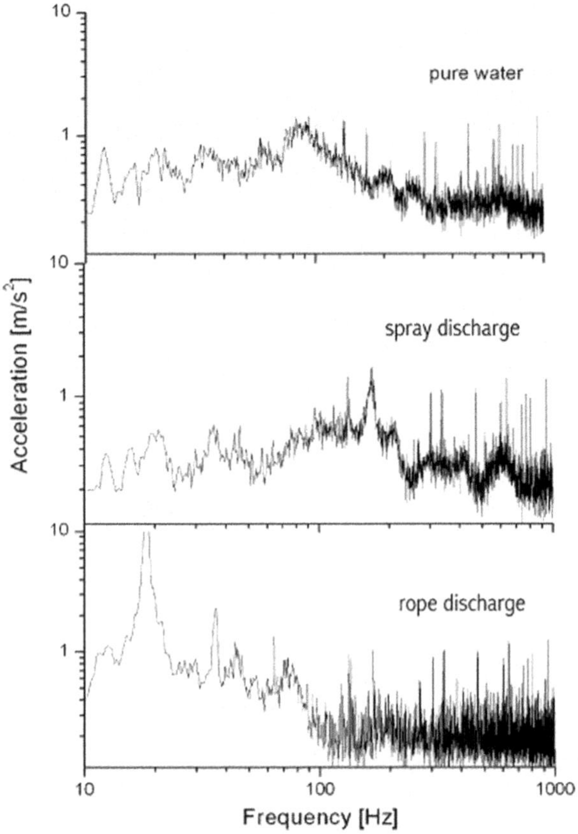

Fig. 11.14 Typical acoustic spectrum of a 150 mm polyurethane hydrocyclone with a vortex finder and an apex diameter of 72 mm, vortex diameter of 29 mm, inlet pressure of 1.6 bar and feed solids content of 400 g/l barytes [6]

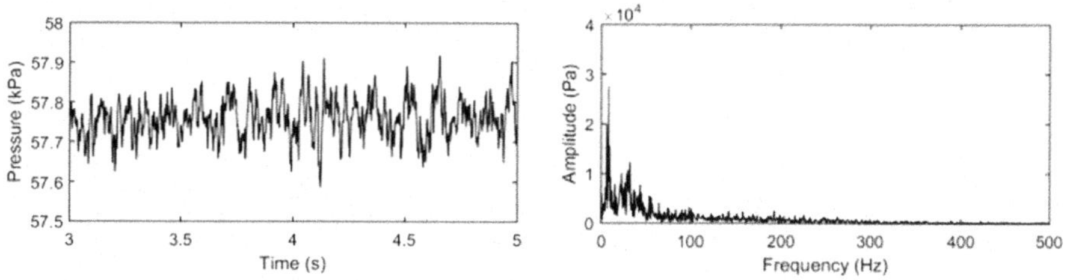

Spray discharge.

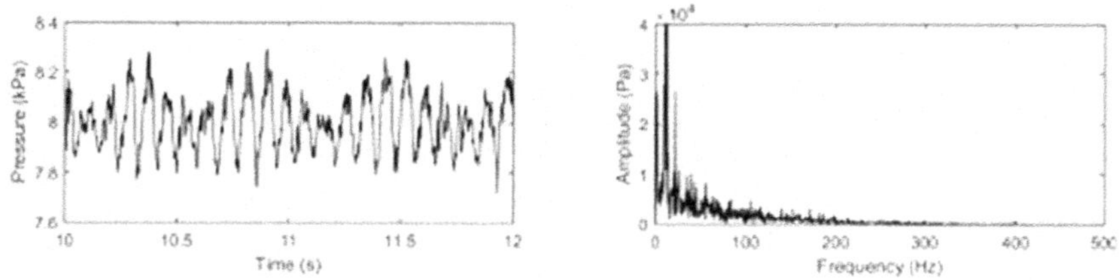

Semi-rope.

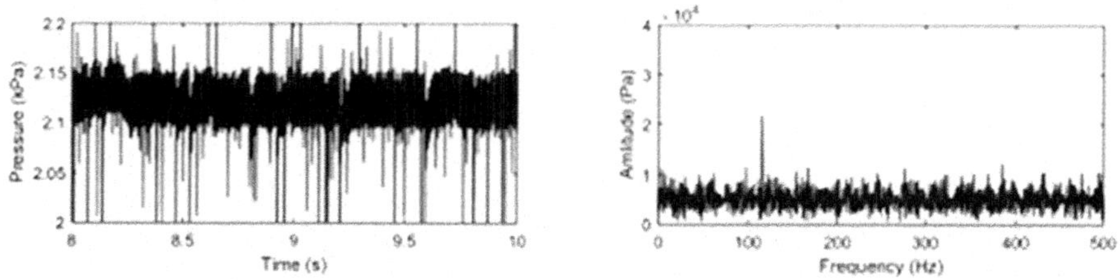

Rope.

Fig. 11.15 Simulated Waveform of the oscillation of the hydrocyclone wall at the apex, in a 75 mm hydrocyclone

References

1. Arellano K (2017) Technical evaluation of the PST in Laguna Seca Concentrator, Escondida Mining. Metallurgical Engineering Thesis, University of Concepción 2017
2. CYCLONEtrac™ PST Versus particle size Tracking System versus traditional particle size measurement techniques. Application to particle size measurement in the overflow flow of hydrocyclones in a mineral concentrator. Cidra Minerals Processing 2015
3. Cirulis D, Dunford S, Snyder J, Bartsch E, Rothman P, Maron R, Newton D, O'Keefe C, Mercuri J (2015) Process optimization using real team tracking of coarse material in individual cyclone overflow streams in Metplant Conference Proceedings, Perth, Australia, 7–8 Sept2015.
4. Meng S, Shi LS, Zhihong L, Jing L (2008) Application of electrical capacitance tomography to the concentration measurement in a cyclone Dipleg. Chin J Chem Eng 16(4):635–639
5. Neesse Th, Schubert H, Graichen K (1991) Practical and theoretical aspects of denso flow separation. Aufbereitungstechnik 32(9):459–472
6. Neesse Th, Schneider M, Golyk V, Tiefel H (2004) Measuring the operating state of the hydrocyclone. Miner Eng 17:697–703
7. Neesse Th, Golyk V, Kaniut P, Reinsch V (2004) Hydrocyclone control in grinding circuits. Miner Eng 17:1237–1240

8. Nesse Th, Schneider M, Dueck J, Golyk V, Buntenbach S, Tiefel H (2004) Hydrocyclone operation at the transition point rope/spray discharge. Miner Eng 17:733–737
9. Pérez D, Cornejo P, Rodríguez C, Concha F (2018) Transition from spray to roping in hydrocyclones. Min Eng 123:71–78
10. Pérez D, Cornejo P, Rodríguez C, Concha F (2020) Transition from spray to roping in hydrocyclones. Min Eng 123:71–84
11. Petersen KRP, Aldrich C, Van Deventer JSJ, McInnes C, Stange WW (1996) Hydrocyclone underflow monitoring using image processing methods. Miner Eng 9:301–315
12. Vakamalla TR, Narashima M (2014) Air-core size measurement of operating hydrocyclone by electrical resistance tomography. Chem Eng Technol 37(5)
13. Van Vuuren J, Aldrich C, Auret L, Bezuidenhoudt C, De Jager C (2010) On monitoring of hydrocyclones by use of image analysis. In: 13th symposium on automation in mining, mineral and metal processing, Cape Town, South Africa, 2–4 Aug 2010
14. Williams RA, Dyakowski T, Xie CG, Luke SP, Gregory PJ, Edwards RB, Gate LF (1995). In: Beck MS et al (eds) Process tomography—implementation for industrial processes. EU/UMIST, Manchester, pp 3–15

Chapter 12
Case Studies

Abstract This chapter presents 6 Case Studies of application of hydrocyclones in the mining industry: Cavex hydrocyclons at Placer Dome La Coipa Chile, at Minsur Perú, at Ok Tedi Paua New Guinea, at El Soldado, Anglo American, Chile. Hydrocyclones optimization at Chuquicamata, SQM Salar and PuCobre Chile. Special hydrocyclones Chile, Economic cost of roping in Minera Escondida, Chile.

12.1 Application of Cavex Hydrocyclones in the Mining Industry

12.1.1 Application of Cavex 650CVX13 Hydrocyclones at Placer Dome's La Coipa Gold Plant in Chile

The La Coipa open pit gold and silver mine is located in northern Chile. The mine employs approximately 440 people and had a projected mine life of 6 years. In 2003, La Coipa was scheduled to produce a total of 177,000 oz of gold and approximately 7.9 million ounces of silver [5] (Fig. 12.1).

Objective
The company wanted to reduce the Mantos de Oro concentrator's P80 cut size below 400 mesh at the overflow product without decreasing the plant capacity. Using a circuit simulation package, the only way to achieve both objectives was to increase separation sharpness. Improving the hydrocyclone classification efficiency was supposed to allow the completion of this task.

The Processing Plant
The mineral processing plant consisted of a pre-crushing circuit, primary crusher, semi-autogenous ball mills and hydrocyclones and a leaching, filtering and washing Merrill-Crowe plant, which is better than a carbon-in-pulp system for high-grade silver materials and the dore refinery. Tailings are filtered to recover excess water as well as residual cyanide and metal credits. A dry tailing disposal system was chosen over a conventional wet tailing impoundment for site-specific environmental reasons.

F. Concha A. and J. L. Bouso A., *Fluid Mechanics Fundamentals of Hydrocyclones and its Applications in the Mining Industry*, Fluid Mechanics and Its Applications 126,
https://doi.org/10.1007/978-3-030-67913-2_12

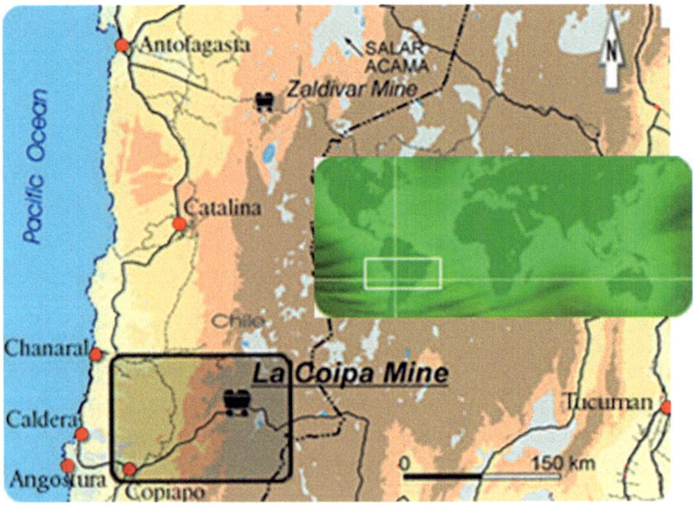

Fig. 12.1 La Coipa mine location in northern Chile

A precipitate retorting facility and a precipitate and calcine handling system were commissioned in April 1998 to replace the nitric acid leach circuit for mercury processing. A new pre-crushing circuit was commissioned in October 1999, increasing the nominal throughput to over 17,000 tons per day.

The grinding circuit has one SAG mill, two ball mills and two clusters of eight D26B hydrocyclones. In this work, the first cluster of eight D26B hydrocyclones was replaced by five CAVEX 650CVX13 hydrocyclones. At the end of the testing, the second cluster was also replaced by five CAVEX 650CVX13 hydrocyclones (Fig. 12.2).

New Hydrocyclone Design

The challenge was to design and fabricate a hydrocyclone with minimum turbulence and maximum residence time. The combination of a new vortex with a new, longer conical section increases residence time, reducing cut size and significantly improving separation sharpness. The new CAVEX inlet head, based on a phenomenological model introduced by Conha et al. [6], not only reduces turbulence intensity, but also yields maximum efficiency. This model was described in detail in Chap. 4, Part III of this book, and there is no need to repeat the description here.

Figure 12.3 illustrates the differences between a traditional hydrocyclone and this new design. The new inlet head curvature allows the use of a very short vortex finder, diminishing the mixing effect in the upper zone of the cyclone. This new hydrocyclone design breaks the paradigm that states "the vortex finder length must be longer than the inlet height at the cyclone entrance," as shown in the figure.

The new CAVEX inlet design reduces turbulence intensity throughout the hydrocyclone and, consequently, reduces equipment wear. Figure 12.4 depicts differences

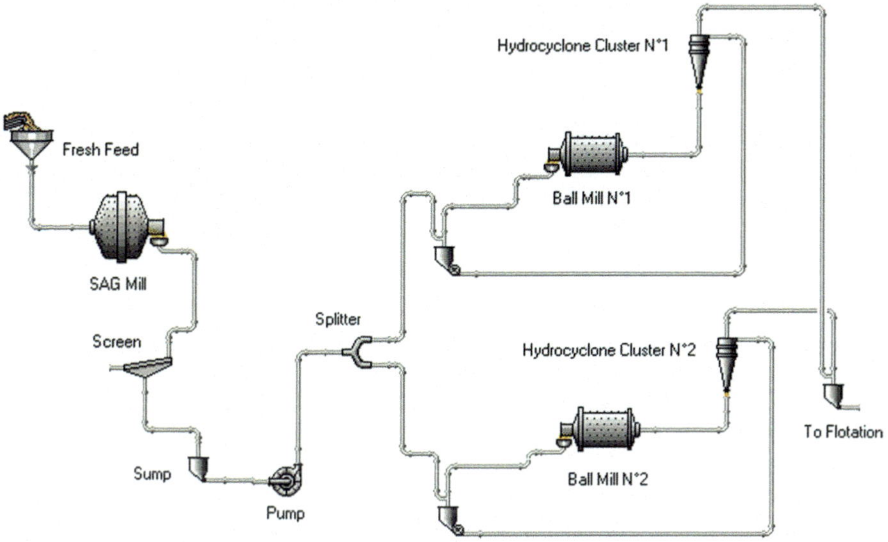

Fig. 12.2 La Coipa grinding-classification circuit

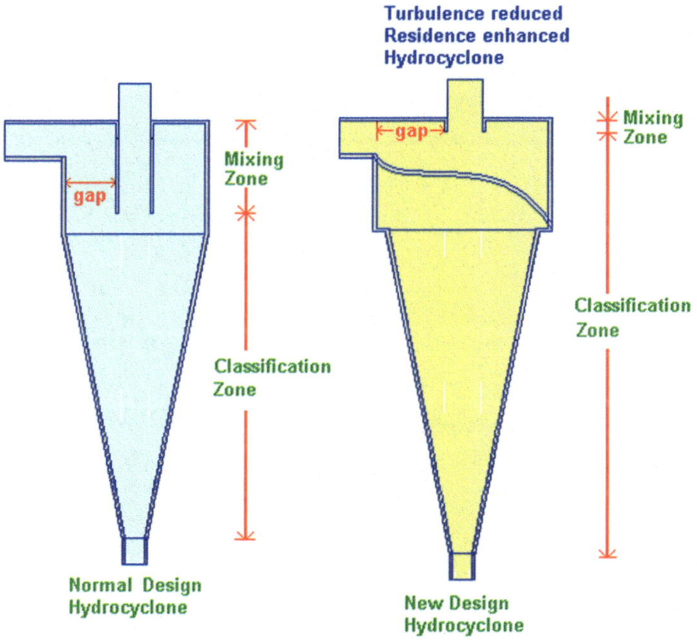

Fig. 12.3 Main geometrical differences between a conventional design and the new Cavex design

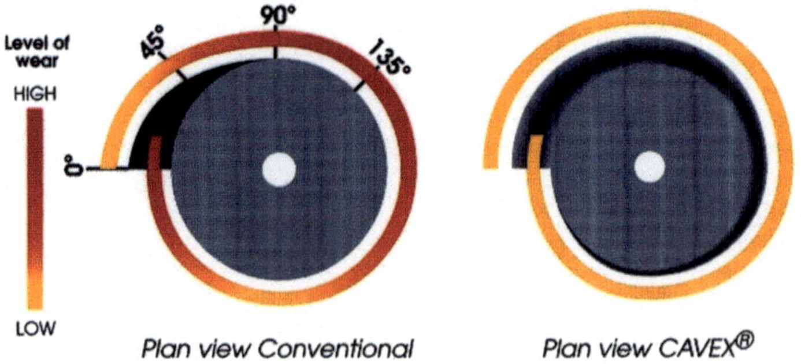

Fig. 12.4 Wear distribution intensity comparison between a conventional and Cavex hydrocyclone

in wear distribution intensity inside the inlet head of a conventional cyclone and this new design.

The above-mentioned changes also increased the flowrate per cyclone, meaning the cyclone can treat a greater volume with less energy consumption and fewer cyclones in the cluster (Figs. 12.5, 12.6 and 12.7).

Results

Several sampling campaigns were conducted for two different hydrocyclones, the Krebs G-max and Cavex, at comparable throughputs. The results, shown in Fig. 12.8,

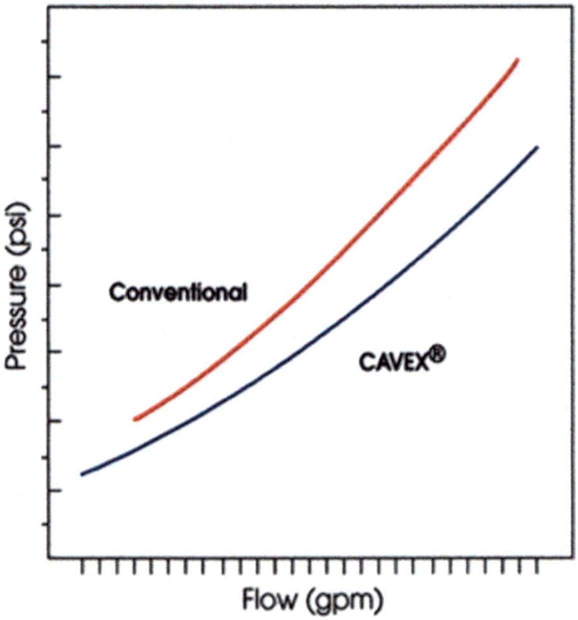

Fig. 12.5 Capacity comparison between conventional and Cavex cyclones

Fig. 12.6 Five CAVEX 650CVX13° installed in cluster 2

Fig. 12.7 Hydrocyclone installation Cavex hydrocyclones installed at La Coipa

indicate that the CAVEX 650CVX13 produces a significantly finer product than the Gmax DS-26 without decreasing circuit capacity. Figure 12.8 shows that the sharpness index and cut size were reduced in the CAVEX hydrocyclone. The La Coipa plant implemented the hydrocyclone change, allowing the production a finer product, decreasing its P80 by 20% and maintaining the same tonnage. In the figure the squares indicate the results for the CAVEX hydrocyclones with a normal apex, triangles CAVEX with a ceramic apex and crosses G-Max 26 Krebs hydrocyclones.

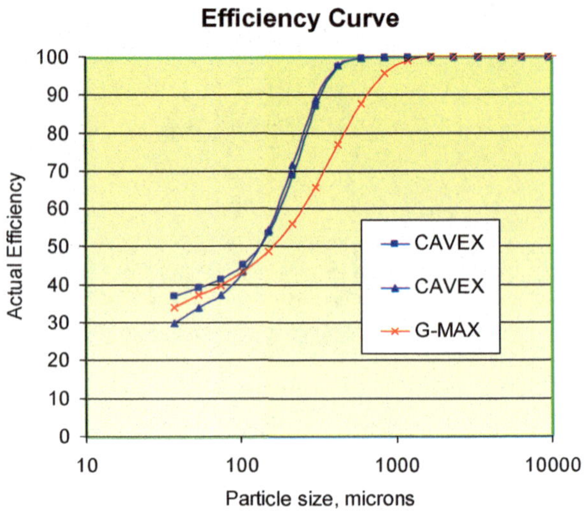

Fig. 12.8 Efficiency curve comparison of new Cavex and G-max at Mantos de Oro

12.1.2 Test of Cavex 400CVX10° Hydrocyclone at Minsur S.A., Puno, Perú

The Minsur S.A. San Rafael mine is located in South America in the southern Peruvian Andes, Antauta/Puno district, at approximately 4,530 m.a.s.l. The ore averages 5.05% tin content and is processed by in a 2 500 t/d concentrator to obtain a final 61.5% tin concentrate with an approximate recovery of 90.6% [2] (Figs. 12.9 and 12.10).

The San Rafael mine had a configuration of three Krebs DS-15 hydrocyclones in its grinding-classification circuit and wanted to optimize the circuit with a target of

Fig. 12.9 Minsur S.A. San Rafael mine location

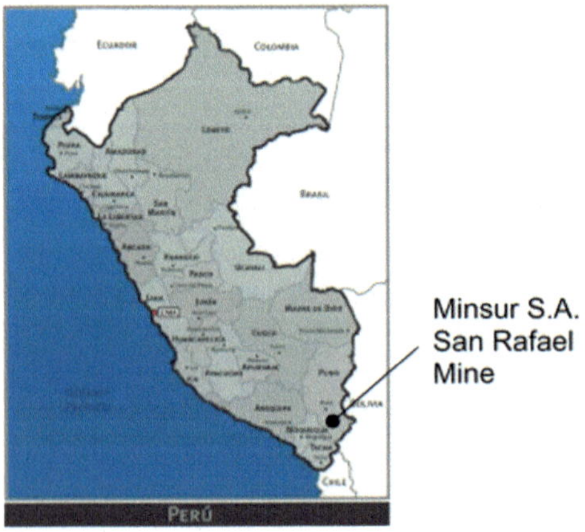

Fig. 12.10 Three-hydrocyclone cluster

5% + 200 mesh at the overflow and a minimum of -200 mesh at the underflow by evaluating the following hydrocyclones: Krebs DS-15, GMax 15–3117 and Cavex 400CVX10.

The results of three tests are given in the table below.

Hydrocyclone brand	N° of Tests					
	1		2		3	
	+ 200 in overflow	– 200 in underflow	+ 200 in overflow	-200 in underflow	+ 200 in overflow	– 200 in underflow
DS-15	66.77	9.6				
Gmax 15	17.9	15.29	4.75	14.6		
Cavex 400	10.4	13.5	8.5	18.3	6.6	14.2

As the Cavex 400CVX10 performed best, the Minsur mine opted to purchase these hydrocyclones for its new cluster and a new classification set point was defined.

12.1.3 Evaluation of Cavex650CVX Hydrocyclone in the Ok Tedi Grinding Circuit in Papua New Guinea (PNG) [4]

The mining/milling operations of the Ok Tedi mine are located 18 km from the PNG/Indonesian border on Mount Fubilan, where average annual rainfall is an incredible 10 m. Operations started in 1984 and mine closure was expected in 2010. Although Ok Tedi is the single largest business contributor to the economy of PNG, in January 2002 BHP Billiton divested its holdings in the mine. Ok Tedi had two grinding circuits, called SAG1 and SAG2, followed by two ball mills in a closed reverse circuit with a cluster of 10 Krebs D26B hydrocyclones each. The plant capacity was 90,000 t/d in 2002, with an average grade of 0.82% copper and 0.8 g/t gold [4].

The concentrator required an increase in capacity without changing the number of hydrocyclones or coarsening the grind. Ten CAVEX 650CVX hydrocyclones were installed in cluster CS1 to compare its results with those of cluster CS2, where the original Krebs D26B models were replaced with DS26 GMAX hydrocyclones (Figs. 12.11 and 12.12).

In 2001, Ok Tedi commissioned JKTech to conduct a survey to compare the performance of the two hydrocyclone clusters. The results showed that:

- Classification in CAVEX hydrocyclones was finer.
- The CAVEX hydrocyclones operated with a lower feed pressure.
- Circulating loads were significantly reduced with the CAVEX cyclones (Fig. 12.13).

Fig. 12.11 SAG2 grinding circuit

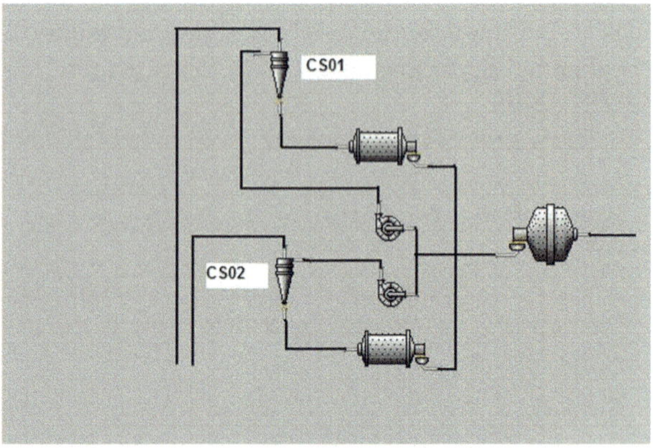

Fig. 12.12 Cavex 650CVX in cluster CS1

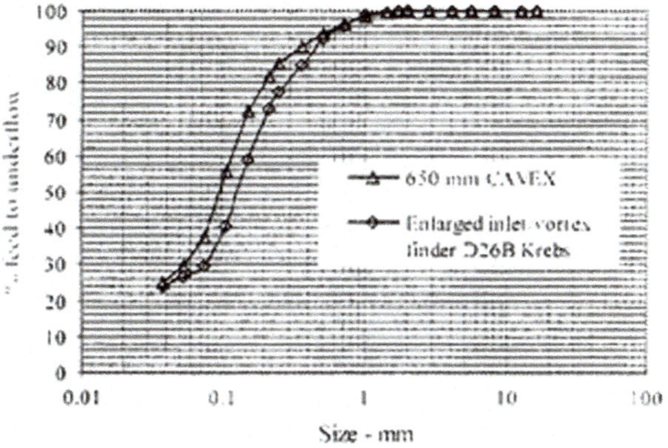

Fig. 12.13 Selectivity functions of both clusters

12.1.4 ReCyclone at El Soldado Copper Mine in Chile

Anglo American's El Soldado mine is a SAG/conventional mill plant located 120 km north of Santiago, Chile (Fig. 12.14).

The deposit consists of numerous isolated ore bodies embedded in trachytes, andesites and tuffs and located in an area 1 800 m long and 800 m wide. The lateral limits of the ore bodies are characterized by abrupt variations in the copper grade. The main mineralogical species in the copper ore processed by the El Soldado operation, in order of importance, are chalcopyrite (40%), bornyte (30%), chalcocite (5%) and pyrite (25%). The two main facilities are the El Soldado mine and the El Cobre concentrator plant (Fig. 12.15).

Fig. 12.14 El Soldado copper mine, Chile

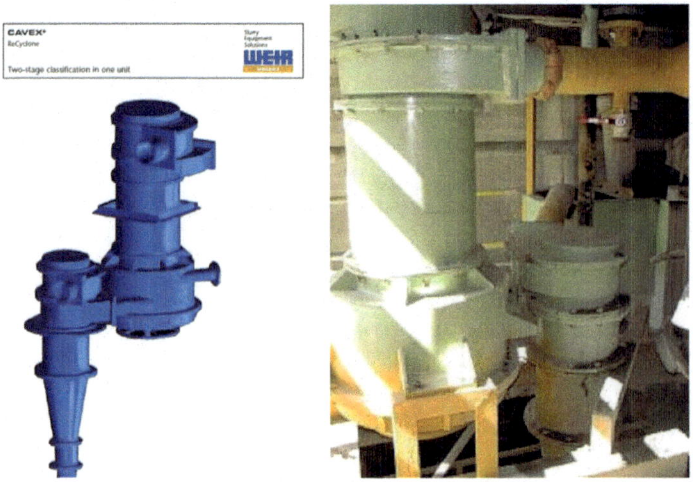

Fig. 12.15 Schematic and photograph of a Cavex ReCyclone

12.2 Hydrocyclone Optimization for the Chuquicama Division of chile's Codelco

There are plans to improve the facilities of Chile's Codelco Norte copper concentrator for the new two ores it is processing. These ores come from the Chuquicamata and Radomiro Tomic mines, which contain very different mineralogies. The challenges presented by this situation were solved by processing these two different minerals in separate plants, with plants A-0 and A-1 treating Radomiro Tomic ores and plant A-2 treating Chuquicamata ores [8].

Sampling around the grinding classification circuit in plant A-2 showed that the conditions under which the circuit was operating were inefficient. Microscopic analysis of the flotation feed in the A2 concentrator indicates that minerals do not need a high liberation to obtain recoveries greater than 85% in the rougher flotation; therefore, the objective of this work was to find the optimum hydrocyclone operating conditions, with particle size restriction of 30–32% + 65 mesh (Fig. 12.16).

The Chuquicamata Division is an open pit copper, molibdenum and gold mine located 15 km north of Calama in the Antofagasta Region, 2 870 m above sea level. It is considered the largest open-pit mine in the world and started production in 1915. In the near future it will be converted to an underground mine (Fig. 12.17).

The Chuquicamata A-2 plant consists of two inverse closed-grinding circuits processing between 58,000 and 63,000 t/d. The only difference between these lines is the type of hydrocyclone. The Fig. 12.18 shows the general flowsheet of the A-2 plant.

Two 32 × 14 foot SAG mills with 34% 5 inch balls rotating at 8 rpm draw an average of 7000 kW. The mill discharges over two double-deck vibrating screens. The 20 mm oversize pebbles are further crushed and return to the SAG mill. The screen

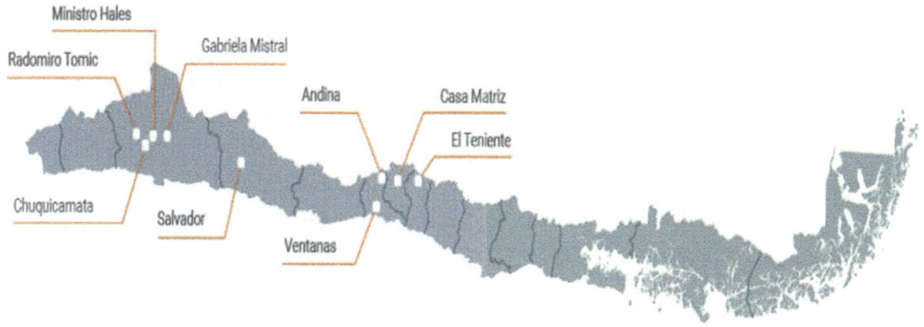

Fig. 12.16 Chile's Codelco mines

Fig. 12.17 Codelco Norte's Chuquicamata open pit mine

undersize pulp is sent to two batteries of five to eight hydrocyclones, one consisting of Krebs D-26 hydrocyclones and the other of Cavex 650CVX hydrocyclones. The hydrocyclone overflow, with a designed 30%-65 mesh, goes to flotation and the underflow feeds the secondary ball mills, which rotate at 13 rpm and have a ball volume level of 32% (Figs. 12.19 and 12.20).

This work was performed in five steps:

1. **Compilation of operational data of grinding and classification for base condition** Sampling point in Fig. 12.18. Results in Table 12.1
2. **Microscopy of the tailings**
   The ore contains 1.2% copper, in several mineralogical species, and 2.41% iron as pyrite. Figure 12.21 shows the mineralogical species distributions.
3. **Circuit balance for normal conditions using JKSimMet software**
   For the base condition the technical results for sections 17-A and 17-B of plant A2 were the following (Table 12.2).

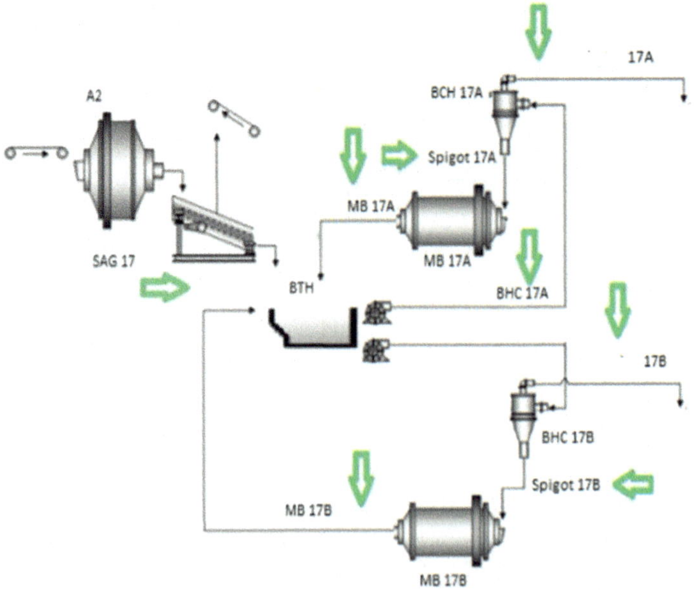

Fig. 12.18 Chuquicamata plant A-2 Sect. 17A flowsheet. Green arrows are sampling points

Fig. 12.19 SAG grinding mills

4. **Technical evaluation of the plant under normal and optimum conditions**
   Simulations were run using base and three additional conditions to find the optimum operating values to achieve maximum capacity with the restriction of 32% + 65 mesh in the hydrocyclone overflow.
   Three simulations were run with different apex/vortex ratios, with information from Sect. 17A of the plant: hydrocyclone diameter 0.666 m, cylindrical part length 0.753 m, feed diameter 0.23 ¿?, cone angle 18°, basic capacity 600 t/h and pressure 8.1 psi.

Fig. 12.20 Secondary grinding circuits

Table 12.1 Operational data for base condition

Operating conditions	Section 17
Hydrocyclone diameter (cm)	66
Hydrocyclone F80 (μm)	1200–1300
Hydrocyclone P80 overflow size (μm)	3002
Hydrocyclone operating pressure (psi)	6–9
Hydrocyclones in operation per mill (N°)	5–8
Solids in hydrocyclone feed (%)	70–72
Mill dimensions (m × m)	5.4 × 7.9
Mill rotation speed (rpm)	13
Recharge ball size (in)	3
Ball mill power (kW)	2800–3100
Feed tonnage to mills (tph)	600

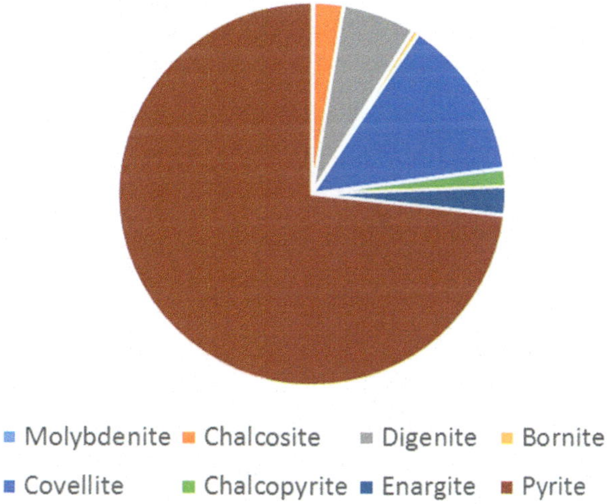

Fig. 12.21 Distribution of mineralogical species

Table 12.2 Balanced grinding-classification circuits, plant A2

Plant	A2	
Tonnage (tph)	1650.2	
Fresh tonnage (tph)	1305.1	
Recirculated tonnage (tph)	345.1	
Water flow to the SAG (m^3/h)	470.6	
Water flow to the pulp feeding tank (m3/h)	833.4	
Section	17-A	17-B
Hydrocyclones feed concentration (%)	70.0	75.9
Hydrocyclones underflow concentration (%)	79.8	77.6
Hydrocyclones overflow cvoncentration (%)	36.1	46.9
Ball mill discharge concentratio (%)	78.2	77.8
BTH F80 feed (μm)	1282.4	1428.7
BHC F80 discharge (μm)	1326.7	1794.1
BHC P80 overflow (μm)	302.7	367.2
Ball mill P80 (μm)	1153.6	1340.6
Ball mill Rotational speed (rpm)	1.2	14.3
CC	613.0	504.9
Hydrocyclones pressure (psi)	8.7	7.9
Ball mill power (kW)	2797.6	2946.5
S.I	0.2	0.1
a (%)	40.0	38.0
Retainede i the overflow (%)	30.1	36.0
Ball mill filling level (%)	27.0	29.0
d50c (μm)	142.5	95.2
Hydrocyclones in operation (number)	6.0	6.0

Section 17-A, plant A2 case	Base	Case 1	Case 2	Case 3
Apex/vortex ratio	0.73	0.51	0.51	0.54
Hydrocyclones number	6	7	7	6
Feed pressure (psi)	8.1	5.7	7.6	8.5
Solids in the feed (%)	70.0	70.0	70.0	66.0
% Solid in the overflow (%)	36.1	38.7	39.1	40.1
% Solid in the underflow (%)	79.8	80.9	79.9	77.0
P80 (μm)	304.7	318.0	325.0	285.1
% + 65 mesh	30.1	32.0	31.9	32.2
Capacity ton/h	600	625	670	650

The simulations of the three cases, with different apex-to-vortex ratios and % solids in the feed, resulted in an increase in circuit capacity of 4.2% for case 1, 11.7% for case 2 and 8.3% for case 3. According to these results the best alternative is case 2, which has the maximum capacity and complies with the particle size restriction.

5. **Economic evaluation**
   An economic evaluation of the three proposed alternatives found that that the net present value was much greater than zero for all three cases, and the highest for case 3.

12.3 Optimization of Hydrocyclones for Brine Recovery in SQM SALAR

SQM SALAR is a Chilean brine company that was founded in 1968 and initially owned jointly by the state of Chile and the Alto Lautaro Niter Company. In 1971 it was nationalized and remained entirely in the hands of the state of Chile until 1988, when it was privatized. The company is located in the Salar de Atacama Region [9].

The Salar de Atacama is a source of underground brine formed by natural leaching from the Andes. The brines, which contain sodium, potassium iodine and lithium, are subjected to solar evaporation and treated separately in plants to produce potassium chloride and sulphate, iodine and lithium carbonate. The plant that produces potassium chloride and sulphate uses the saturated brine to transport the solid salts and perform the entire grinding, classification, flotation and solid–liquid separation process. This work is related to the recovery of the brine with hydrocyclones in the solid–liquid separation circuit (Fig. 12.22).

The concentrator plant consists of four processes: (1) grinding the material from the evaporation ponds to release the material of interest, (2) flotation of the ground material to separate the potassium chloride (sylvite) from sodium chloride (halite), (3) leaching of salts to dissolve sodium chloride contaminants and (4) solid–liquid separation in two steps, (a) filtration to obtain a final product with 5% humidity and a filtrate of brine containing a small amounts of potassium chloride; (b) hydrocycloning to separate the low-Cl solids and the maximum amount of brine. The hydrocyclone underflow goes to the tailing filters, where the cake is sent to the salt deposit and the brine filtrate is recovered. The main function of the hydrocyclones is to recover most of the brine, with the restriction of 55% solids in the discharge and less than 8% solids in the overflow. The objective of this work was to find the best conditions for the hydrocyclone performance to optimize brine recovery [10] was used to simulate the experimental.

Hydrocyclone Type and Operational Data
The hydrocyclones used in SQM were Warman 15 inch diameter Cavex 400CVX models, which are shown in Fig. 12.23 below.

Fig. 12.22 Atacama Salar in the Antofagasta Region of Chile

Fig. 12.23 Hydrocyclone rack in SQM

Two sampling campaigns were made; the first over three months, during which 8 samples were taken to obtain representative experimental data with which to calibrate the CIMM mathematical model, and the second over one month to validate the model. The samples were analyzed for flow rates, particle size, concentration of the feed, underflow and overflow and solid and brine densities. The results are shown in Table 12.3 and Fig. 12.24.

Table 12.3 Experimental data

Design variables		Operating variables	
Number of hydrocyclones	4	Solid density	2.14 t/m^3
Hydrocyclone diameter (in)	14.96	Saturated brine density	1.23 t/m^3
Height (in)	52.1	Feed rate per unit	123 t/h
Feed diameter (in)	4.33	Volume flow rate per unit	391.59 m^3/h
Vortex diameter (in)	3.94	Pressure	24 psi
Apex diameter (in)	3.15	% solids in the feed	23.0
Apex-to-vortex ratio	0.80	% solids in underflow	45.3
		% solids in overflow	5.7

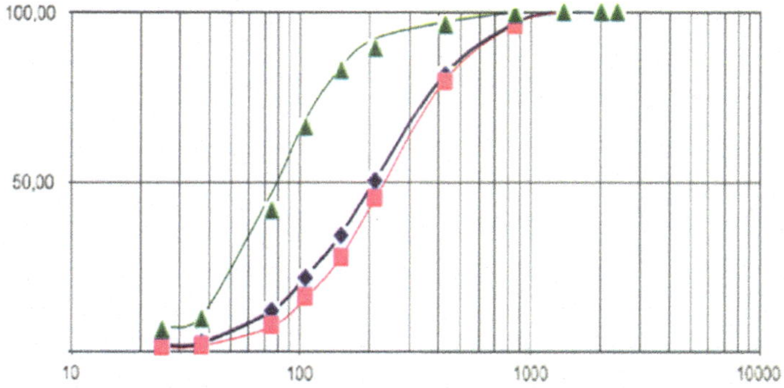

Fig. 12.24 Experimental and simulated particle size of the feed, overflow and underflow with the Molycop Tools model and data from one sample

From these results and the other experimental data obtained from the sampling campaign, average CIMM model parameters were calculated and are shown in Table 12.4.

Simulation and Optimization

(a) **Effect of feed flow rate and concentration on brine recovery**

The following data were used for the simulation (Table 12.5):

With the feed volume flow rate varied and the solids capacity maintained, the brine recovery in the overflow increases from 22 to 82% and the underflow concentration from 50.65 to 56.44% (Fig. 12.25).

Table 12.4 Average value of CIMM parametersf

α_1	α_2	α_3	α_4	λ
14.29	17.33	17.97	0.87	1.86

Table 12.5 Design and operating variables for feed flow

Design variables		Operating variables	
Number of hydrocyclones	4	Solid density	2.14 t/m^3
Hydrocyclone diameter (in)	14.96	Saturated brine density	1.23 t/m^3
Height (in)	52.05	Feed rate per unit	106.35 t/h
Feed diameter (in)	4.33	Volume flow rate per unit	370; 400;430; 475 m^3/h
Vortex diameter (in)	3.94	Pressure	24 psi
Apex diameter (in)	2.76		
Apex-to-vortex ratio	0.80		

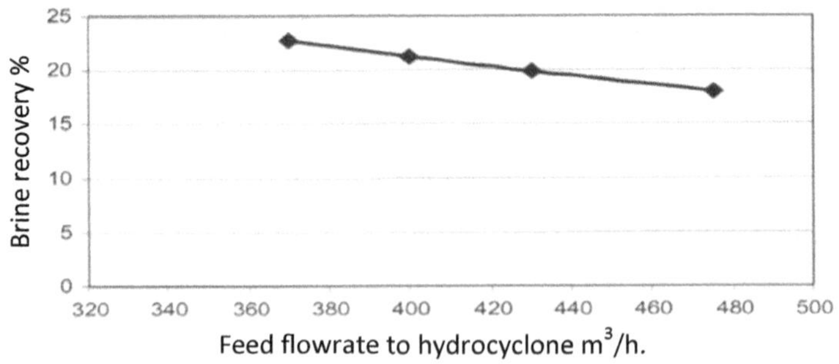

Fig. 12.25 Brine recovery at the underflow with changes in the feed concentration

(b) **Apex-to-vortex ratio**

The following data were used for the simulation (Table 12.6):

With an increase in the apex-to-vortex ratio, the recovery of brine in the overflow increases from 11 to 36%, with underflow concentration decreasing from 56.44 to 41.52% (Fig. 12.26).

(c) **Number of hydrocyclone**

Finally, the number of hydrocyclones was increased from 4 to 5 (Table 12.7).

Table 12.6 Design and operating variables for vortex ratio

Design variables		Operating variables	
Number of hydrocyclones	4	Solid density	2.14 t/m^3
Hydrocyclone diameter (in)	14.96	Saturated brine density	1.23 t/m^3
Height (in)	52.05	Feed rate per unit	136.53 t/h
Feed diameter (in)	4.33	Volume flow rate per unit	475 m^3/h
Vortex diameter (in)	3.94	Pressure	24 psi
Apex diameter (in)	2.76; 3.15;3.54	% solid in the feed	21.26
Apex to vortex ratio	0.70; 0.80; 0.90		

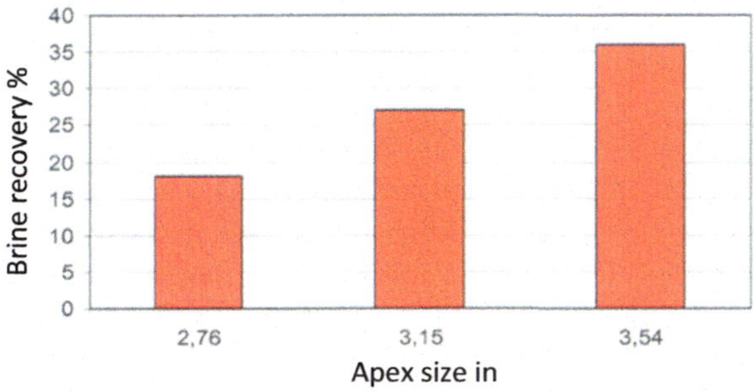

Fig. 12.26 Brine recovery at the underflow with changes in the apex-to-vortex ratio

Table 12.7 Design and operating flow for number of hydrocyclones

Design variables		Operating variables	
Number of hydrocyclones	4; 5	Solid density	2.14 t/m^3
Hydrocyclone diameter (in)	14.96	Saturated brine density	1.23 t/m^3
Height (in)	52.05	Feed rate per unit	136.53 t/h
Feed diameter (in)	4.33	Volume flow rate per unit	475 m^3/h
Vortex diameter (in)	3.94	Pressure	24 psi
Apex diameter (in)	2.76	% solid in the feed	21.26
Apex-to-vortex ratio	0.70		

With the number of hydrocyclones increased in one unit, the recovery of brine in the overflow decreases from 82 to 78%.

(d) Best design and operational variables

According to the results of this work, the following are the best conditions (Table 12.8).

Table 12.8 Design and operating variables for best design and operation

	Feed	Underflow	Overflow
No. of hydrocyclones = 4			
Apex-to-vortex ratio = 0.70			
t/h	136.53	118.36	18.17
Brine m3/h	411.20	74.28	663.92
Pulp t/h	642.31	209.73	432.58
Pulp m^3/h	475.00	129.59	345.41
% solids by weight	21.26	56.44	4.20

Fig. 12.27 Grinding-classification system at Valle Central concentrator

12.4 Special Types of Hydrocyclones

12.4.1 Grinding Circuits

Minera Valle Central, Rancagua, Chile, process the tailings of Minera El Teniente Codelco the largest copper underground mine in the word. Four lines of a first stage of 750 mm (30 in.) and 500 mm (20 in.) diameter hydrocyclones clusters, receive the full tailings from the mine to perform a classification of the coarser and heavy particles. These hydrocyclones batteries are gravity fed. The underflow of this stage is the fed to a second stage of 8 lines of a closed-circuit grinding circuit with one ball mill each and 400 mm diameter hydrocyclones (Fig. 12.27).

12.4.2 Coarse Classification (Fractioning)

The Cerro Negro plant owned by Minera del Pacifico Mining Company (CAP Mineria) in Copiapó Chile, has an iron ore treatment processing plant using magnetic separation The final product is classified in two fractions, coarse and fines, with short flat bottom hydrocyclones 750 mm (30 in) diameter in diameter hydrocyclones. These

Fig. 12.28 Coarse classification at Minera del Pacífico

hydrocyclones perform a coarse cut size around 600 μm, which can only be achieved with these types of cyclones with a short cylindrical section. Changes in cut size can be easily achieved increasing or diminishing its length and the number of cylinder units.

The underflow with the coarse fraction of these hydrocyclones is sent to a dewatering screen and the fine overflow to a thickener. In this way the blocking of the thickeners are avoided plant and the environmental impact is reduced (Fig. 12. 28).

12.4.3 Desliming/Classification

In copper plants tailings disposal, a classification is often made to produce two fractions, a coarse and fine fraction. The coarse fraction is used to build the tailings dam. At same time that the fine fraction, with most of the liquid, is sent to the tailing pool for a further settling and better water recovery.

The Chilean El Soldado Anglo American copper plant located at Calera, Chile (north of Santiago), classify the final tailings with a cluster of 500 mm diameter hydrocyclones, obtaining a coarse fraction (underflow) with a minimum content of finer particles, smaller than 74 μm (200 mesh), which is pumped with special piston pump to a large distance to the tailing pond (Fig. 12.29).

Fig. 12.29 Tailing dam at El Soldado Anglo American Copper plant

12.4.4 Thickening

Cementos Polpaico in La Calera, Chile is one of the few cement production operations in the world using a wet process. The tailings dam is built with 250 mm diameter (10 in.) flat bottom hydrocyclones manufactured in casted polyurethane. These hydrocyclones are very light making it ease to moved along the upper part of the wall optimizing the wall construction (Fig. 12.30).

Fig. 12.30 Construction of the tailing dam at Cementos Polpaico

Fig. 12.31 Mantos copper mining tailings treatment plant

12.4.5 *Classification/Thickening*

"Hydro-drain" is a new technology for tailings filtration, based on small hydrocyclones, 100 mm diameter, in combination with high frequency dewatering screens. The hydrocyclones achieve a very fine cut size to recover most particle as possible from the dewatering filtrate, to reduce the amount of fine particle being sent to the settling stage with thickeners. This process minimize the environmental impact, reducing the tailings dump size but also diminishing the fresh water consumption.

The Mantos Copper mining company, in Antofagasta, Chile, into the Atacama Desert was one of the first mines introducing this technology which let them to recover around 70% of its tailings volume in a solid form, reducing and recycling the water used in their process (Figs. 12.31 and 12.32).

12.4.6 *Densifications with Hydrocyclones*

Densification in hydrocyclones involves the removal of most of the liquid from the pulp achieving highest possible concentration of solids form the hydrocyclone underflow. If all the solids are recovered, the process becomes a clarification operation.

Fig. 12.32 Discharge of the dry tailings high frequency dewatering screens

As solids recovery is always limited, the overflow current is usually cloudy and the clarification is partial.

To perform densification, the geometry of a hydrocyclone must be elongated with a low cone angle, usually between 5° and 20° for hydrocyclones with diameters between 10 and 500 mm (Fig. 12.33).

In addition to a low cone angle, extended cylindrical sections are often used to increase the length of the hydrocyclone, resulting in a longer pulp residence time during its passage through the hydrocyclone and consequently a greater recovery of solids in the discharge underflow.

Fig. 12.33 Two rigs of 100 mm and 500 mm densification hydrocyclones

12.4.7 Desliming Hydrocyclones

Feed to flotation, magnetic separation, gravimetric separation and filtration processes must have the lowest possible fine particle content. The removal of these particles can be accomplished through a desliming operation. The resulting cutting size is different depending on the concentration process used. In flotation, for example, feeds with considerable amount of fine particles absorb a great quantity of very expense reagents resulting in low recovery, desliming is recommended.

The geometry of the desliming hydrocyclones is similar to that of thickening hydrocyclones, with the only difference that in this case the cutting size, usually between 10 and 100 μm, is vital (Fig. 12.34).

Fig. 12.34 Hydrocyclones in a 10 microns desliming operation for flotation

12.4.8 Degritting

In industrial mineral treatment plants such as that process kaolin, a refined product requires multi-stage classification therefore these stages are combined to obtain the fine product, with wash-up stages of the coarse fraction. Fine particles retained in the underflow flows (short-circuit) are recovered to obtain cleaner coarse fractions, improving the fine particles recovery at same time. In general, small diameter hydrocyclones between 40 mm (16 in.) and 100 mm (40 in.) are used, due to the small cut size (10–20 μm) of these operations.

Some kaolin minerals have quartz and feldspar sands, mica and other mineralogical species as by-products, so that treatment plants have hydrocyclones stages with the combination of desliming and washing. The process where a small amount of coarse particles is rejected from a flow with fine particle in suspension is called degritting (Fig. 12.35).

12.5 Optimizing the Grinding-Classification Circuit at Pucobre S. A.'s Medium-Sized San José Copper Concentrator

Punta Del Cobre S. A. (Pucobre) is a Chile-based mining company engaged in the exploration and exploitation of mineral resources, primarily copper in the Atacama Region of Chile. The company's facilities comprise two plants: the San Jose

Fig. 12.35 Multi-stage classification with small hydrocyclones

Fig. 12.36 Punta del Cobre S.A. San José plant

plant, which produces copper concentrates, and the Biocobre plant, which produces cathodes. Through its subsidiaries, the company also operates in Canada [7].

The concentrator is located 20 km from the Chilean city of Copiapó and belongs to the Candelaria mining company. The company processes 10 820 t/d of chalcopyrite copper ore with an average grade of 1% copper, producing 361 t/d of a 28% copper concentrate (Fig. 12.36).

The grinding-classification section of the plant consisted in two sections, one with a 13′ × 18′, 1600 HP motor, ball mill with four hydrocyclones and a second 16.5′ × 18′, 5200 HP motor, ball mill with four Cavex hydrocyclones (Figs. 12.37 and 12.38).

The company decided to install new Cavex 650 hydrocyclones and wanted to find the optimum operation conditions for the plant, with a restriction of 52% −200 mesh in the product. Based on grinding-classification circuit sampling and the use of Molycop Tools (2005), for result reconciliation and circuit simulation, the following were the conditions before and after the study [7].

Results

Operating conditions	Original operation	Optimized operation
Dominion ball mill	5.0 m × 8.5 m	5.0 × 8.5 m (16.5 × 28.0 in)
Maximum power (kW)	3 900	3 896
Operational work index (kWh/t)	24.29	18.97

(continued)

(continued)

Operating conditions	Original operation	Optimized operation
Ball size (in)	6.35	6.35; (2.5)
% ball volume	33.5	33.5[a]
% solid ball mill discharge	82	82
% fraction of critical speed	78.84	78.74
Solid capacity (t/h)	315	340
% -1/4 in feed size	93.6	93.6
% -200 mesh product particle size	24.2	24.2
% solids in discharge	81.71	82
Three Cavex 650 hydrocyclones		
Diameter (cm;in)	56.3;26	56.3;26
Pressure (psi)	10.2	11.2
Apex-to-vortex ratio	0.43	0.56
% -200 mesh feed particle size	24.2%	24.2%
% -200 mesh overflow particle size	49.83%	52.17%
% -200 mesh underflow particle size	19.23%	16.75%
Volume flow (m^3/h)	815	1 252
% solids in the feed	77.38	53
% solids in the overflow	60	40
% solids in the underflow	81.71	87.37
Circulating ratio	402	375.5
Mineral density (t/m^3)	2.90	2.90

[a]Optimum ball volume was 45%, but it had overflow mill restriction.

The target of 52% −200 mesh was obtained by changing the hydrocyclone feed concentration from 77.3 to 53% solids and the apex/vortex ratio from 0.43 to 0.56.

12.6 Economic Cost of Roping in Hydrocyclones: Evaluation of the Performance of the PST Instrument in Minera Escondida's Laguna Seca Plant, Chile

In 2016 a campaign was carried out to evaluate the operation of the new PST particle size analysis instrument installed in the hydrocyclones of the comminution-classification systems in one of the largest copper mines in Chile. Although the objective of the campaign was not a study of roping, the fact that roping occurred allowed the reasons for its occurrence and the associated economic costs to be established [1].

The plant processed porphyry copper, with an average of 143,000 t/d. The grinding stage consisted of a SAG mill followed by four ball mills in a closed circuit with 4

Fig. 12.37 Dominium Ball Mill

Fig. 12.38 Cavex hydrocyclone

clusters of 19 hydrocyclones Krebs gMax of 33 in (839 mm) in cm-diameter each, with a product overrun of 80% minus 145 μm (Figs. 12.39, 12.40 and 12.41).

To balance the load in the 19 hydrocyclones in each rig, the hydrocyclones were organized into four quadrants in each of the batteries, as shown in the Fig. 12.42

During the operation of the plant, there were several episodes of hydrocyclone roping with an increase in the particle size in the overflow. During these episodes, the pressure and feed concentration remained constant and the only parameter that changed was the size distribution of the ore, a smaller than usual Cu/Fe fraction,

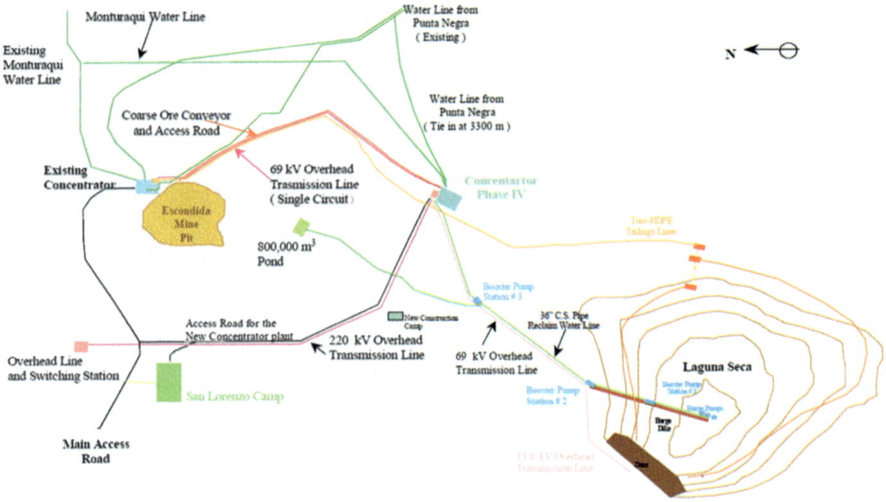

Fig. 12.39 Escondida Los Colorados and Laguna Seca mines and concentrators (Phase IV)

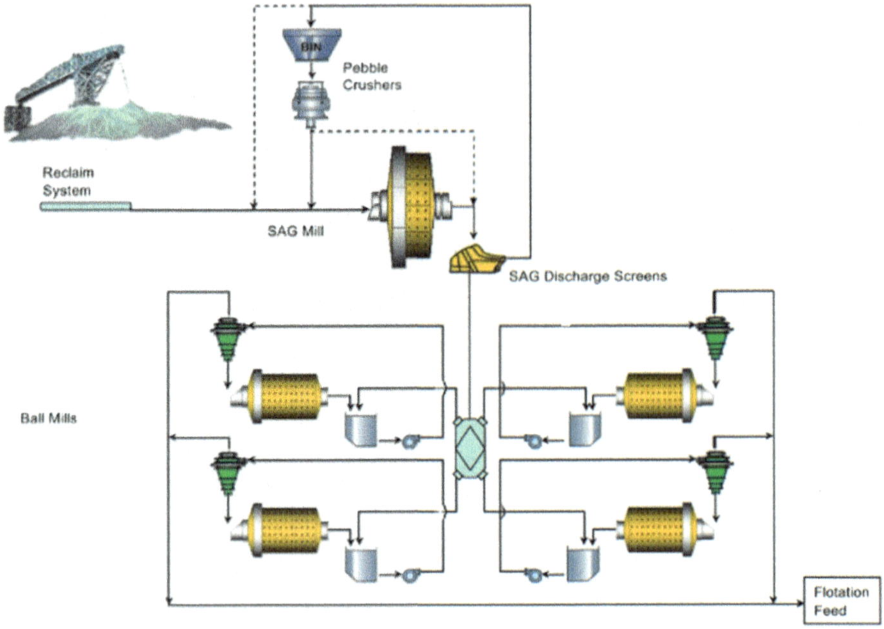

Fig. 12.40 Flowsheet of the grinding section of the concentrator

Fig. 12.41 One of four hydrocyclone batteries

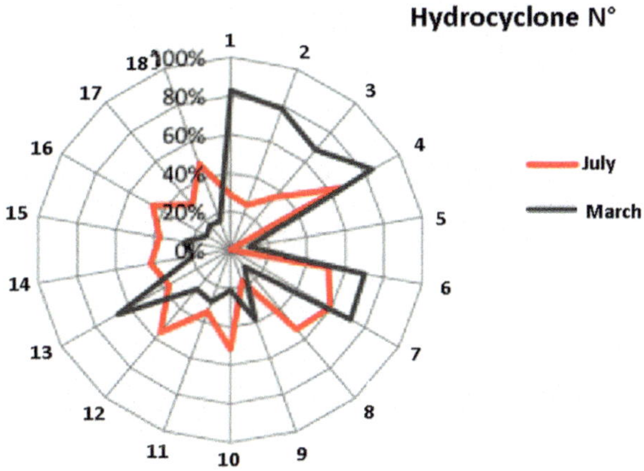

Fig. 12.42 Hydrocyclone distribution before and after optimization

which produced roping. These episodes lasted for about one-hour total and 12 min each (Fig. 12.43).

The consequences of such events for plants are catastrophic for both the equipment and production. In this case, a massive embankment affected mainly the central axes of the rougher cells because the coarse material remained at the bottom of the flotation cell, affecting its normal functioning; in the most critical events the axes of the cells may even break. To evaluate the effects on production, data from executive reports were complemented with data on the metallurgical parameters of the week in which the embankment occurred (Fig. 12.44).

The above figure shows that one of the most affected parameters is copper recovery. The larger particle sizes affected flotation, diminishing the effectiveness of flotation and causing the embankment of the rougher flotation cells. To solve the

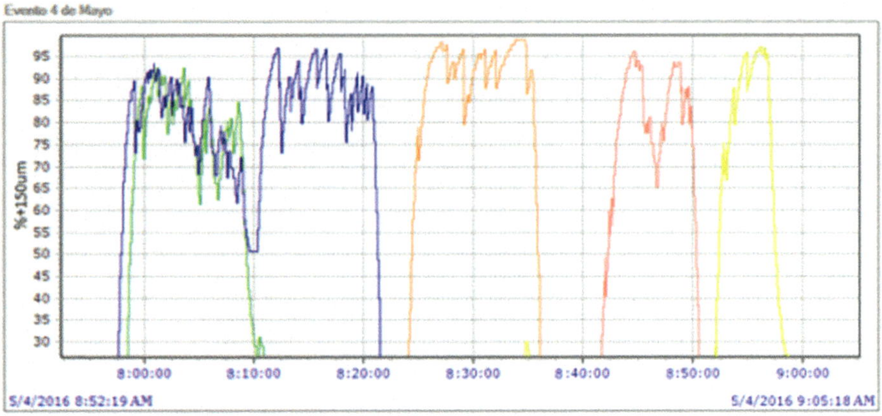

Fig. 12.43 Visualization of the increase in particle size in the overflow due to massive roping

N° of hydrocyclones in roping	5
Hydrocyclones	4, 8, 17, 18 and 19
MSAG detention time	3:24:00
Affected equipment	M-013 embanked, FT-207 shaft cut, cell 2016 embanked, column FT-312 embanked.
Events duration	1:01:09
Estimated fine Cu loss	328 (t)
Estimated fine Cu los (lb) [kg]	723.117 (lb); 328.000 (kg)
Loss	US$ 1.516.376

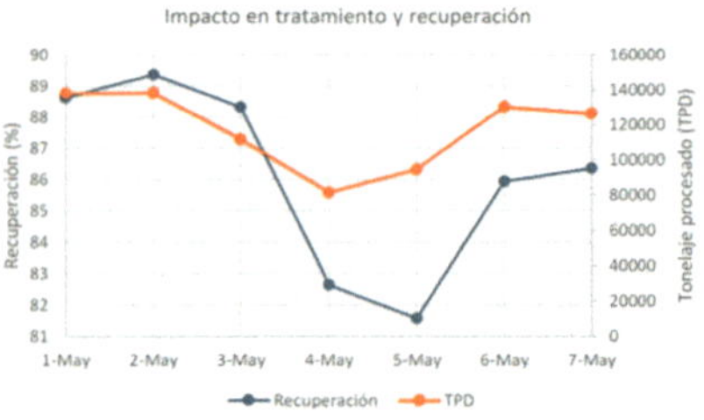

Fig. 12.44 Embankment condition and impact on the throughput and concentrate recovery

problem, the ore feed rate in the grinding circuit had to be decreased, reducing plant capacity, which resulted in the loss of more than 300 t of fine copper and more than 1.5 million dollars.

12.7 Hydrocyclones in Industrial Minerals and Other Industries

12.7.1 Fine Silica Grinding for Fiber Cement Manufacturing

The use of asbestos in the manufacture of fiber-cement was banned approximately 50 years ago. Asbestos was generally replaced by a mixture of ground calcium carbonate and ground silica along with cellulose. Fiber-cement companies developed ultrafine grinding processes for silica and limestone to replace asbestos, usually by wet grinding of the silica and dry grinding of the calcium carbonate. The grinding degree is of the order of 80% finer than 38 microns (400 mesh), equivalent to about 4000–5000 Blaine (g/cm^2). For comparison, the fineness of the cement is about 2500–3000 Blaine.

Some typical installations all using hydrocyclones PU casted.

- Pizarreño (Chile): 3 Closed Grinding Circuits. Classification at 38 microns with 3 Eral hydrocyclones model, PP025101 of 250 mm + 1 grinding circuit with 4 Eral hydrocyclones model PP015081, 150 mm.
- Eternit Argentina: 1 Closed grinding circuit at 38 microns with 1 Eral Hydrocyclone model PP025101 de 250 mm.
- Eternit Perú: 1 Closed grinding circuit at 38 microns with 1 Eral hydrocyclone model PP025101 de 250 mm.
- Eternit Colombiana: 1 Closed grinding circuit at 38 microns with 1 Eral hydrocyclone model PP025101 de 250 mm.
- Etex Colombia: 1 Closed grinding circuit at 38 microns with 1 Eral hydrocyclone model PP025101 de 250 mm.
- Nigerite: 3 Closed grinding circuits at 38 microns with 1 Eral hydrocyclone, each, model PP025101 de 250 mm (Fig. 12.45).

Fig. 12.45 Quartz sand closed grinding circuit, Eternit Perú

12.7.2 *Desliming, Classification and Refining of Silica Sand and Kaolin*

In most deposits, kaolin is accompanied by quartz and feldspar. The kaolin particles are much finer, generally below 20–45 microns (600–325 mesh) and therefore the process of obtaining kaolin is referred to wet separation by means of small diameter hydrocyclones, between 40 and 100 mm. The separation is usually carried out in several stages so that kaolin of various qualities (fineness) can be produced for use mainly in ceramics and paper. In sedimentary deposits with the presence of coarse quartz, cyclones of larger diameters, 250–500 mm for example, are used to produce silica sands for use in glass and foundry production.

Some installations of this type all using hydrocyclones casted in PU.

- Sumicol: 4 Eral hydrocyclones model SR040102 de 400 mm + 2 Eral Hydrocyclones model PP025101 de 250 mm + 12 Eral hydrocyclones model PP010041 + 68 ERAL hydrocyclones model PP050050 de 50 mm + 34 Eral hydrocyclones model PP040040 de 40 mm
- Arevi: 3 Eral Hydrocyclones model PP050101 de 500 mm + 30 Eral Hydrocyclones model PP050050 de 50 mm + 68 Eral Hydrocyclones model PP040040 de 40 mm (Fig. 12.46).

Fig. 12.46 Hydrocyclone' clusters in Kaolin process. Sumicol, Colombia

Fig. 12.47 Arenas Patagónicas: 4 Eral Hydrocyclones model SR075151 de 750 mm + 3 Eral Hydrocyclones model PP050101 de 500 mm + 3 Eral Hydrocyclones model PP040101 de 400 mm + 4 Eral Hydrocyclones model PP025101 de 250 mm

12.7.3 Desliming and Classification of Silica Sand for Fracking

The production of oil and gas through the technique of "Fracking" requires the use of very specific sands to achieve high production wells. The sands used must be well rounded, with very narrow grain sizes and absolutely clean. The most used sizes are 0.85/0.42 mm (20/40 meshes), 0.42/0.21 mm (40/70 meshes) and 0.21/0.11 mm (70/140 meshes). The production of these sands is quite complex and requires multiple classification stages using screens, hydroclassifiers and hydrocyclones.

A typical installation shown in the Fig. 12.47.

12.7.4 Quartz Sand Washing and Classification for Glass Production

The production of glass and crystal requires the use of very high-quality silica sands with fine grain sizes, generally 0.15–0.6 mm. Foundries also use quartz sands of

Fig. 12.48 Complete planta of Silica sand process. Migrin Las Piedras, Chile

similar particle size. The production plants for this type of sand use hydrocyclones of different diameters, generally between 100 and 500 mm. In addition to hydro-classifiers and gravity concentration equipment (spirals), magnetic separation, and even closed grinding circuit stages (ball mills) with hydrocyclones, some of them in Flat Bottom configuration.

Some plants producing glass sand, using hydrocyclones, all casted in PU.

- Migrin Las Piedras: 6 Eral hydrocyclones model PP040903 de 400 mm
- Migrin El turco: 5 Eral hydrocyclones model PP050903 de 500 mm + 10 Eral hydrocyclones model PP025101 de 250 mm
- Cristalerias Toro: 3 Eral hydrocyclones model PP025903 de 250 mm
- Dorila: 3 Eral hydrocyclones model PP025903 de 250 mm + 20 Eral hydrocyclones model PP005050 de 50 mm (Fig. 12.48).

12.7.5 Washing and Desliming of Sand for Construction and Fine Sand Recovery

The production of high resistance concrete requires the use of high quality sands to achieve the required strength, while maintaining a reduced consumption of cement.

Fig. 12.49 Hydro-Dewatering plant for construction sand. Itaquareia, Brasil

Nowadays the production of fine aggregates sands, is done by hydrocyclone washing plants.

A list of important producers of sands for the production of concrete are Itaquareia, Grupo Argos, Grupo BSA, Grupo Petreos, Grupo Holcim, Unicon, Cemex.

More than 50 Eral hydrocyclones model SR075151 de 750 mm, model SR050101 de 500 mm, model SR040101 de 400 mm and model PP025101 de 250 mm. Hydrocyclones Steel-Rubber lined and full casted in PU (Fig. 12.49).

12.7.6 Desliming and Classification of Dredging Sand

Production of construction sand and other uses from dredging rivers and lagoons requires effective washing systems to be able to produce quality sand, regardless of the area where dredging is taking place.

The use of hydrocyclones to pre-wash the sands during the dredging process, followed by a final washing of the dredged sands at the time of disembarkation, is a very interesting application that allows the achievement of quality sands with high efficiency. In this type of application, large hydrocyclones are used, with diameters of 750, 1000 and 1200 mm and generally with a flat bottom configuration.

Fig. 12.50 Eral hydrocyclones model SR 100151, 1000 mm in Dredger

SanPedro / Blinki installation with 3 Eral hydrocyclones model SR120903 of 1200 mm + 2 model SR100903 of 1000 mm + 6 model SR075903 of 750 mm. All of them Flat Bottom configuration (Figs. 12.50 and 12.51).

12.7.7 Heating Oil Deggriting

Removal of small number of coarse particles in solutions at high pressure and temperature, by means of special hydrocyclones built in alloy steels (Fig. 12.52).

12.7.8 Alumina Classification

In the Bayer process for producing alumina from bauxite, thousands of hydrocyclones are used in numerous applications around the world. The hydrocyclones used must be of a very special construction given the thousands of hydrocyclones that are used in numerous applications around the world in the Bayer process for producing alumina from bauxite. The hydrocyclones used must be of a very special construction given the high temperature and corrosion conditions of the slurry to be classified.

Fig. 12.51 Screening, Washing and Dewatering plant with Eral Hydrocyclones of 750 mm and 1000 mm diameter

Fig. 12.52 Akavortex Gritbox Hydrocyclones

Fig. 12.53 Installations: **a** Bauxilum, Venezuela. **b** Sao Luis, Brasil. **c** Alumina Australia, **d** Stade Germany,: More tan 600 Akavortex Hydrocyclones model RWS4122 de 200 mm

Therefore, hydrocyclones must have an external protective capsule to prevent any type of spillage (Fig. 12.53).

12.7.9 Wolfram Mineral Ore Treatment

In concentration plants producing heavy minerals such as tin, tungsten, titanium, zirconium, etc., hydrocyclones are used in various stages of the process: grinding circuits, thickening before gravity concentration stage, classification, etc. There are numerous different types of hydrocyclones used depending on the process (Fig. 12.54).

12.7.10 Complex Sulphides, Cu, Pb, Zn, Closed Grinding Circuit

Perhaps the best-known application of hydrocyclones in mineral processing is the classification in grinding circuits. Sometimes the ores are of very difficult treatment, as is the case with complex sulphide ores containing copper, lead, zinc and gold. In these cases the grinding stages must be very fine and consequently the hydrocyclones

Fig. 12.54 Los Santos, Tungsten process plant, Spain: Six Eral hydrocyclones model PP025903 of 250 mm, Flat Bottom configuration, ahead of spirals concentration bank

required in the grinding circuits must be of small diameter, especially in regrind circuits, where sizes up to 12 microns (1200 mesh) must be ground. Because of this, clusters of a large number of small hydrocyclones are required (Fig. 12.55).

12.7.11 Phosphate Classification

Phosphate is one of the highest producing minerals in the world. One of the richest areas is the Sahara desert in Africa. The application of hydrocyclones in the treatment of phosphate minerals employs hundreds of hydrocyclones worldwide. The main work is the combined classification/concentration by removing the fine silica and clay particles that accompany the phosphate fractions. At the Bucraa plant in the former Spanish Sahara, one of the largest in the world, 1800 t/h, phosphate concentration is carried out exclusively by means of classification with hydrocyclones, modifying its cut-off size between 90 and 110 microns, depending on the area of the deposit being exploited (Fig. 12.56).

Fig. 12.55 Andaluza de Piritas, Spain: Closed Grinding Circuit at 38 micras (400 mesh) and 12 micras (1200 mesh): 18 Eral hydrocyclones model PP0500101 de 500 mm + 48 Hydrocyclones model PP020081 de 200 mm + 74 Eral hydrocyclones model PP010081 de 100 mm

Fig. 12.56 Bucraa, Spanish Sahara: Phospate Classificacion at 90 μm and 110 μm: 72 Akavortex Hydrocyclones RWN 6118, 300 mm

12.7.12 Gold Process

Often alluvial construction aggregates deposits contain small amounts of free gold, which is usually concentrated in sand fractions. A curious application of hydrocyclones is to use them in a pre-washing stage of the sands, so that the washed sand fraction is sent to a stage of gravimetric concentration by means of concentration spirals or centrifugal concentrators. The hydrocyclones must carry out the combined work of desliming the fine fractions harmful to the sand destined for the manufacture of concrete, at the same time as they must recover the greatest possible quantity of free gold particles contained in the sand to be sent to the gravimetric concentration stage prior to the final storage of the washed sand (Fig. 12.57).

Fig. 12.57 Sorigué, Spain: Cluster with five Eral hydrocyclones model SR075151, 750 mm, in Gold pre-concentration ahead centrifugal concentrators

References

1. Arellano, K., Espinoza, P., Sanchez, J. and Betancourt, F. Evaluation of the performance of the PST instrument in the Planta Laguna Seca Plant of Escondida. Metallurgical Engineering Thesis, University of Concepción (2017).
2. BHP, Evaluation of CAVEX Hydrocyclone in the Ok Tedi Grinding Circuit, Seventh Mill Operators Conference, Kalgooorlie, October 2000.
3. Bouso, J.L. Internal information
4. Bouso, J.L. Hydrocyclone practice, (jlbouso@eralchile.com.)
5. Castro, O, Application of a new generation of Hydrocyclones at La Coipa: Placer Dome Gold Plant in Chile, 1st International Symposium on Hydrocyclones, Mineral Engineering International, Hydrocyclones 03, Mount Nelson Hotel, Cape Town, South Africa, September 24–25, 2003.
6. Concha F, Barrientos A, Muñoz A, Castro O (1996) A phenomenological model of a hydrocyclone. In: Hydrocyclone 96 conference, Cambridge, UK, 2–4 April 1996.
7. Parga G, Soto C, Concha F (2007) Optimization of the Cavex 650 hydrocyclones in section 2 grinding-classification in the start-up of the San José plant. Engineering thesis, Metallurgical Engineering, University of Concepción (in Spanish)
8. Peña MA, Moyano G, Betancourt F (2017) Technical and economical evaluation of the modification of the operating conditions of the hydrocyclones in Codelco Chile, Chuquicamata Division using JKSimMet. Metallurgical Engineering thesis, University of Concepción
9. Rojas PF (2012) Mathematical modeling and optimization of a hydrocyclone for the recovery of brine at SQM SALAR. Engineering thesis, Metallurgical Engineering, University of Concepción, Chile (in Spanish)
10. Sepúlveda J (2005) Moly-Cop Tools-2, Software for the assessment and optimization of grinding-classification circuit performance

Printed by Printforce, the Netherlands